Methods in Molecular Biology

Series Editor
John M. Walker
School of Life and Medical Sciences
University of Hertfordshire
Hatfield, Hertfordshire, UK

For further volumes:
http://www.springer.com/series/7651

For over 35 years, biological scientists have come to rely on the research protocols and methodologies in the critically acclaimed Methods in Molecular Biology series. The series was the first to introduce the step-by-step protocols approach that has become the standard in all biomedical protocol publishing. Each protocol is provided in readily-reproducible step-by step fashion, opening with an introductory overview, a list of the materials and reagents needed to complete the experiment, and followed by a detailed procedure that is supported with a helpful notes section offering tips and tricks of the trade as well as troubleshooting advice. These hallmark features were introduced by series editor Dr. John Walker and constitute the key ingredient in each and every volume of the Methods in Molecular Biology series. Tested and trusted, comprehensive and reliable, all protocols from the series are indexed in PubMed.

Manuscript Instructions https://drive.google.com/file/d/1K4Az5vpAFVzYadJ3jKx2-QSo7Dqa_SNA/view?usp=sharing

Doubled Haploid Methods in Cereals

Edited by

Ana María Castillo

Department of Genetics and Plant Breeding, Estación Experimental de Aula Dei, Consejo Superior de Investigaciones Científicas (EEAD-CSIC), Zaragoza, Spain

Begoña Echávarri

Department of Genetics and Plant Breeding, Estación Experimental de Aula Dei, Consejo Superior de Investigaciones Científicas (EEAD-CSIC), Zaragoza, Spain

María Pilar Vallés

Department of Genetics and Plant Breeding, Estación Experimental de Aula Dei, Consejo Superior de Investigaciones Científicas (EEAD-CSIC), Zaragoza, Spain

Humana Press

Editors
Ana María Castillo
Department of Genetics and Plant Breeding
Estación Experimental de Aula Dei
Consejo Superior de Investigaciones Científicas
(EEAD-CSIC)
Zaragoza, Spain

Begoña Echávarri
Department of Genetics and Plant Breeding
Estación Experimental de Aula Dei
Consejo Superior de Investigaciones Científicas
(EEAD-CSIC)
Zaragoza, Spain

María Pilar Vallés
Department of Genetics and Plant Breeding
Estación Experimental de Aula Dei
Consejo Superior de Investigaciones Científicas
(EEAD-CSIC)
Zaragoza, Spain

ISSN 1064-3745 ISSN 1940-6029 (electronic)
Methods in Molecular Biology
ISBN 978-1-0716-5244-2 ISBN 978-1-0716-5245-9 (eBook)
https://doi.org/10.1007/978-1-0716-5245-9

This Humana imprint is published by the registered company Springer Science+Business Media, LLC, part of Springer Nature.
The registered company address is: 1 New York Plaza, New York, NY 10004, U.S.A.

Preface

Cereal crops are fundamental to the global food supply, and safeguarding their availability for a growing population in the face of climate-change uncertainties remains one of the most urgent challenges for the scientific community today. In this context, promising solutions for developing new cereal varieties that can maintain both productivity and quality are emerging from methods that accelerate breeding programs.

Doubled haploid (DH) technology is particularly noteworthy among these methodologies, as it enables the rapid production of completely homozygous lines within a single generation. Breeding programs based on DH technology constitute a substantial advancement over conventional approaches, enhancing selection accuracy and intensity, thereby reducing the breeding cycle to approximately 4–6 years for the development of new varieties or pure lines for hybrid production. The integration of DH technology with advanced breeding tools, including marker-assisted, genotyping-by-sequencing, speed breeding, transgenic and genome editing approaches, as well as machine learning and AI, will enable breeders and scientists to fully exploit its potential.

Although recent updates on double haploid (DH) production in cereals were published in an earlier volume (Double Haploid Technology, Seguí-Simarro, 2021), the demand for more efficient and reliable DH systems continues to grow. Rapid methodological advances make it essential to revisit, update, and broaden existing protocols. This volume responds to that ongoing need by providing timely developments that support continued progress in cereal DH technology. We present a collection of general chapters addressing key questions in cereal breeding and DH production, complemented by detailed and up-to-date protocols for generating DH lines for breeding programs. Additionally, we provide methodologies for morphological, biochemical, molecular, and cytogenetic characterization that support these protocols.

The opening chapter (Chap. 1) addresses current and future challenges in cereal breeding, as well as the technological tools available to accelerate progress and increase breeding efficiency. The review of recent findings on the transcriptional dynamics of ME in barley and wheat provides an integrated overview of molecular mechanisms underpinning DH production, introducing new perspectives in Chap. 2. Albinism in cereal ME continues to limit the efficiency of DH production. Chap. 3 explores its underlying causes, with emphasis on plastid function, gene regulation, and chloroplast development, and discusses potential strategies to reduce its incidence.

The protocol chapters cover both in vivo and in vitro methods for producing DH plants. In in vivo approaches, haploids are generated through either interspecific or intergeneric (wide) crosses, or through intraspecific pollination using haploid-inducer lines. In wide crosses, loss of the male chromosomes and endosperm abortion make in vitro embryo rescue necessary to obtain haploid plants. Chap. 4 and 5 present protocols for producing bread wheat haploids via interspecific crosses with maize, while Chap. 6 details a complementary approach using *Imperata cylindrica*. Additionally, Chap. 7 covers the hybridization of bread wheat with barley leading to haploid or hybrid embryos, and details molecular cytogenetic techniques to determine chromosome composition. Chap. 13 addresses haploid induction by intraspecific pollination using haploid-inducer lines, which is widely used in maize.

In vitro methods involve the developmental reprogramming of gametic cells towards an embryogenic pathway in response to stress treatments, followed by an in vitro culture phase to produce embryos and ultimately haploid or DH plants. Among the available in vitro methods, microspore embryogenesis, via anther or isolated microspore culture, has demonstrated higher efficiency and potential compared to gynogenesis. Step-by-step efficient protocols for anther culture are presented for bread wheat (Chap. 8), durum wheat (Chap. 10), barley (Chap. 11), and both indica and japonica rice (Chap. 14 and 15), while microspore culture protocols are provided for bread wheat (Chap. 9), barley (Chap. 12), triticale (Chap. 16), and rye (Chap. 17). In addition, Chap. 12 describes methodologies to examine changes in cellular organization and for localizing key molecular determinants through immunofluorescent assays within the context of microspore embryogenesis.

Recent advances in gene-editing technologies have further expanded the potential of DH production. Chap. 18 details the generation of transgene-free, homozygous edited wheat lines via intergeneric crosses between maize and genome-edited wheat, and Chap. 19 describes the production of rice haploid plant production by the knockout of an egg-specific gene.

This volume provides a comprehensive and authoritative guide to methodologies for cereal DH production. By presenting up-to-date, reproducible approaches, it aims to inspire and empower researchers and breeders to develop, refine, and implement DH systems across diverse cereal species. We sincerely thank the authors for their dedication and invaluable contributions, which have advanced not only the scientific understanding of DH technology but also its practical application in modern cereal breeding.

Zaragoza, Spain — *Ana María Castillo*
Zaragoza, Spain — *Begoña Echávarri*
Zaragoza, Spain — *María Pilar Vallés*

Contents

Contributors

MOHAMMAD ALI • *Institute of Genetics and Biotechnology, Hungarian University of Agriculture and Life Sciences, Gödöllő, Hungary*

SERGIO G. ATIENZA • *Department of Plant Breeding, Instituto de Agricultura Sostenible, Consejo Superior de Investigaciones Científicas (IAS-CSIC), Córdoba, Spain*

MELTEM BAYRAKTAR • *John Innes Centre, Department of Crop Genetics, Norwich Research Park, Norwich, Norfolk, UK; Department of Genetic and Bioengineering, Faculty of Engineering and Architecture, Kırşehir Ahi Evran University, Kirsehir, Türkiye*

SANTOSH KUMAR BISHNOI • *ICAR-Indian Institute of Wheat and Barley Research, Karnal, Haryana, India*

PRASANNA M. BODDUPALLI • *International Maize and Wheat Improvement Center (CIMMYT), Nairobi, Kenya*

SUE BROUGHTON • *Department of Primary Industries and Regional Development, South Perth, Australia*

ANA M. CASAS • *Department of Genetics and Plant Breeding, Estación Experimental de Aula Dei, Consejo Superior de Investigaciones Científicas (EEAD-CSIC), Zaragoza, Spain*

MARIECLAIRE CASTELLO • *Department of Primary Industries and Regional Development, South Perth, Australia*

ANA MARÍA CASTILLO • *Department of Genetics and Plant Breeding, Estación Experimental de Aula Dei, Consejo Superior de Investigaciones Científicas (EEAD-CSIC), Zaragoza, Spain*

VIJAY CHAIKAM • *International Maize and Wheat Improvement Center (CIMMYT), Nairobi, Kenya*

FRANCISCO J. CIUDAD • *Instituto Tecnológico Agrario de Castilla y León, Junta de Castilla y León, ITACyL, Valladolid, Spain*

MARÍA ASUNCIÓN COSTAR • *Department of Genetics and Plant Breeding, Estación Experimental de Aula Dei, Consejo Superior de Investigaciones Científicas (EEAD-CSIC), Zaragoza, Spain*

EWA DUBAS • *The Franciszek Górski Institute of Plant Physiology, Polish Academy of Sciences, Kraków, Poland*

BEGOÑA ECHÁVARRI • *Department of Genetics and Plant Breeding, Estación Experimental de Aula Dei, Consejo Superior de Investigaciones Científicas (EEAD-CSIC), Zaragoza, Spain*

SAKINEH FARHADI-TOOLI • *Department of Tissue and Cell Culture, Agricultural Biotechnology Research Institute of Iran (ABRII), Agricultural Research, Education and Extension Organization (AREEO), Karaj, Iran*

ALISON M. R. FERRIE • *Aquatic and Crop Resource Development Research Centre, National Research Council Canada, Saskatoon, SK, Canada*

DAOLIN FU • *State Key Laboratory of Wheat Improvement, College of Agronomy, Shandong Agricultural University, Taian, China; Spring Valley Agriscience Co., Ltd., Jinan, China*

PATRICIA FUSTERO • *Department of Genetics and Plant Breeding, Estación Experimental de Aula Dei, Consejo Superior de Investigaciones Científicas (EEAD-CSIC), Zaragoza, Spain*

MONIKA GAJECKA-LESIK • *Institute of Biology, Biotechnology and Environmental Protection, Faculty of Natural Sciences, University of Silesia, Katowice, Poland*

ABDELBAGI MUKHTAR ALI GHANIM • *Plant Breeding and Genetics Laboratory, Joint FAO/IAEA Division of Nuclear Techniques in Food and Agriculture, Department of Nuclear Sciences and Applications, International Atomic Energy Agency, Vienna, Austria*

PATRICIA GIRALDO • *Department of Biotechnology-Plant Biology, Universidad Politécnica de Madrid (UPM), Madrid, Spain*

XIZHEN GUAN • *State Key Laboratory of Wheat Improvement, College of Agronomy, Shandong Agricultural University, Taian, China; Spring Valley Agriscience Co., Ltd., Jinan, China*

SADIYE HAYTA • *John Innes Centre, Department of Crop Genetics, Norwich Research Park, Norwich, Norfolk, UK*

ERNESTO IGARTUA • *Department of Genetics and Plant Breeding, Estación Experimental de Aula Dei, Consejo Superior de Investigaciones Científicas (EEAD-CSIC), Zaragoza, Spain*

NURIA ISACH • *Department of Genetics and Plant Breeding, Estación Experimental de Aula Dei, Consejo Superior de Investigaciones Científicas (EEAD-CSIC), Zaragoza, Spain; Department of Pharmacology, Faculty of Medicine and Dentistry University of Valencia, Valencia, Spain*

MIHÁLY JANCSÓ • *Research Center for Irrigation and Water Management, Institute of Environmental Sciences, Hungarian University of Agriculture and Life Sciences, Szarvas, Hungary*

JULIE KILLEN • *Department of Primary Industries and Regional Development, South Perth, Australia*

ANDRÁS KIS • *Institute of Genetics and Biotechnology, Hungarian University of Agriculture and Life Sciences, Gödöllő, Hungary*

VIKTOR KORZUN • *Global Lead Scientific Affairs, KWS SAAT SE & Co. KGaA, Einbeck, Germany*

MONIKA KRZEWSKA • *The Franciszek Górski Institute of Plant Physiology, Polish Academy of Sciences, Kraków, Poland*

CSABA LANTOS • *Cereal Research Non-Profit Ltd., Biotechnology Laboratory, Szeged, Hungary*

THEANO B. LAZARIDOU • *University of Western Macedonia–School of Agricultural Sciences, Florina, Greece*

YUEQIANG LENG • *Department of Plant Pathology, Microbiology, Biotechnology, North Dakota State University, Fargo, ND, USA*

LI LIU • *Department of Primary Industries and Regional Development, South Perth, Australia*

LEOCADIO MARTINEZ • *International Maize and Wheat Improvement Center(CYMMIT), Apdo, Mexico, DF, Mexico*

CHRISTOPHER MCMULLAN • *Department of Primary Industries and Regional Development, South Perth, Australia*

AIMAR NAVARRO • *Department of Genetics and Plant Breeding, Estación Experimental de Aula Dei, Consejo Superior de Investigaciones Científicas (EEAD-CSIC), Zaragoza, Spain*

JOHN OCHIENG • *International Maize and Wheat Improvement Center (CIMMYT), Nairobi, Kenya*

DHARAM PAL • *ICAR-Indian Agricultural Research Institute, Regional Station, Tutikandi Centre, Amartara Cottage, Shimla, India*

MADHU PATIAL • *ICAR-Indian Agricultural Research Institute, Regional Station, Tutikandi Centre, Amartara Cottage, Shimla, India*

JÁNOS PAUK • *Cereal Research Non-Profit Ltd., Biotechnology Laboratory, Szeged, Hungary*

JUNHUA PENG • *Spring Valley Agriscience Co., Ltd., Jinan, China*

YOLANDA PÉREZ-PÉREZ • *Pollen Biotechnology of Crop Plants Group, Margarita Salas Center of Biological Research (CIB), CSIC, Madrid, Spain*

DÁVID POLGÁRI • *Institute of Genetics and Biotechnology, Hungarian University of Agriculture and Life Sciences, Gödöllő, Hungary; HUN-REN Centre for Agricultural Research, Martonvásár, Hungary*

KALLOL KUMAR PRAMANICK • *ICAR-Indian Agricultural Research Institute, Regional Station, Tutikandi Centre, Amartara Cottage, Shimla, India*

ELENA PRATS • *Department of Plant Breeding, Instituto de Agricultura Sostenible, Consejo Superior de Investigaciones Científicas (IAS-CSIC), Córdoba, Spain*

PEDRO REVILLA • *Department of Plant Production, Misión Biológica de Galicia, Consejo Superior de Investigaciones Científicas (MBG-CSIC), Pontevedra, Spain*

LÁSZLÓ SÁGI • *HUN-REN Centre for Agricultural Research, Martonvásár, Hungary*

ADÉL SEPSI • *HUN-REN Centre for Agricultural Research, Martonvásár, Hungary*

MEHRAN E. SHARIATPANAHI • *Department of Tissue and Cell Culture, Agricultural Biotechnology Research Institute of Iran (ABRII), Agricultural Research, Education and Extension Organization (AREEO), Karaj, Iran; Plant Breeding and Genetics Laboratory, Joint FAO/IAEA Division of Nuclear Techniques in Food and Agriculture, Department of Nuclear Sciences and Applications, International Atomic Energy Agency, Vienna, Austria*

CE SHI • *State Key Laboratory of Hybrid Rice, College of Life Sciences, Wuhan University, Wuhan, China*

MENG-XIANG SUN • *State Key Laboratory of Hybrid Rice, College of Life Sciences, Wuhan University, Wuhan, China*

TÍMEA SZALÓKI • *Research Center for Irrigation and Water Management, Institute of Environmental Sciences, Hungarian University of Agriculture and Life Sciences, Szarvas, Hungary*

IWONA SZAREJKO • *Institute of Biology, Biotechnology and Environmental Protection, Faculty of Natural Sciences, University of Silesia, Katowice, Poland*

ÁRPÁD SZÉKELY • *Research Center for Irrigation and Water Management, Institute of Environmental Sciences, Hungarian University of Agriculture and Life Sciences, Szarvas, Hungary*

SAMIRA TAJEDINI • *Department of Tissue and Cell Culture, Agricultural Biotechnology Research Institute of Iran (ABRII), Agricultural Research, Education and Extension Organization (AREEO), Karaj, Iran; Plant Breeding and Genetics Laboratory, Joint FAO/IAEA Division of Nuclear Techniques in Food and Agriculture, Department of Nuclear Sciences and Applications, International Atomic Energy Agency, Vienna, Austria*

MARYAM TAVAKOLI • *Department of Tissue and Cell Culture, Agricultural Biotechnology Research Institute of Iran (ABRII), Agricultural Research, Education and Extension Organization (AREEO), Karaj, Iran*

PILAR S. TESTILLANO • *Pollen Biotechnology of Crop Plants Group, Margarita Salas Center of Biological Research (CIB), CSIC, Madrid, Spain*

ISABEL VALERO-RUBIRA • *Department of Genetics and Plant Breeding, Estación Experimental de Aula Dei, Consejo Superior de Investigaciones Científicas (EEAD-CSIC), Zaragoza, Spain; Certest Biotec S.L. San Mateo de Gállego, Zaragoza, Spain*

MARÍA PILAR VALLÉS • *Department of Genetics and Plant Breeding, Estación Experimental de Aula Dei, Consejo Superior de Investigaciones Científicas (EEAD-CSIC), Zaragoza, Spain*

IOANNIS N. XYNIAS • *University of Western Macedonia–School of Agricultural Sciences, Florina, Greece*

SHENGMING YANG • *Cereals Crops Improvement Research Unit, Edward T. Schafer Agricultural Research Center, USDA-ARS, Fargo, ND, USA*

HAI YING YUAN • *Aquatic and Crop Resource Development Research Centre, National Research Council Canada, Saskatoon, SK, Canada*

XUECHENG ZHANG • *State Key Laboratory of Hybrid Rice, College of Life Sciences, Wuhan University, Wuhan, China*

ZIFU ZHAO • *State Key Laboratory of Hybrid Rice, College of Life Sciences, Wuhan University, Wuhan, China*

SHAOBIN ZHONG • *Cereal Disease Laboratory, USDA-ARS, St. Paul, MN, USA; Department of Plant Pathology, Microbiology, Biotechnology, North Dakota State University, Fargo, ND, USA*

IWONA ŻUR • *The Franciszek Górski Institute of Plant Physiology, Polish Academy of Sciences, Kraków, Poland*

Part I

General Chapters

Chapter 1

Current Challenges in Cereal Breeding

Ernesto Igartua, Elena Prats, Sergio G. Atienza, Pedro Revilla, Francisco J. Ciudad, Patricia Giraldo, Ana M. Casas and Ana María Castillo

Abstract

Over the last 150 years, plant breeding has delivered increasingly productive cereal varieties. This growth rate is hardly sustainable, as new factors need to be considered and will affect breeding programs. Climate change threatens agricultural production, and crops must be adapted to the new conditions. In addition, the environmental footprint of agriculture must be taken into account in breeding activities. This chapter reviews the current and expected challenges affecting cereal breeding, including increasing productivity, developing resilient cultivars, improving quality, and delivering varieties for a sustainable agriculture. The rich catalog of tools available to speed up and increase the efficiency of cereal breeding are also reviewed: a wider use of natural genetic diversity through planned prebreeding initiatives; high-throughput genotyping and phenotyping methods; genetic modification; and acceleration of breeding cycles by speed breeding or doubled haploid technologies. Developing strategies for the successful alignment of these tools for cereal breeding will ensure its key role in supporting food safety worldwide.

Keywords Cereals, Plant breeding, Doubled haploid, Climate change, Sustainability

1 Introduction

Plant breeding has been very successful in delivering high quality and high yielding varieties to meet the needs of producers and consumers. First, for about 10,000 years, it was an empirical discipline, based on trial and error, with little theoretical support. Despite this, it produced feats like the domestication of crop species, as illustrated by the comparison between today's maize cob and the tiny ear of its ancestor, teosinte. This success helped to establish agriculture as the main means of food production, supporting population growth throughout modern history. Over the last 100–150 years, plant breeding has been enriched by a strong theoretical background in genetics, molecular biology, mathematics, and physics. As an integrative discipline, it harnesses and applies developments in many areas of science. Cereal plant breeding benefited from this era of

Ana María Castillo et al. (eds.), *Doubled Haploid Methods in Cereals*, Methods in Molecular Biology, vol. 3029,
https://doi.org/10.1007/978-1-0716-5245-9_1,

discovery, resulting in great progress. The yield of the major staple crops (rice, wheat, maize) increased on average by 52.6 kg/ha in 50 years (1960–2010). It is generally accepted that half of this increase is due to agronomic practices and the other half to genetic improvement [1]. This bright perspective should not hide some shadows. This growth rate is hardly sustainable. Climate change threatens agricultural production, and the degradation of the environment and the biodiversity crisis must be considered in breeding activities.

1.1 Are the Challenges in Plant Breeding the Same as They Have Always Been?

In short, yes, but new challenges have emerged. Traditionally, plant breeding has aimed to increase crop production to ensure food security and economic growth. This goal involved controlling other species that fed on the crops or thrived on the resources provided for the crops. These challenges persist today and are likely to persist for the foreseeable future. Nowadays, however, the sustainability of healthy and safe food production has become a top priority.

The remarkable increase in crop production has been necessary to feed a world population that has nearly tripled in size, from 3 billion to over 8 billion, in just over six decades. Although there is enough food to feed the world, the inequal distribution makes it essential to continue working to increase production. This is needed at both ends of the agricultural production spectrum. At one end, plant breeders must strive to reduce the yield gap caused by biotic and abiotic stresses and resource constraints. At the other end, plant breeding must break through production ceilings on areas of highly intensive agriculture with abundant inputs (termed "sustainable intensification"). In both cases, the goals will be achieved using the tools of nature, the amazing genetic diversity of plant species but also the tools of human ingenuity to modify and improve natural processes.

Climate change has introduced new challenges. In the past, plant breeding worked for stable target areas, with stable soils and climates. However, this stable scenario is changing dramatically. In the short term, plant breeding programs will go on as usual. Variety development in cereal breeding programs is extremely rapid (5–10 years), with current improvements in doubled haploid (DH) production and speed breeding reducing this time even further. Current breeding programs also select rapidly and effectively the best genetic combinations aided by powerful methods like genomic selection and high-throughput phenotyping. In this context, field trials within each program will address climate change trends over such a short period. However, in the medium term, the accumulation of subtle but gradual shifts in climate factors (like temperature, rainfall patterns, or the frequency of extreme weather events), that may seem minor in the short term, can alter growing conditions in ways that may challenge current elite varieties, which may not be able to respond, due to the exhaustion of genetic variance. New conditions will require widening the search for solutions outside the elite germplasm. Fortunately, a huge catalog of cereal genetic diversity is available in gene banks and research organizations [2].

Another source of new challenges for plant breeding is the transition of most of our food production systems to sustainable ones. We have suffered from a lack of integration between the study of agricultural and natural ecosystems, with these scientific disciplines developing with their backs turned to each other. The impact on the environment and the cost of environmental damage must be factored when developing new varieties that are more efficient in the use of resources such as water and fertilizers. Resources such as groundwater and soil are finite. Soil degradation is a problem in all agricultural systems [3], and the health of aquifers is threatened by depletion from excessive irrigation and contamination from fertilizers. Agriculture inevitably causes environmental disturbances that, if not properly managed, can lead to long-term environmental damage [4]. Sustainability is now at the forefront of societal concerns and must be considered in any activity that affects food production. Trade-offs between agricultural production and environmental degradation must be carefully considered, and plant breeding must be sensitive to these concerns.

Another source of new targets for breeding is the interaction of the crop with other organisms, particularly microorganisms. There is already indisputable evidence of the intense dialogue between the plant and the microbiota present at the rhizosphere [5] and the phyllosphere [6], affecting plant responses to biotic and abiotic stresses and nutrient acquisition. These responses depend on the genetics of the variety. Therefore, the recruiting capacity of beneficial microorganisms may become a new breeding target.

A recently proposed breeding goal, mainly in tree species, is carbon farming. Many researchers believe that this goal should also be explored in annual crops [7]. The economic output of a cereal crop is no longer limited to grain and straw but also includes the small part of the root residue that remains in the soil, increasing the soil carbon budget. Although only a small fraction enters the stable soil carbon pool, the scale of cereal cropping makes any small improvement in this respect a potentially relevant source of carbon sequestration, with the added benefit of improving soil health. Strategies to improve dry matter allocation to the roots, increase the depth of root penetration, and modify root structure may be part of breeding programs in the near future. In addition, these strategies can support other breeding targets that can positively impact overall yield and strengthen resilience particularly in marginal soils, where improved root systems can enable crops to thrive under conditions of low fertility and limited water availability.

Finally, the big data era is posing new challenges for all researchers and, particularly to an integrative science such as plant breeding.

The data deluge is a mixed blessing. It has to be tamed and made usable to live up to its promise of new discoveries. Genomic, proteomic, metabolomic, phenomic, environmental data must be combined to allow integrated analyses. This is especially relevant to facilitate the use of artificial intelligence approaches. The importance of recording metadata, and using the FAIR principles in all breeding activities must be implemented by all researchers.

1.2 What are the Current and Future Breeding Targets in Cereal Breeding Programs?

1.2.1 Increasing Productivity

Increasing crop production should remain a standing objective of cereal breeders. Advances in many realms of science are coming to fruition. Achieving better photosynthetic performance seems within reach, either by exploiting existing genetic variation in traits such as the Rubisco catalytic activity or by genetic modification of photosynthetic pathways genes [8]. Any increase in cereals yield must come from an increase of assimilates accumulated in the grain. It has long been widely debated whether cereal yields are source-limited or sink-limited [9]. There may not be a single answer to that question. Recent gains in small grain cereal yields were generally caused by increases in sink strength [10]. However, the trade-offs between yield components and between all plant organs are complex, with strong interactions between sources and sinks. As stated by Murchie et al. [11]: "If yields are to be increased, especially in species with a high harvest index, it is essential that the capacity of the source is optimized for the sink, and vice versa." They provide a "wiring diagram" to study breeding strategies taking into account the dynamic relationship between plant organs and tissues trade-offs [11]. Further improvements in cereal productivity would benefit from multidisciplinary approaches involving not only breeders but also physiologists and agronomists.

1.2.2 Developing Resilient Cultivars

The development of resilient varieties has a dual objective. One is to reduce the yield gap (the difference between actual and potential yields). In addition, resilience should protect against dramatic yield losses in the presence of strong stress events. Resilience can be achieved through improvement in these areas:

- *Abiotic stress tolerance*. This is an area where research is extremely active worldwide. Climate change will cause a number of stresses to become more frequent, with more extreme events and more geographically widespread. Heat will become more common everywhere, while drought, waterlogging, and frost will increase regionally. There are many examples of cereal genotypes that offer good promise for increased stress tolerance. While breeding efforts for different abiotic stresses will have specific targets, many of these stresses induce a common physiological effect: oxidative stress. When plants are subjected to various environmental stress conditions, the balance between the production of reactive oxygen species and their detoxification through antioxidant mechanisms is disturbed. Enhancing the plant's antioxidant capacity to combat oxidative stress has thus become a crucial breeding target, as it helps to sustain photosynthetic activity and improve resilience to multiple stress factors [12].

- *Biotic stress tolerance.* Climate change is causing shifts in the incidence and distribution of pests and diseases, leading to significant production losses. This area is experiencing a great impulse due to the insights gained from considering resistance mechanisms as part of the plant immune system [13]. The durability of resistance is crucial for sustainable crop production, as many commonly used types of resistance, such as those based on single major genes, are often short-lived in the field. These single-gene resistances, while initially effective, are susceptible to being overcome by virulent pathogen strains, particularly in high-selection pressure environments. Multiresistant varieties can be developed through pyramiding multiple resistant alleles, with the aim to confer durable resistance that keep pathogen populations from evolving into more aggressive strains. However, achieving truly durable resistance requires a deep understanding of the mechanisms underlying resistance responses, as some are inherently more stable over time [14]. For example, nonhypersensitive defense responses, structural avoidance traits, and mechanisms that prevent pathogen compatibility—such as blocking haustorium formation—often provide more enduring protection than race-specific resistance. Additionally, integrating traits that hinder the pathogen's ability to establish compatibility can reduce infection success and limit the pathogen's capacity to evolve counter-resistance strategies. Breeding programs that target these more complex resistance pathways, alongside traditional gene pyramiding, are essential for developing robust cultivars that maintain their resistance over time and under the unpredictable impacts of a changing climate. This approach is only feasible thanks to marker assisted selection.

The challenge ahead, however, is to test the responses of plants to combinations of stresses that will become the norm in the near future. The combination of abiotic stresses causes a very different response in plants compared to single stresses [15]. Although the combination of abiotic stresses has been considered in recent decades, the combination of abiotic with biotic stresses is likely to play an important role eventually, as the resilience of plants against any stress may be compromised when a combination of abiotic×abiotic or abiotic×biotic stresses is produced. Importantly, the signaling pathways that plants activate for resistance to biotic and abiotic stresses often share common nodes, presenting both opportunities and challenges for breeding [16]. Identifying these shared molecular targets may enable breeders to achieve synergistic effects across multiple stress responses or avoid potential antagonisms that could impair resilience. In addition, activating resistance to pathogens often triggers physiological changes—related to stomatal function and photosynthetic processes—that can inadvertently increase susceptibility to drought or other abiotic stresses [17]. Therefore, understanding the molecular basis of resistance responses and assessing the variability in these responses across genotypes is essential. To this end, we should develop advanced phenotyping methods to integrate physiological and biochemical criteria to optimize selection for tolerance to abiotic stress. A major challenge is to develop low-cost methods that allow the combined application of stresses under field conditions for breeding programs. In this context, multidisciplinary research involving agronomists, physiologists, and breeders would be needed.

- *Leveraging GxE.* The approaches to solve the genotype by environment riddle has dominated cereal breeding research for decades. A third dimension, management, has been incorporated into the equation, GxExM. Agricultural systems are rapidly shifting towards sustainability, at least in Europe, and will have to be considered globally as well. Breeders must define their target environments based on commonalities of environmental features but also including management aspects. In this sense, GxE studies allow breeders to select varieties that perform well across the average conditions of a broad mega-environment while also identifying genotypes that are particularly suited to specific local conditions [18]. Further analysis may also reveal the main climate and edaphic characteristics driving the performance of the best-adapted genotypes, providing critical insights for defining breeding targets [19]. When GxE studies are combined with genomic data, they can offer insights into the genetic plasticity of different genotypes, revealing markers that explain a higher percentage of phenotypic variance within average mega-environments, as well as those that are significant only under particular or extreme environmental conditions [20]. Identifying these environmentally sensitive markers is essential in the context of climate change, as it enables breeders to select genotypes capable of performing well under increasingly variable and extreme conditions. This approach not only enhances resilience but also supports precision breeding, allowing for improved targeting of genetic adaptations suited to the complex climate scenarios of the future.

- *Stability*. In the current scenario, environments are becoming more uncertain, more variable, with increasing year-to-year fluctuations. Both interseason and intraseason variations strain the adaptive capacity of crops. In this context, stability, an old tool in the breeders' toolkit deserves more attention. Large field trial networks and state-of-the-art statistical methods to analyze them and model genetic gains are needed to ensure progress in stability. Combining these trials with GxE analyses and genomic data allows breeders to identify varieties that maintain high performance across a range of conditions while pinpointing specific traits or markers associated with resilience to extreme or unique stresses. Such approaches are essential to develop cultivars that perform consistently, buffering against climatic extremes and the multistress environments of the future.

1.2.3 Improving Quality

Climate change not only reduces total crop production. It also threatens the production of cereals with the quality demanded by society. Therefore, quality must always be kept as a main target in cereal breeding programs, particularly in wheat and rice. Cereals are the main source of carbohydrates for humans and animals worldwide and, for a large part of humanity, they are an important source of protein, lipids, minerals, and fiber. In addition, cereals can provide antioxidants, such as carotenoids or anthocyanins, making them functional foods with high added value. Some varieties are also sources of vitamins such as folic acid, vitamin A, or provitamin E in maize [21]. Furthermore, sensory quality is a key aspect of cereal value that discriminates between the final use of the varieties, such as wheat for bakery or pasta, or maize for bread, talo, tortilla, polenta, or gofio. Consumer-driven quality traits, such as flavor, color, aroma, and nutritional value are becoming increasingly important. Additionally, new industrial processes, such as frozen doughs, require specific end-use quality parameters. Breeding for improved/differential quality will be also helpful for farmers and industry. The development of specific products targeting gourmet markers may increase the income of the farmers and even compensate for lower yields in some cases. The success of tritordeum is a good example. Tritordeum is commercialized as the Golden cereal due to its high grain carotenoid content. Consumers value this trait along with its distinctive flavor, and thus, they are willing to pay higher for a product with differential quality. Similar strategies can be used in other cereals to increase the income of farmers ensuring that agriculture constitutes a profitable activity.

There are nutrition gaps in many areas of the world where cereals are the main staple. Therefore, enriching cereal grains in mineral or protein content (biofortification) are efficient ways to significantly decrease these gaps, as has been proposed for iron and zinc [22]. On the other hand, cereals like oats and barley are gaining ground as food, increasingly valued by consumers due to evidence-based health claims [23, 24]. Gluten content is of great importance in wheat, rye, and barley, where it is responsible for the rheological quality of bakery products. Recently, however, gluten reduction has also become a breeding target, due to health concerns. In some cases, the scientific evidence is clear (coeliac disease) whereas in others the causal link between certain conditions and compounds like gluten and FODMAPs must still be established [25]. On the other hand, gluten-free cereals, such as maize or rice, are suitable for gluten-intolerant people.

Another important feature of grain quality, also related to food safety, is mycotoxin contamination. These fungal contaminants cause serious health problems in animals and people. The Mycotoxin Awareness Network and FAO recognize more than 300 mycotoxins, and breeders must work to reduce their presence in food and feed grains.

1.2.4 Delivering Varieties for a Sustainable Agriculture

Current research trends clearly advocate for the need to expand the boundaries of plant breeding from the crop to the agroecosystem. The effects of crops on the agroecosystem must be factored in cereal breeding programs. Besides optimizing the direct economic output, breeders must be aware of an efficient and sustainable use of resources like soil, water, and nutrients. The main goals in this respect are:

- *Increasing water use efficiency (WUE).* This is an important challenge given the lack of available water in large parts of the world and the increasing frequency and intensity of drought and heat episodes. Research has repeatedly found a trade-off between WUE and transpiration and total water use by the crop [26]. Hence, it was proposed to use the magnitude "efficient use of water" as the target for breeders aiming at raising production under drought conditions. Sources of drought tolerance are being identified among accessions adapted to arid environments, their mechanisms of tolerance to water scarcity should be investigated, using efficient methods of phenotyping at large scale, and more efficient methods of breeding for drought tolerance must be designed with those resources [27]. An area of specific interest to drought tolerance is the ability of plant roots to extract water from the soil. The development of new phenotyping methods and the growing interest in improving water and nutrient acquisition efficiency are driving root research in various directions [28]. Root research applied to plant breeding is still in its infancy but holds great potential for the future.
- *Fertilizer use efficiency.* Reduction of synthetic fertilizers is another issue at the forefront of societal demands, for environmental and regulatory reasons. Nitrification is a crucial process in the nitrogen cycle. Enhancing the biological nitrification inhibition (BNI) capacity of cereal crops offers significant potential

to develop next-generation varieties that naturally suppress/regulate nitrification [29]. This approach has the potential to reduce nitrogen leakage to underground waters and N_2O and N_2 gas emissions, thereby providing a dual protective effect on the environment. Given that current global cereal yields rely heavily on environmentally harmful inorganic fertilizers, improving nitrogen use efficiency (NUE) is essential for sustainable agriculture. The introgression or over-expression of genes that enhance NUE in cereal crops offers a promising approach to developing varieties capable of maintaining high yields with reduced nitrogen input, promoting long-term agricultural sustainability [30].

- *Plant–microorganism interaction.* Plant breeders are adopting the view that crop plants do not live in isolation. Plants provide many niches for proliferation of microorganisms that can form complex coassociations with plants and have important roles in enhancing the crop productivity and health [31]. Specific compounds exudated by cereal roots have been linked to the capacity to recruit soil microorganisms affecting plant defenses and even herbivore performance [32]. The production of exudates is under genetic control [33]. This evidence has triggered the interest to explore and exploit plant–microbiota interactions at the gene level and apply them to plant breeding for a sustainable crop production, less dependent on agrochemicals.
- *Management systems.* Crop management is shifting towards practices more respectful with the environment, particularly soil conservation. Local regulatory efforts, like the ecoschemes promoted by the EU Common Agricultural Policy contribute to the trend towards a more resource-conserving agriculture. The target environment for cereal breeders must now include management type, leading to varieties specifically suited to different management systems like conservation agriculture or organic agriculture. This perspective is causing renewed interest in traditional varieties, characterized by local adaptation, often under low-input conditions [34]. In addition, European Regulation 2018/848, which recognizes the need to have varieties adapted to organic farming, opens the door to the interest of using traditional varieties in this type of agriculture [35]. Other management alternatives aiming at increasing sustainability of agroecosystems are intercropping and regenerative agriculture. Both will impact the type of varieties better suited to each condition. Intercropping cereals with legumes or other crops can improve soil and crop nutrient status without external inputs and reduce herbicide use [36]. Regenerative agriculture puts the emphasis on soil health, often including the use of cattle in crop lands, by caring for the root-soil dialogue and its influence on the rhizosphere microbiota [37].

1.3 Which are the Tools to Deliver Future-Ready Varieties?

We will summarize the main tools in the breeder's toolbox, in terms of their impact on the breeder's genetic gain equation $R_t = \mathbf{i}\mathbf{r}\sigma_A/y$, where R_t is genetic gain over time, i is selection intensity, r is selection accuracy, σ_A is the genetic variance, and y is the years per breeding cycle. The following aspects are already affecting (or will affect) parts of the breeder's equation:

- *A wider use of germplasm banks* will increase the genetic variance (σ_A). One clear goal of cereal breeding is to widen the genetic base of the crops. Germplasm banks are the treasure trove of plant breeding. Some accessions are low-hanging fruits, with possible direct use, waiting for a reappraisal of their value under current societal demands. However, to find the hidden gems in the genomes, researchers must explore carefully among the thousands of accessions, with the use of smart mining techniques. Fortunately, the democratization of sequencing and genomics, coupled with increasingly efficient methods of gene delivery, are facilitating this daunting task [38]. Wild relatives, either stored in germplasm banks or in situ, are also an important resource for future breeding. Their use in breeding requires more effort than other resources such as landraces. However, they have proven to be very valuable, particularly for transferring resistance to many biotic stresses. The same can be expected for abiotic stresses, quality traits or even yield. The development of prebreeding material, such as introgression lines in elite backgrounds, will be key to transferring potential traits that may be needed in the medium term to new varieties adapted to emerging challenges.
- *Planning prebreeding initiatives* will allow the deployment of new diversity to increase the genetic variance *(σA)* available for breeding programs. This is an area where public–private collaboration is strategic [39, 40]. Where the private sector is the main responsible for delivering genetic innovations to farmers, there is a risk that breeding activities, with a necessarily short-term vision, will fail when confronted with large environmental shifts. These situations will occur as a result of the accumulation of small changes. This threat calls for a strategic vision. In several countries, public–private prebreeding programs have been set in place, aiming at exploring the germplasm and creating trait-focused prebreeding populations. What type of traits should be targeted in prebreeding programs? This depends on the expected climate in a timeframe that coincides with the delivery of prebreeding activities, between 10 and 30 years. Target identification and ideotyping (suggesting trait combinations fit to perform well in specific situations) are two areas of active research. The climate of the future is a moving target. We must rely on climate models to predict the environments that crops will encounter in the future. How will current varieties

perform under future conditions? This is an area of intersection between climate modeling, functional crop growth modeling, and genomic prediction, which is just starting to be explored [41, 42]. However, it is likely that it will shape the activity of plant prebreeders seeking to improve the selection accuracy ($\boldsymbol{r}$) for their future target environments.

- *High-throughput genotyping methods* allow for the rapid and efficient characterization and selection of genetic diversity. These methods, together with modern computing capacity, open the possibility of performing selection attending mainly to the genotype. Genomic selection, is already widely used in plant breeding, although methodological developments, mostly addressing a judicious use of genotype by environment interaction and population structure effects, are continuously being proposed. The main outcome of genomic selection is a higher selection intensity ($\boldsymbol{i}$), using larger populations of individuals. Selection accuracy is barely improved in crops but is compensated by the increase in population size.
- *Genetic modification*, either through transgenics or gene editing, will optimize selection accuracy ($\boldsymbol{r}$) by addressing genes with a known direct effect on the trait of interest. This technology could lead to the introgression of genes from related species, thereby increasing genetic variance ($\boldsymbol{\sigma_A}$), down-regulating or knocking-out of certain genes, and accelerating the development of plants with desirable traits ($\boldsymbol{y}$). Genome editing has been used to achieve important agronomic and quality traits in cereals [43]. An example of this is the development bread wheat with low gliadin content [44]. Challenges in this area include the development of tissue culture-independent transformation systems, genotype-independent regeneration systems, the discovery of more efficient "molecular scissors," and the development of base-editing or prime-editing technologies that allow for more precise insertion and reduce the potential for off-target editing [43, 45].
- *High-throughput phenotyping* methods can be used in plant breeding in a similar way to genomic prediction. However, in general, phenotyping activities aim at improving the prediction of the economic yield (in cereals, usually, grain weight) using faster and cheaper methods than measuring yield itself. Even with well-developed methods, phenotyping cannot compete with the speed and scale of genotyping for predictions. Phenotyping activities are essential to discover simple traits that are important in determining yield in target environments, thus increasing selection accuracy ($\boldsymbol{r}$).
- *DH technology* allows complete homozygosity for all loci to be reached in one generation vs the 6–8 generations of

self-pollination required by traditional breeding methods. Selection accuracy ($\boldsymbol{r}$) in breeding programs based on DH is significantly increased due to the absence of dominant effects, thereby facilitating the selection of favorable traits. Selection intensity ($\boldsymbol{i}$) can also be enhanced since the desired alleles are not lost in subsequent generations. Consequently, the time required for the development of new varieties or inbred lines is reduced by 4–5 years (reducing $\boldsymbol{y}$). One constraint of DH is the limited rounds of meiotic recombination, especially when applied to F1 or F2, which could lead to a reduction of genetic variance ($\boldsymbol{\sigma_A}$). However, this could be partially addressed by creating large populations of DH. On the other hand, DH lines fully meet the criteria of distinctness, uniformity, and stability which is required for obtaining plant variety protection [46, 47].

The use of DH is also highly relevant to genetic mapping, and QTLs analysis, as the homozygosity and immortality characteristics of DH populations allows a more accurate and effective phenotyping in early generations of the breeding cycle. The integration of DH with marker assisted selection is also a good approach for gene pyramiding, gene stacking and introgression of genes. Similarly, combining genomic resources and genomic selection with DH technology provides new opportunities to increase predictive accuracy, genetic gain ($\boldsymbol{R_t}$), and reduce $\boldsymbol{y}$ [48, 49]. In addition, the application of transgenesis or editing technologies, as well as mutagenesis to haploid cells enables the rapid fixation of the new genetic variability. Other uses of DH have been: prediction of the potential value of a cross in pre-breeding studies, G×E studies, and conservation of the genetic diversity in untapped germplasm and landraces [49, 50].

DH are being used in barley, wheat, maize, rice, and triticale breeding programs. However, new challenges to increase the efficiency of the protocols are: the development of more genotype-independent protocols to increase the embryogenesis rate and to reduce the albinism in in vitro methods; the identification of effective haploid inducer lines and phenotypic markers to discriminate haploid and diploid seeds in maize and other cereals; the reduction of genotype specificity and the identification of wild-relative species that could lead to a high percentage of high quality haploid embryos in wide crosses. Finally, the identification of a nontoxic and effective induction agent for genome doubling is mandatory to overcome one of the major bottlenecks in most species, regardless of the DH method used.

- *Speed breeding technique*, which led to the reduction of y, is the latest improvement to respond quickly to emerging needs. Up to six generations have been obtained in wheat and barley by using fully-enclosed, controlled-environment growth-chambers with an extended photoperiod (22 h light/2 h night) [51]. This methodology is particularly useful when combined with a marker-assisted selection strategy, as it allows rapid deployment and pyramiding of resistance genes to new pathogens or races not previously present in a target environment.

Cereal breeding will continue to be a strategic activity to support food safety policies worldwide. Agricultural productions depend on delivering locally adapted, productive, and quality cultivars that respond to the many challenges ahead. This will be achieved by cereal prebreeders and breeders with the access to the broad catalog of multidisciplinary tools available to ensure genetic progress, and with the increasingly deeper understanding of the biological processes involved in plant production.

Acknowledgments

All authors acknowledge funding from the Research Network on Resilient and Quality Cereals for Spanish Food Security (Red CERES, RED2022-134922-T) funded by the Agencia Estatal de Investigación, Ministerio de Ciencia, Innovación y Universidades, Spain. EEAD-CSIC authors funded by grant A08_23R (Government of Aragón).

References

1. Fischer RA, Byerlee D, Edmeades GO (2014) Crop yields and global food security: will yield increase continue to feed the world? ACIAR Monograph No. 158. Australian Centre for International Agricultural Research, Canberra. xxii + p 634
2. Schulthess AW, Kale SM, Liu F et al (2022) Genomics-informed prebreeding unlocks the diversity in genebanks for wheat improvement. Nat Genet 54:1544–1552. https://doi.org/10.1038/s41588-022-01189-7
3. Ferreira CS, Seifollahi-Aghmiuni S, Destouni G et al (2022) Soil degradation in the European Mediterranean region: processes, status and consequences. Sci Total Environ 805:150106. https://doi.org/10.1016/j.scitotenv.2021.150106
4. Horrigan L, Lawrence RS, Walker P (2002) How sustainable agriculture can address the environmental and human health harms of industrial agriculture. Environ Health Perspect 110:445–456
5. Babalola OO, Emmanuel OC, Adeleke BS et al (2021) Rhizosphere microbiome cooperations: strategies for sustainable crop production. Curr Microbiol 78:1069–1085. https://doi.org/10.1007/s00284-021-02375-2
6. Zhan C, Matsumoto H, Liu Y, Wang M (2022) Pathways to engineering the phyllosphere microbiome for sustainable crop production. Nat Food 3:997–1004. https://doi.org/10.1038/s43016-022-00636-2
7. Jansson C, Faiola C, Wingler A et al (2021) Crops for carbon farming. Front Plant Sci 12:636709. https://doi.org/10.3389/fpls.2021.636709
8. Furbank RT, Sharwood R, Estavillo GM et al (2020) Photons to food: genetic improvement of cereal crop photosynthesis. J Exp Bot 71:2226–2238. https://doi.org/10.1093/jxb/eraa077
9. Reynolds M, Foulkes J, Furbank R et al (2012) Achieving yield gains in wheat. Plant Cell Environ 35:1799–1823. https://doi.org/10.1111/j.1365-3040.2012.02588.x
10. Reynolds MP, Slafer GA, Foulkes JM et al (2022) A wiring diagram to integrate physiological traits of wheat yield potential. Nat Food 3:318–324. https://doi.org/10.1038/s43016-022-00512-z

11. Murchie EH, Reynolds M, Slafer GA et al (2023) A 'wiring diagram' for source strength traits impacting wheat yield potential. J Exp Bot 74:72–90. https://doi.org/10.1093/jxb/erac415
12. Mishra N, Jiang C, Chen L et al (2023) Achieving abiotic stress tolerance in plants through antioxidative defense mechanisms. Front Plant Sci 14:1110622. https://doi.org/10.3389/fpls.2023.1110622
13. Jones JD, Staskawicz BJ, Dangl JL (2024) The plant immune system: from discovery to deployment. Cell 187:2095–2116. https://doi.org/10.1016/j.cell.2024.03.045
14. Park RF, Boshoff WHP, Cabral AL et al (2022) Breeding oat for resistance to the crown rust pathogen *Puccinia coronata* f. sp. *avenae*: achievements and prospects. Theor Appl Genet 135:3709–3734. https://doi.org/10.1007/s00122-022-04121-z
15. Atienza SG, Faccioli P, Perrotta G et al (2004) Large scale analysis of transcripts abundance in barley subjected to several single and combined abiotic stress conditions. Plant Sci 167:1359–1365. https://doi.org/10.1016/j.plantsci.2004.07.006
16. Peck S, Mittler R (2020) Plant signaling in biotic and abiotic stress. J Exp Bot 71:1649–1651. https://doi.org/10.1093/jxb/eraa051
17. Sánchez-Martín J, Montilla-Bascón G, Mur LAJ et al (2016) Compromised photosynthetic electron flow and H_2O_2 generation correlate with genotype-specific stomatal dysfunctions during resistance against powdery mildew in oats. Front. Plant Sci 8:1660. https://doi.org/10.3389/fpls.2016.01660
18. Millet EJ, Kruijer W, Coupel-Ledru A et al. (2019) Genomic prediction of maize yield across European environmental conditions. Nat Genet 51:952–956. https://doi.org/10.1038/s41588-019-0414-y
19. Canales FJ, Montilla-Bascón G, Gallego-Sánchez LM et al (2021) Deciphering main climate and edaphic components driving oat adaptation to Mediterranean environments. Front Plant Sci 12:780562. https://doi.org/10.3389/fpls.2021.780562
20. Rispail N, Montilla-Bascón G, Sánchez-Martín J et al (2018) Multi-environmental trials reveal genetic plasticity of oat agronomic traits associated with climate variable changes. Front Plant Sci 9:1358. https://doi.org/10.3389/fpls.2018.01358
21. Revilla P, Alves ML, Andelković V et al (2022) Traditional foods from maize (*Zea mays* L.) in Europe. Front Nutr 8:1235. https://doi.org/10.3389/fnut.2021.683399
22. Balk J, Connorton JM, Wan Y et al (2019) Improving wheat as a source of iron and zinc for global nutrition. Nutr Bull 44(1):53–59. https://doi.org/10.1111/nbu.1236
23. EFSA Panel on Dietetic Products, Nutrition and Allergies (NDA) (2010) Scientific opinion on the substantiation of a health claim related to oat beta-glucan and lowering blood cholesterol and reduced risk of (coronary) heart disease pursuant to Article 14 of Regulation (EC) No 1924/2006. EFSA J 2010 8:1885. https://doi.org/10.2903/j.efsa.2010.1885
24. EFSA Panel on Dietetic Products, Nutrition and Allergies (NDA) (2011) Scientific opinion on the substantiation of a health claim related to barley beta-glucan and lowering of blood cholesterol and reduced risk of(coronary) heart disease pursuant to Article 14 of Regulation (EC). EFSA J 9(No 1924/2006):2470. https://doi.org/10.2903/j.efsa.2011.2470
25. Dionne J, Ford AC, Yuan Y et al (2018) A systematic review and meta-analysis evaluating the efficacy of a gluten-free diet and a low FODMAPs diet in treating symptoms of irritable bowel syndrome. Am J Gastroenterol 113(9):1290–1300. https://doi.org/10.1038/s41395-018-0195-4
26. Blum A (2009) Effective use of water (EUW) and not water-use efficiency (WUE) is the target of crop yield improvement under drought stress. Field Crops Res 112(2–3):119–123. https://doi.org/10.1016/j.fcr.2009.03.009
27. Vadez V, Grondin A, Chenu K et al (2024) Crop traits and production under drought. Nat Rev Earth Environ 5:211–225. https://doi.org/10.1038/s43017-023-00514-w
28. Lynch JP (2022) Harnessing root architecture to address global challenges. Plant J 109(2):415–431. https://doi.org/10.1111/tpj.15560
29. Subbarao GV, Kishii M, Bozal-Leorri A et al (2021) Enlisting wild grass genes to combat nitrification in wheat farming: A nature-based solution. Proc Natl Acad Sci U S A 118:e2106595118. https://doi.org/10.1073/pnas.2106595118
30. Wu K, Wang S, Song W et al (2020) Enhanced sustainable green revolution yield via nitrogen-responsive chromatin modulation in rice. Science 367:6478. https://doi.org/10.1126/science.aaz2046
31. Trivedi P, Leach JE, Tringe SG et al (2020) Plant–microbiome interactions: from community assembly to plant health. Nat Rev Microbiol 18:607–621. https://doi.org/10.1038/s41579-020-0412-1
32. Hu L, Robert CAM, Cadot S et al (2018) Root exudate metabolites drive plant-soil feedbacks on growth and defense by shaping the rhizosphere microbiota. Nat Commun 9:2738. https://doi.org/10.1038/s41467-018-05122-7

33. Escudero-Martinez C, Coulter M, Alegria Terrazas R et al (2022) Identifying plant genes shaping microbiota composition in the barley rhizosphere. Nat Commun 13:3443. https://doi.org/10.1038/s41467-022-31022-y
34. Pascual L, Ruiz M, López-Fernández M et al (2020) Genomic analysis of Spanish wheat landraces reveals their variability and potential for breeding. BMC Genom 21:122. https://doi.org/10.1186/s12864-020-6536-x
35. Thanopoulos R, Negri V, Pinheiro de Carvalho MAA et al (2024) Landrace legislation in the world: status and perspectives with emphasis in EU system. Genet Resour Crop Evol 71:957–997. https://doi.org/10.1007/s10722-023-01824-0
36. Jensen ES, Carlsson G, Hauggaard-Nielsen H (2020) Intercropping of grain legumes and cereals improves the use of soil N resources and reduces the requirement for synthetic fertilizer N: a global-scale analysis. Agron Sustain Dev 40:5. https://doi.org/10.1007/s13593-020-0607-x
37. Mooney SJ, Castrillo G, Cooper HV et al (2024) Root–soil–microbiome management is key to the success of regenerative agriculture. Nat Food 5:451–453. https://doi.org/10.1038/s43016-024-01001-138
38. Mascher M, Schreiber M, Scholz U et al (2019) Genebank genomics bridges the gap between the conservation of crop diversity and plant breeding. Nat Genet 51:1076–1081. https://doi.org/10.1038/s41588-019-0443-6
39. Cheng S, Feng C, Wingen LU et al (2024) Harnessing landrace diversity empowers wheat breeding. Nature 632:823–831. https://doi.org/10.1038/s41586-024-07682-9
40. Yuan Z, Rembe M, Mascher M et al (2024) Capitalizing on genebank core collections for rare and novel disease resistance loci to enhance barley resilience. J Exp Bot 75:5940–5954. https://doi.org/10.1093/jxb/erae283
41. Paleari L, Li T, Yang Y et al (2022) A trait-based model ensemble approach to design rice plant types for future climate. Glob Chang Biol 28:2689–2710. https://doi.org/10.1111/gcb.16087
42. Yang Y, Wilson LT, Li T et al (2022) Integration of genomics with crop modeling for predicting rice days to flowering: a multi-model analysis. Field Crops Res 276:108394. https://doi.org/10.1016/j.fcr.2021.108394
43. Matres JM, Hilscher J, Datta A et al (2021) Genome editing in cereal crops: an overview. Transgenic Res 30:461–498. https://doi.org/10.1007/s11248-021-00259-6
44. Sánchez-León S, Marín-Sanz M, Guzmán-López MH et al (2024) CRISPR/Cas9-mediated multiplex gene editing of gamma and omega gliadins: paving the way for gliadin-free wheat. J Exp Bot 75:erae376. https://doi.org/10.1093/jxb/erae376
45. Ahmar S, Hensel G, Gruszka D (2023) CRISPR/Cas9-mediated genome editing techniques and new breeding strategies in cereals–current status, improvements, and perspectives. Biotechnol Adv 108248:43. https://doi.org/10.1016/j.biotechadv.2023.108248
46. Maqbool MA, Beshir A, Khokhar ES (2020) Doubled haploids in maize: development, deployment, and challenges. Crop Sci 60(6):2815–2840. https://doi.org/10.1002/csc2.20261
47. Patial M, Chauhan R, Chaudhary HK et al (2022) Au-courant and novel technologies for efficient doubled haploid development in barley (*Hordeum vulgare* L.). Crit Rev Biotechnol 43(4):575–593. https://doi.org/10.1080/07388551.2022.2050181
48. Weyen J (2021) Applications of doubled haploids in plant breeding and applied research. In: Segui-Simarro JM (Ed.)Eds. Doubled haploid technology. Methods in molecular biology. Humana, New York, 23–39. https://doi.org/10.1007/978-1-0716-1315-3_2
49. Hale B, Ferrie AM, Chellamma S et al (2022) Androgenesis-based doubled haploidy: past, present, and future perspectives. Front Plant Sci 12:751230. https://doi.org/10.3389/fpls.2021.751230
50. Singhal IRK, Meena VK, Saroj R et al (2023) Doubled-haploid technology in maize (Zea mays L.) and its practical implications in modern agriculture. In: Sharma D, Singh S, Sharma SK, Singh R (Eds.)Eds. Smart plant breeding for field crops in post-genomics era. Springer, Singapore, 195–219. https://doi.org/10.1007/978-981-19-8218-7_6
51. Watson A, Ghosh S, Williams MJ et al (2018) Speed breeding is a powerful tool to accelerate crop research and breeding. Nat Plants 4:23–29. https://doi.org/10.1038/s41477-017-0083-8

Chapter 2

Molecular Insights into Microspore Embryogenesis: Transcriptomic Dynamics in Barley and Wheat

María Pilar Vallés, Begoña Echávarri, Isabel Valero-Rubira and Ana María Castillo

Abstract

The development of efficient protocols for doubled haploid (DH) production in cereals has long been a major focus of plant biotechnology research. Most efforts have concentrated on in vitro microspore embryogenesis (ME) systems, which provide a powerful route for the rapid generation of homozygous lines. To optimize these systems, numerous empirical, trial-and-error approaches have been employed to evaluate the effects of genotype, stress treatments, culture media composition, and mother-plant growth and culture conditions on the efficiency of DH production. These strategies have been essential, given the limited understanding of the cellular and molecular mechanisms underlying ME. However, recent advances in high-throughput transcriptomic technologies, particularly RNA sequencing (RNA-seq), have substantially expanded our insight into the regulatory networks involved in cellular reprogramming and developmental transitions during ME. This review summarizes the most recent findings on the transcriptional dynamics of ME in barley and wheat, aiming to provide comprehensive insights into the molecular mechanisms involved, thereby facilitating the development of more effective strategies for DH production.

Keywords Cereals, Developmental reprogramming, Microspore embryogenesis, Stress response, Early embryogenesis, Transcriptomics, RNA-seq

1 Introduction

Microspore embryogenesis (ME) is an attractive example of cellular reprogramming, in which microspores, upon exposure to stress, adapt to adverse environmental conditions by altering their developmental trajectory to ensure survival and continued growth. In this process, microspores escape their default gametophytic pathway and instead initiate a sporophytic development, forming embryos, embryo-like structures (ELS), or callus [1]. This reprogramming ultimately results in the generation of haploid or DH plants, either through spontaneous chromosome doubling or via the application of chromosome doubling agents (for review, see [2]). DH plants

Ana María Castillo et al. (eds.), *Doubled Haploid Methods in Cereals*, Methods in Molecular Biology, vol. 3029,
https://doi.org/10.1007/978-1-0716-5245-9_2,

have become a valuable tool in plant breeding and genetics, as they enable the rapid fixation of genetic variability by achieving complete homozygosity within a single generation [3].

The first successful report of anther culture, described in *Datura innoxia* Mill. [4], opened new avenues in plant biotechnology. Since then, numerous protocols for ME via both anther and isolated microspore culture have been established across a wide range of plant species (for review, see [5, 6]). However, these protocols have not always resulted in successful plant regeneration, and when regeneration is achieved, efficiency is often too low for routine use or large-scale applications.

Following the first reports of DH production in barley [7] and wheat [8], both species have become model systems for studying ME in cereals [9]. The highest regeneration efficiencies are obtained with microspores at the mid-late to late uninucleate stage subjected to a temperature shock (cold or heat), carbohydrate starvation, or a combination of stress and chemical inducers. Genotype dependency and the frequent occurrence of albinism remain major obstacles ([10], see Chap. 3). Additionally, the requirement for ovary co-culture and the low frequency of spontaneous chromosome doubling represent further challenges for ME, particularly in wheat [2, 11, 12]. The book in which this chapter appears provides a comprehensive overview of the current state of ME protocols and their successful application in both barley and wheat (see Chaps. 8–12).

The optimization of ME protocols has traditionally relied on trial-and-error approaches due to limited insight into the cellular and molecular processes involved. Parallel efforts have aimed to identify key regulatory elements governing ME induction and early embryogenic development. Progress in this field has been accelerated by advanced gene function analysis tools, which have facilitated the exploration of molecular pathways driving cellular reprogramming.

Early studies employed differential hybridization screening to compare gene expression between young microspore-derived embryoids and mature pollen in wheat [13, 14], as well as between induced and non-induced microspores in barley [15]. These analyses identified a limited subset of differentially expressed genes associated with ME induction and the initial stages of embryogenic development. Early functional genomics studies employing macroarray and microarray technologies facilitated large-scale analysis of expressed sequence tags (ESTs), primarily in barley, which benefited from the most advanced genomic resources available at the time. However, these studies were limited by incomplete sequencing of the barley genome and the restricted gene representation on the arrays. Focusing on mannitol-induced stress treatment, Maraschin et al. [16] utilized a 1.4k macroarray derived from a cDNA library encompassing the first 15 days of seed development [17] to analyze gene expression in isolated microspores. On the other hand, Muñoz-Amatriaín et al. [18] and Sánchez-Díaz [19] utilized the commercial Barley1

GeneChip microarray (22k), which contains transcripts from 84 cDNA libraries representing a broad range of developmental stages, as well as responses to abiotic and biotic stress treatments [20], to conduct the analyses on anther culture. These studies partially elucidated stress-responsive mechanisms involved in ME, revealing that the broad transcriptional changes observed are consistent with those reported in the model species *Brassica* [21–24], suggesting the existence of conserved regulatory patterns across species.

While advances in high-throughput transcriptomics have greatly accelerated the analysis of transcriptome landscapes, the use of bulk RNA sequencing (RNA-Seq) in ME in cereals remains limited. Two pioneering studies by Seifert et al. [25] in wheat and Bélanger et al. [26] in barley provide early insights into transcriptomic changes in isolated microspores during stress response and early embryogenic development in highly responsive cultivars. Significantly, these studies highlighted the epigenetic regulation of the process, emphasizing the role of chromatin remodeling in cellular reprogramming.

In recent years, the integration of genome sequencing technologies with advances in high-resolution transcriptome analysis via RNA-seq has markedly improved our ability to study complex biological systems. Although the available literature on ME in cereals remains limited—comprising a single publication on barley [27], and a master´s thesis and a doctoral dissertation on wheat [28, 29]—these studies have nonetheless introduced innovative analytical approaches. They have explored the differential responsiveness of cultivars and the role of inductive treatments in ME, such as the application of the histone deacetylase inhibitor trichostatin A (TSA). These transcriptional studies have contributed to clarifying the complexity of the mechanisms involved, identifying novel regulatory components, and validating previously characterized elements.

This review presents the current state of knowledge on the transcriptional landscape of ME in barley and wheat. To provide a comprehensive overview, we synthesize the available transcriptomic data. Transcriptomic analyses are typically aligned with discrete stages of ME; however, the timing of transcriptional changes may not strictly correspond to the protocol-defined stages. Despite these limitations, this review aims to contribute to a deeper understanding of the complex reprogramming processes that microspores undergo in response to stress, ultimately resulting in embryogenic development.

2 Transcriptomic Complexity Dynamics During Microspore Embryogenesis

In both barley and wheat, analyses of microspore transcript abundance and diversity show that the majority of genes are transcribed across all stages of ME. However, significant differences in transcript profiles are identified, reflecting the combined influence of stress-inducing treatments and the distinct phases of the emerging developmental program [25–29]. While the transition from the uninucleate microspore to binuclear pollen is associated with a reduction in transcript abundance [29], ME-inductive stress treatments consistently lead to a substantial increase in the number of transcripts [25, 27, 29], most notably during the early phase of stress exposure [29]. In contrast, during the first hours of culture, coinciding with the initiation of embryogenic development, transcript abundance declines compared to the end of the stress treatment [25, 28, 29]. However, a continuous increase in transcript number throughout the ME process has also been reported by Belanger et al. [26]. Interestingly, regarding stage-specific gene expression, the highest number of uniquely expressed genes is consistently recorded during the early stages of culture, although the overall numbers remained low [25–27].

Transcript diversity also undergoes significant changes throughout stress treatment and subsequent culture stages. While the majority of genes belong to the very low or low expression categories, the most pronounced changes occur among medium- and highly expressed genes [27, 29], likely reflecting transitions associated with the activation of specialized cellular processes.

To further elucidate transcriptional dynamics during ME, comparative analyses have been conducted to identify differentially expressed genes (DEGs) at different time points within a developmental stage, between developmental stages, and across induction treatments and cultivars. Notably, a substantial proportion of genes were not differentially expressed between stages within a stress treatment (76.5%) [26] or between treatments (62.3%) [28]. Although notable discrepancies exist among studies, they are likely attributable to the differing experimental conditions.

Comparative analyses of stage transitions during ME have yielded conflicting results regarding which transition involves the most significant differential changes in DEG numbers. Early studies pointed to the initiation of embryogenic development as the stage with the most pronounced transcriptional shifts [25]. In contrast, Belanger et al. [26] reported comparable changes between stress induction and embryogenesis initiation. More recent studies, however, highlight stress treatment as the primary factor responsible for the highest number of DEGs [19, 28, 29]. Furthermore, within the stress treatment, early responses induce more pronounced transcriptional changes than those observed in prolonged stress. At the earliest stages of culture, the number of DEGs remains relatively low compared to levels observed at the end of stress [29]. However, as embryogenic development progresses, a marked increase in DEG number becomes evident [28].

Considerable discrepancies exist among studies regarding the number of DEGs that are up- or downregulated during the transitions from non-stressed uninucleate to stress-treated microspores, and to microspores at the culture stage. Muñoz-Amatriaín et al. [18] show that most DEGs are downregulated during the stress transition, a pattern also observed across both transitions by Seifert et al. [25]. Similarly, Sánchez-Díaz [19] reported a predominance of downregulated DEGs during the stress transition, followed by upregulation after culture initiation. In contrast, other studies have described a higher number of upregulated genes compared to downregulated ones in both transitions [16, 26–29].

Other studies have examined transcriptional differences associated with cultivar responses or stress treatments. Notably, the cultivar with the highest embryogenic response showed a greater number of genes transcribed across all developmental stages, with the most pronounced differences occurring during the transition from culture initiation to later stages of embryogenic development [27]. Regarding inductive treatments, the application of TSA, known to enhance ME efficiency [30, 31], was generally associated with only minor transcriptomic changes between treated and untreated samples across all stages. Nevertheless, TSA led to an increase in transcript abundance either at the end of the stress treatment [29] or during the early hours of culture [28].

3 The Transcriptomic Signature of the Uninucleate Microspore: Implications for Microspore Embryogenesis

The uninucleate microspore representsmicrospore represents a critical stage in the transition from sporophytic to gametophytic development, marked by the occurrence of pollen mitosis I (PMI). Despite its central role in pollen development, this stage remains one of the least studied [32].

In the context of ME, the transcriptional profile of the uninucleate microspore has been relatively understudied and is typically considered only as the baseline for induction. However, a key aspect of microspore reprogramming involves the silencing of gametophytic developmental regulators that are active in uninucleate microspores, including genes involved in cell cycle progression and pollen development [33]. Accordingly, transcriptional studies have shown that genes downregulated in response to stress and maintaining low expression throughout the ME process include those encoding proteins associated with chromatin organization, such as nucleosome assembly factors, as well as components of the translational machinery, including ribosomal proteins and fibrillarin [16, 26]. In addition, genes encoding storage proteins (*Hv11SGSS*) and cell wall modifiers (*HvGGL*) are also downregulated, suggesting a redirection of energy and cellular metabolism [27]. Furthermore, the significant downregulation of EIL, ARF, and Aux/IAA transcription factors (TFs) is consistent with broad reprogramming [18].

In wheat, genes involved in sugar metabolism and the deregulation of pollen development are also downregulated during ME, with "chloroplast RNA processing" and "transmembrane transport" identified as the most significantly enriched Gene Ontology (GO) terms [29]. Notably, key genes associated with pollen development, such as *TaYAO*, *TaNFD6-A*, *TabZIPF1*, and *TaAGO802*, exhibit a more rapid downregulation in the highly responsive cultivar [34]. However, the mannitol combined with TSA treatment results in delayed downregulation of some of these genes, consistent with TSA's interference in their transcriptional silencing, which may be modulated by histone deacetylase TaHDT1 [34].

Interestingly, the downregulation of the GO term "embryo development", and in particular the repression of the *SERK1* gene, known to be involved in embryogenesis, indicates that these genes may also play roles in male gametophyte development [25]. Likewise, the downregulation of stress-related GO terms such as "oxidation-reduction process", "response to abiotic stimulus", "osmotic stress", and "pollen dehydration" [18, 25] may reflect the presence of a stage-specific protective mechanism in the uninucleate microspores, which are particularly sensitive to environmental stress.

Studies have also identified additional patterns among downregulated genes, suggesting the involvement of diverse regulatory mechanisms. A limited number of transcripts exhibit progressive downregulation during both stress treatment and the initiation of embryogenic development. These transcripts are primarily associated with catalytic activities, including the generation of precursor metabolites and the metabolism of lipids and nucleic acids, which are required to maintain cellular dedifferentiation [25, 28]. Another group of genes, only marginally affected by stress but strongly downregulated at culture initiation, is enriched in GO terms such as "protein metabolic process", "proteolysis involved in cellular processes", "Golgi apparatus", and "proteasome core complex". This pattern suggests the specific targeted degradation of gametophytic cell fate-associated proteins, coinciding with the initiation of embryogenic development [25].

Another distinct transcriptional pattern is characterized by an initial repression followed by upregulation upon culture initiation. In barley, for example, 46.2% of TFs showed this profile [35]. Genes following this pattern are enriched in GO terms related to nuclear processes, likely associated with cell cycle reactivation, including "nucleic acid metabolic process", "nuclear transport", "cytokinesis", "chromatin and chromosome organization", and "gene silencing" [26, 28, 29]. Notably, this group includes genes homologous to members of the rice minichromosome maintenance protein complex (MCM) [26, 28].

To the best of our knowledge, only two studies have directly compared the transcriptomes of uninucleate microspores and bicellular pollen within the context of ME [16, 29]. Many transcripts repressed by ME-inductive stress—particularly those involved in sucrose, lipid, and nucleic acid metabolism, as well as storage of metabolic reserves—are also repressed during normal gametophytic development [29]. Similarly, genes associated with biosynthetic activity and cell division at the initiation of mitosis are highly downregulated [16]. These findings support the hypothesis that there is an overlap in gene expression requirements between the gametophytic and embryogenic developmental pathways [16].

Recent studies have emphasized the critical role of both the transcriptional and epigenetic states of uninucleate microspores as key determinants of genotypic responsiveness to ME [27, 34]. In wheat, the most responsive cultivar displays the highest expression of early microspore developmental genes, such as *TaMS1* and *Ta-2B-LBD16*, indicating that a transcriptionally less advanced state favors reprogramming [34]. Similarly, microspores from the most responsive barley cultivar show significant enrichment in chromatin organization, DNA replication, translation, and protein catabolism. Notably, this includes the upregulation of genes involved in ubiquitin-dependent proteolysis (*HvFBPs*) and redox homeostasis (*HvKAT*, *HvAKR*, and *HvCOX*) [27].

Consequently, careful consideration must be given to the initial physiological and molecular state of the microspore prior to induction, as well as to the interactions between the gametophytic and embryogenic developmental programs.

4 Microspore Stress Responses: From Survival to Reprogramming

In barley and wheat, ME is commonly induced by exposing microspores to low temperatures or osmotic shock combined with nutrient deprivation by incubation in mannitol (see Chaps. 8, 10, and 11). In recent years, combining these stress treatments with specific compounds has significantly enhanced ME efficiency in cereals [30, 31], (see Chaps. 16 and 17). The success of these induction protocols relies on achieving a delicate balance between eliciting a sufficient stress response, preserving microspore viability, and promoting developmental reprogramming. Although treatment efficiency, genotype responsiveness, and induction kinetics vary among systems, the underlying mechanisms are likely to be common to all of them.

The first transcriptomic studies in wheat and barley ME primarily focused on comparing the transcriptional profiles of microspores before and after the inductive stress treatment [16, 18]. Subsequent studies, however, suggest that distinct expression profiles also emerge at different stages of stress treatment [29, 36, 37]. Moreover, there is a partial overlap between transcriptional changes that constitute a general stress response and those associated with adaptive mechanisms leading to developmental reprogramming, which are also present at culture initiation [27–29]. These complexities may confound transcriptomic analyses and should be carefully considered in future research.

The stress response can be divided into several phases, beginning with the perception of stress signals arising from alterations in cellular homeostasis largely mediated by reactive oxygen species (ROS) [38].

In ME, the redox status of microspores during stress treatment significantly influences induction efficiency, requiring an optimal balance between ROS generation and protection against oxidative damage [39–41]. Furthermore, it has been suggested that each species and genotype has a specific ROS threshold necessary for reprogramming, conditioned by the physiological state of microspores and the type of stress applied [39]. Induction strategies, including exogenous ROS scavengers, have been shown to improve ME efficiency [40, 42, and references therein]. Therefore, a well-coordinated antioxidant system, including both enzymatic and non-enzymatic (low-molecular-weight) antioxidants, as well as components of the ascorbate-glutathione cycle, is crucial for effective ME.

Changes in the expression of genes related to oxidative homeostasis were first reported in barley, including activation of GST genes involved in glutathione redox regulation [15]. GSTs are among the earliest antioxidant-related genes activated by mannitol treatment in both barley and wheat, although minor differences in expression have been observed between cultivars differing in ME response [36, 37]. Transcriptomic analyses have produced conflicting results regarding the role of GSTs in ME, likely due to the diverse stress responses of the GST gene superfamily [43]. For example, high concentrations of mannitol has been shown to both upregulate and downregulate different GST genes simultaneously [18], whereas a lower concentration consistently upregulates them [26]. It has been hypothesized that GST gene activation may reflect a general stress response rather than specific association with ME induction [16]. Conversely, studies comparing near-isogenic barley lines suggested a positive correlation between GST expression and ME efficiency [44]. Additionally, GST genes exhibited heterogeneous expression profiles, with peaks occurring at different stages (Vallés, unpublished data).

Recent transcriptome studies have confirmed the GO enrichment for oxidative stress-related terms in response to both cold and mannitol treatments in wheat [25, 28, 29]. Comparisons across genotypes and induction treatments further indicate that higher ME efficiency is associated with elevated expression of genes involved in ROS regulation, including *NQR*, *NDA1*, *cytochrome P450*, *Cysteine-rich peptides* (*CRPs*), and *ribA* [27, 44].

Stress signaling involves various second messengers, including calcium, which plays a key role in early stress perception and response. In ME induction, the addition of calcium—either alone or in combination with with n-butanol—has been shown to enhance ME efficiency in barley and wheat [37, 45–48]. However, the precise role of calcium throughout the ME process remains unclear [49]. In wheat, cold treatment has been associated with an increase in GO terms linked to calcium-dependent protein serine/threonine kinase activity, calmodulin binding, and calmodulin-dependent protein kinase activity. Additionally, the *PARP* gene, known to be activated by calcium in response to oxidative stress, has been identified in stressed microspores [28]. Moreover, expression of over 24 calmodulin or calmodulin-related genes peaks at the end of mannitol treatment (Vallés, unpublished data).

Stress significantly alters metabolic pathways to meet energy demands, access resources, and produce metabolites [50]. Biochemical, proteomic, and metabolomic analyses have been used to investigate the metabolic changes associated with ME-inducing treatments in cereals. Enhanced ME efficiency involves a metabolic shift from photosynthesis-related anabolism to catabolic pathways that mobilize stored carbon sources (starch, fatty acids, and glutamine), meeting increased energy demands and maintaining protein quality [51–53]. In wheat, recent integrative analysis of metabolomic and transcriptomic data has confirmed that energy production, primarily sustained on starch degradation, is essential in the stress response [54].

Transcriptomic data are largely consistent with proteomic and metabolomic findings, revealing the downregulation of starch biosynthesis genes, associated with the arrest of gametophytic development, and the induction of genes involved in sugar hydrolysis and starch degradation [16, 18]. Numerous genes have been identified, including those encoding maltase, GAPD, CWI, and FBA, with particular emphasis on β-amylase and vacuolar invertase [16, 18, 28]. Although starch biosynthesis genes such as *SSI*, *SBEII*, and *GLGC* have been implicated, their differential expression may reflect the presence of microspores accumulating starch prior to cell death, and/or the high incidence of albinism [18, 28]. Comparative studies across barley lines also confirm that ME efficiency is associated with active degradation of carbohydrates and lipids, resulting in increased sugar availability. Key genes include those encoding β-D-xylosidase, α-galactosidase, lipase, and sucrose transporters [44]. Moreover, the highly efficient mannitol plus TSA treatment led to an additional enrichment of genes related to carbohydrate metabolism, particularly within the "polysaccharide catabolic process" category [29]. Genes involved in energy metabolism have also been identified, including those encoding oxidoreductases, peptide transporters, enolase, and alcohol dehydrogenase [27, 44]. Interestingly, in wheat, differences between early and late responses to mannitol treatment are observed with respect to secondary metabolites [29]. The early response involves targeted degradation, likely contributing to resource recycling, metabolite homeostasis, and initial stress adaptation, whereas prolonged treatment induces the synthesis of secondary metabolites implicated in stress signaling and the regulation of intracellular pH.

The functional regulation of primary metabolites and plant growth regulators (PGRs) is closely interconnected to coordinating stress responses and developmental reprogramming [50]. In cereals, extensive research on ME has highlighted the role of auxins, citokinins, gibberellins, and abscisic acid (ABA), and the importance of maintaining their homeostasis [55, 56 and references therein]. While auxins and cytokinins have been consistently shown to play key roles, the involvement of ABA remains uncertain [57].

ABA initiates signaling cascades that regulate numerous processes and increase microspore viability [55] and references therein]. Recent studies propose that a genotype-specific ABA threshold may be critical for ME induction, with low ABA-mediated responses linked to reduced efficiency in winter wheat [57]. Transcriptomic analyses exhibit generally low expression of genes involved in ABA biosynthesis, degradation, and signaling [19]. After mannitol treatment, ABA biosynthetic genes are not significantly upregulated [18], though targeted analysis of *AO5a* and *ABA8'OH* genes confirms *de novo* ABA synthesis [19]. ABA signaling TFs also show increased expression, but with distinct dynamics in mannitol treatment: *ABI1b* peaks at mid-treatment, while *ABI5* remains elevated at later stress stages [19, 29]. Also, positive regulators such as *PKABA1* are upregulated, while negative regulators (*ROP6*, *PP2C*, *CIPK3*, and *SAD*) are downregulated [18]. Additionally, ABA-responsive genes, including those encoding LEA proteins and ADH3, are upregulated during stress [16, 18, 27]. Nonetheless, no significant differences in *ABI5* or *ADH3* expression were observed between cultivars with differing ME responses [27, 34].

Although most studies on the role of auxins in ME have focused on early stages of embryogenic development, their involvement during stress treatments has also been explored. While cold treatment has minimal impact on auxin levels, mannitol stimulates auxin biosynthesis and accumulation [57]. However, a balanced IAA/cytokinin ratio is essential, as excessive auxin accumulation may lead to embryo morphological abnormalities [56–58]. Interestingly, genes involved in auxin biosynthesis (*TaYUC9-D1*), transport (*AUX1*), and signaling (*SAUR*) are upregulated by cold or mannitol treatments [28, 29]. Notably, mannitol combined with TSA leads to increased expression of these genes and other auxin-related genes, including *TaYUC11D*, *TaGH3-3.11*, *NIT-6B-2*, *ILL-7A*, and *Ta-2B-LBD16*, which are involved in auxin biosynthesis, homeostasis, and signaling pathways [29, 34]. These results highlight the critical role of auxin regulation in successful ME induction.

In the context of protein biosynthesis, mechanisms ensuring proper folding and degradation of damaged or misfolded proteins are directly associated with the efficiency of ME [53]. The ubiquitin–proteasome system represents the primary pathway for selective protein degradation in plants, supported by various peptidases such as cysteine (CPs), serine (SPs), aspartic (APs), and metalloproteases (FtSH), which play essential roles in stress adaptation [59]. In barley and wheat, stress treatment upregulates genes encoding proteasome subunits (20S and 26S), assembling factors (RPT4 and PSMD10), and ubiquitin ligases (UB-E3) that mediate protein targeting for degradation. Moreover, several genes encoding proteases, including CPs and APs, CATHB2, CBP1, APM1 and FtSH, are also induced

[16, 18, 26, 28]. However, comparative analyses between optimal and suboptimal treatments, or cultivars with contrasting responses, suggest that only specific changes mediated by the 26S proteasome are associated with the acquisition of ME competence [16, 44].

Processes associated with the ubiquitin–proteasome system, such as autophagy and plant senescence, contribute to nutrient recycling as part of a pro-survival strategy. However, autophagy plays a dual role, either promoting cell survival or triggering programmed cell death (PCD) [60]. In ME, many microspores fail to achieve a proper balance between stress response and survival after stress treatment and during the first hours of culture. Stress treatment increases caspase-3-like activity and cell death [61]. Notably, the application of specific compounds that inhibit apoptosis, autophagy, and cysteine protease (CP) activity after stress treatment enhances ME efficiency [61–66]. Accordingly, transcriptomic analysis data confirm the activation of autophagy and PCD processes in response to stress, as revealed by the upregulation of autophagy-related genes (*HvATG5* and *HvATG6*), cathepsin-like genes (*HvPap-1* and *HvPap-6*), PCD regulators (*BI-1*), and *Poly(ADP-ribose) Polymerases* and LSD TF family [16, 28, 29, 44, 64]. In addition, low-responding cultivars exhibit higher expression of the autophagy *TaATG18fD* or the senescence-related *SAG102* genes suggesting that excessive activation of degradation and senescence pathways may hinder the acquisition of embryogenic competence [34, 44].

Proteins associated with stress tolerance, including small signaling peptides, redox regulators, and cell wall remodeling enzymes, have been implicated in stress response during ME [67]. Among these, the heat shock HSP90 is considered exclusively related to stress response, whereas HSP70 is likely involved in triticale ME induction [52]. Transcriptomic studies further reveal that most *HSPs* and *HSFs* are constitutively expressed after mannitol treatment [18], while genes encoding HSP70, HSP20, and an HSF are upregulated in cold treatment [26, 28]. Notably, expression of *HSC70* and *HSP81-2* correlates positively with the number of green plants produced, suggesting a potential role of specific HSPs in enhancing ME efficiency [44].

A potential role of pathogenesis-related (PR) proteins in ME induction has been proposed after their identification among the molecules secreted in maize embryogenic cultures [68]. In barley, mannitol treatment upregulates several PR genes, including *PDF2.4*, a basic endochitinase, *PR-1*, and *MLO-1* [18]. Notably, the expression of *PR-2*, *ELI3-1*, *osmotin-like proteins*, and a *non-specific LTP* correlates with ME efficiency [44]. In addition, genes encoding PR proteins (Barwin, PR4, glucanase, PR1, and PR10) are strongly stress-induced, though with species- and genotype-specific patterns [36, 37]. Among these, PR4, which is likely involved in the hydrolysis of arabinogalactan proteins (AGPs), exhibits the highest level of induction. Recent studies have shown that the addition of exogenous

AGPs to the stress treatment significantly enhances the efficiency of green plant regeneration, supporting their direct role in barley ME induction [69]. Additional genes associated with membrane and cell wall modification have been identified in wheat, including *Cer1* and *TaMS*1, which are involved in cuticular wax biosynthesis and pollen exine development, respectively [25, 28, 34].

When exposed to stress, cells must establish a balance between growth and development versus survival. During stress treatments, many microspores re-enter the cell cycle, progressing through different phases or even undergoing nuclear division with dynamics that vary depending on the specific treatment [70]. These results contrast with transcriptomic analyses, which report no significant differences in the expression of core cell cycle genes [16, 18, 28]. However, a GO term related to the cell cycle, including a subunit of the anaphase-promoting complex (HvAPC/C), is activated during stress treatment in barley [27]. Interestingly, in wheat, genes involved in DNA replication, transcription, and chromatin organization, including histone H3 genes, are downregulated at early stress and reactivated later, resulting in non-significant differences at the end of a mannitol treatment. Notably, the activation of these genes is delayed until the early culture stage under combined mannitol plus TSA treatment [29].

Epigenetic mechanisms regulating chromatin structure and DNA accessibility for transcription are also altered by ME-inducing treatments, with DNA hypomethylation and histone hyperacetylation being the most commonly associated changes. Several studies have explored the use of epigenetic modulators to improve ME efficiency, either during or immediately after treatment, following the work of Li et al. using TSA in *Brassica* [71]. In cereals, various strategies have been assayed using histone deacetylase inhibitors such as TSA, suberoylanilide hydroxamic acid (SAHA), and sodium butyrate [30, 31, 72–75], as well as the histone methyltransferase G9a inhibitor, such as BIX [76], and the DNA methyltransferase inhibitor 5-azacytidine (AzaC) [77, 78]. More recently, additional compounds have been tested, including Aurora kinase inhibitor II (AUKI-II), which affects histone H3 phosphorylation [79].

Consistent with these observations, cold treatment in wheat upregulated genes associated with DNA methylation, histone deacetylation, and microRNA processing, including *DCL1*, *DRM1*, *HDA6*, and *HDA19* [25]. Similarly, in a high-responding barley cultivar, GO enrichment analyses revealed upregulation of terms related to histone lysine methylation, and genes involved in chromatin remodeling are upregulated [27]. Interestingly, supplementing mannitol treatment with TSA in wheat delays activation of genes associated with chromatin modification, including those involved in histone demethylation (*PcG*, *JMJ*, and SDG family members) and DNA methylation (*CMT* and *MET1*), until the early culture stage [29]. Moreover, analysis at two time points during mannitol treatment identified two distinct chromatin-associated gene networks:

an early-acting linked to auxin signaling and cell wall modification, including *TaHDA15*, *TaHAG3*, *TaHAM*, and *TaDRM3*; and a late-acting associated with ABA and ethylene signaling and autophagy, comprising *TaHDA18*, *TaHAC1*, *TaHAC4*, and *TaSDG1a-7A* [34].

Stresses can trigger both specific and common transcriptional responses, which are regulated by upstream signaling pathways mediated by TFs. In barley, stress treatments activate a wide range of TF families, distinguishing two functional groups: one linked to stress responses, comprising *ICE1*, *ATAF2-like*, *NAC*, *WRKY33*, *WRKY46*, *AtbZIP60*, and *HSF7*; and another associated with developmental regulation, including *BARD1*, *CURLY LEAF*, *ERF*, *YABBY5*, and *ZIML2* [18, 27, 28]. Notably, *HvICE1*, a key TF in stress response, is also upregulated during the early culture stage in barley, suggesting that stress response mechanisms remain active at this stage [27].

In wheat, early response to mannitol treatment is marked by enrichment of TF families involved in cellular reorganization through calcium/calmodulin signaling, programmed cell death, and developmental regulation (CAMTA, LSD, CPP). In contrast, at the end of treatment, Dof and DBB TF families, associated with diverse biological processes, are the most highly enriched [29]. Interestingly, several members of the AP2/ERF TF family, which are associated with cell proliferation and embryogenesis induction, are strongly upregulated during mannitol plus TSA treatment, along with members of MADS and BZR families, which participate in diverse regulatory functions [29].

Unraveling the complexity of the mechanisms that govern the microspore response to stress treatment remains a major challenge for ME induction. Recent efforts to establish an integrative model of ME induction have revealed new connections, such as the mobilization of β-glucans to fuel glycolysis and tricarboxylic acid (TCA) cycle [80]. Nevertheless, additional elements identified through transcriptomic studies could further expand our understanding and contribute to a more accurate and comprehensive model.

5 Establishment and Progression of the Embryogenic Pathway

The removal of stress and further exposure to culture conditions trigger developmental reprogramming toward the sporophytic pathway, ultimately leading to embryogenesis. At this stage, most microspores remain as enlarged uninucleate cells, exhibiting generally similar morphologies, with differences primarily in cytoplasmic organization and cell wall structure. Among these, the star-like morphology is one of the most distinctive form [81, 82].

The first microspore division is typically symmetrical; however, ME can also be initiated through an asymmetric division, in which the vegetative-like cell exhibits predominant proliferative activity [83]. As a result, microspores show a heterogeneous developmental response, with up to nine distinct early structure types observed within the same culture [83, 84], although only a small fraction progress into well-developed embryos [81]. Consequently, early morphological features, such as star-like structures or symmetric versus asymmetric divisions, are not reliable predictors of embryogenic potential [41]. Therefore, most studies focus on the multicellular structures released from the exine, known as proembryos or ELS, which are considered as markers of differentiation. Nonetheless, recent transcriptomic analyses targeting the earliest culture stages, from microspore to multicellular structures enclosed within the exine, have provided new insights into microspore reprogramming as early as 3 h after culture initiation [28].

Developmental shifts are characterized by intensified cellular activity, particularly the reactivation of the cell cycle, which demands high energy and nutrient availability to sustain DNA replication, protein synthesis, and cell wall formation [85]. Transcriptomic analyses consistently reveal an enrichment of GO terms associated with the cell cycle, transcription, and translation in early culture. Interestingly, genes associated with these processes are already transcriptionally active in microspores prior to stress, as part of the preparatory program for the PMI. This expression is transiently downregulated during stress but reactivated at early culture stages [26, 29]. Among the genes identified, MCM family members such as *PRL*, which regulate DNA replication, are strongly upregulated under inductive conditions, suggesting a central role in ME [25, 26, 28]. Additionally, genes encoding components of the origin recognition complex (ORC), cyclin A, S-phase checkpoint regulators such as MRE11 and RB, and E2F, as well as genes involved in chromosome organization and DNA conformational changes, are activated following effective inductive treatments in wheat [28]. Remarkably, after mannitol treatment, genes associated with these processes are already activated, whereas after the most effective mannitol plus TSA treatment, their activation is delayed until the culture phase [29].

Microtubules play critical roles in cell division, polarity establishment, cell wall formation, and intracellular transport. Their fragmentation and stabilization lead to new configurations essential for ME induction [82]. Enrichment of GO-terms such as "microtubule binding", "cytokinesis by cell plate formation", and "cytoskeleton" has been described at early stages in barley and wheat [25, 27, 28].

Chromatin dynamics during the cell cycle play a key role in regulating cell identity transitions by modulating chromatin accessibility and organization [85]. In barley, a new pattern of chromatin modifications is established within the first 24 h of microspore culture, indicative of cell fate transition [72]. The reduced DNA methylation observed at the end of stress treatment is maintained in two to four cells structures but progressively increases at later developmental stages [86]. Moreover, the application of the DNA methyltransferase inhibitor AzaC during the first days of culture enhances embryo induction in barley and triticale [77, 78]. However, during periods of rapid cell division, the maintenance or even enhancement of DNA methylation is essential for genome stability, transposon silencing, and transcriptional regulation. Transcriptomic analyses indicate that, at the early culture stage, de novo DNA methylation occurs via the canonical RdDM pathway, as evidenced by upregulation of *DCL3* and *AGO4/5/6* genes, while epigenetic silencing is maintained after DNA replication through the upregulation of key DNA methyltransferases, including MET1, CMT3, and DRM2 [25, 27]. Importantly, the most effective mannitol plus TSA treatment results in the upregulation of *MET1* and *CMT* genes at the culture stage [29]. Nonetheless, global DNA methylation levels may reflect the overall epigenetic state of cultured tissues rather than the embryogenic development itself [87].

Histones and their post-translational modifications play a central role in proliferating cells and are closely associated with cell cycle progression, especially during the S-phase [85]. In ME, the expression of histone genes is upregulated during the early stages of culture in both barley and wheat [26, 29]. As described in the previous section, histone hyperacetylation is critical immediately before or at culture initiation. Transcriptomic studies have revealed that genes encoding histone acetyltransferases (HvHMT, HvHAT) and histone deacetylases (HD2) are upregulated during early culture, whereas other deacetylases (HDA6 and HDA19) are downregulated [25, 27]. In addition, genes involved in histone methylation are also upregulated, including those encoding histone methyltransferases (NSD3, SUVH1, SUVR5, HvHMT-SETDB1-A, and TaSDG54), histone demethylases (TaJMJ1 and TaJMJ9), and proteins associated with the Polycomb Repressive Complex 2 (MSI1 and HSL1) [27–29]. Collectively, these data suggest the existence of a dynamic regulatory network of histone-modifying enzymes that maintains cellular homeostasis and modulates developmental cues during the early ME. In addition, enrichment of histone phosphorylation has been observed at 48 h of culture in wheat, corresponding to chromosome condensation during the M phase [28]. In this context, inhibition of histone phosphorylation during mannitol treatment has been associated with an increased frequency of chromosome doubling in wheat [79].

The rapid growth and differentiation occurring during the early stages of embryogenic development require tightly regulated remodeling of the cell wall components, including AGPs, expansins, callose, and pectins (see the reviews [88, 89 and references therein]). The importance of AGPs in ME is highlighted by their detection in the conditioned medium of barley and maize microspore cultures [68, 90], as well as their role in promoting ME efficiency in wheat and barley [69, 91]. Specifically, supplementation of the culture medium with gum arabic alleviates stress, facilitates the androgenic switch, and promotes efficient energy management [92, 93]. Genes associated with cell wall composition and remodeling are upregulated in culture, including those encoding AGPs and related homologous proteins, such as *ECA1*, *ENOD93*, *PcL*, *TaAGP31-like*, and *FLA26* [15, 19, 94]. Interestingly, FLA26 has been linked to the ovary's role in enhancing ME efficiency in wheat [12]. Other genes involved in cell wall remodeling, such as *β-glucosidases*, *HvEH2*, and *HvGT*, are also upregulated, with *HvGT* showing notably high expression in the most responsive barley cultivar [27, 28]. In addition, in wheat, mannitol plus TSA treatment enriches GO terms related to cell wall organization during culture [29, 31]. However, recent studies in *Brassica napus* report no changes in cell wall structure or composition associated with ME efficiency [95].

To support the accumulation of cell wall precursors, essential nutrients and increased energy demands while maintaining redox homeostasis, a metabolic readjustment is required during this stage. In barley, the upregulation of mitochondrial respiratory chain genes, such as *HvCA* and *HvBCP*, reflects a high energy demand [27]. Genes involved in carbohydrate transport, glycolysis/gluconeogenesis (*HvFBA*), and amino acid uptake (*HvAAP*) are also upregulated after an efficient treatment with cold plus TSA in wheat, or in highly responsive barley cultivars [27, 28]. In wheat, however, an alternative metabolic state has been observed, characterized by the upregulation of the *GBSSI* gene, associated with starch and glycan biosynthesis, and the downregulation of genes involved in general metabolism and mitochondrial function [28]. Integrated transcriptomic and metabolomic data in wheat further highlight the importance of starch accumulation at this stage, evidenced by the upregulation of genes encoding proteins involved in starch synthesis, energy production via the TCA cycle, sugar metabolism (e.g., *UGDH*, *TPS*), amino acid biosynthesis, and nitrogen assimilation, particularly *GDH* [54].

Curiously, a broad set of genes encoding enzymes involved in secondary metabolism is upregulated at this stage. These include key components of the phenylpropanoid (*HCT1*, *PAL2*, *CCR1*, and *4CL3/4*), the lipoxygenase pathway (homologs of *LOX6/8/4*), and the jasmonic acid biosynthesis (*AOS1/2*, *AOC*, *ACOX2/3*, and *OPR7/12*)pathways [19, 26]. Notably, in the most responsive barley cultivar, *HvHCT* is specifically upregulated [27]. Recent studies have emphasized the multifunctional roles of secondary metabolites, which not only contribute to defense but also regulate growth and development by acting as hormone mimics or antioxidants [96].

Antioxidant compounds such as glutathione and its precursor glutamine have been used as supplements in culture media to improve ME efficiency [97, 98]. More recently, novel small-molecule antioxidants have produced similar results [66]. However, although *GST* is among the first genes linked to early embryogenesis in wheat, its expression does not correlate with embryogenic competence, suggesting that its role may be primarily associated with stress signaling

[15]. Nevertheless, several GST genes (*GSTF2*, *GSTA2*, and *HvGST*), together with genes involved in antioxidant defense (*HvDEHs*, *HvNDHQ*, and *HvSDR*) and in γ-aminobutyric acid (GABA) biosynthesis, such as *HvGAD* and *HvP*, are upregulated during culture, with some of them exhibiting genotype-dependent expression patterns [27, 39, 94]. Consistently, a cold plus TSA treatment appears to mitigate oxidative stress and hypoxic conditions more efficiently during early culture [28].

Signaling molecules, including peptides and hormones, are key regulators of developmental transitions and are likely critical during the early stages of culture. Several signaling-related genes are upregulated during barley ME, including the CRP genes *ECA1* and *ECLTP* [15], which are also expressed in the egg cell and the proembryo of wheat [99]. Similarly, a marker of early culture stages in wheat (*EcMt*) and its barley homolog (*HvMT*) exhibits high expression, particularly in highly responsive cultivars [14, 27]. Early markers of zygotic reprogramming, such as *EA* and *TaTPD1-like* [100], as well as genes involved in very-long-chain fatty acid biosynthesis, and a candidate gene mediating the ovary-inductive effect on embryo development (*FER*), are upregulated during early embryogenesis [12, 19, 94, 101]. Moreover, components of the MAPK signaling cascade, including *YDA* and the YDA-related *IRAK*, also exhibit elevated expression [28]. Differential expression of CRPs and other signaling-related genes has been reported across genotypes. In the highly responding cultivars, *HvCHIs* genes are strongly expressed at early culture stages in barley [27], whereas *Tad1*, *CHI3*, *TPD1-like*, and *TAA1b* are induced at later stages in wheat [29, 94]. In contrast, stress-related genes, including *KUN1*, *PR* (*BBI*), *MEGL*, and *PR60*, show low expression in a high-responding wheat genotype [19, 27].

Hormone levels, dynamics, and signaling are critical determinants of the transition to microspore embryogenic development, with auxin, cytokinin, abscisic acid, ethylene, and brassinosteroids all playing major roles [26]. In barley, auxin accumulation begins at multinucleate structures and continues throughout embryogenesis, coinciding with the activation of polar auxin transport. This process is associated with the expression of auxin biosynthesis genes (*TAA1*, *TAR2*) and the auxin efflux transporter *HvPIN1-like* [65]. Transcriptional analyses have indicated that auxin early response genes (*AERGs*), which are rapidly induced in response to changes in auxin levels [102] and include *Aux/IAA*, *GH3*, and *SAUR*, are specifically activated during culture initiation ([26], Vallés, unpublished data). In wheat, a highly inductive treatment with mannitol plus TSA results in elevated expression of genes involved in auxin synthesis (*YUC*), transport (*PIN*), and signaling (*ARFs* and *LBD*) during the early stages of culture [34]. In this context, two regulatory modules (miR167–*ARF8* and miR1130–*ZIFL1*) have been implicated in the regulation of auxin accumulation in barley microspores [103]. Overall, these findings emphasize the intricate, multilayered regulation of auxin homeostasis and response involving novel molecular players within specific and interconnected regulatory networks.

Brassinosteroids (BRs) have also emerged as significant regulators in ME. As demonstrated in several species, changes in BR concentrations can markedly influence embryogenesis efficiency and embryo quality (for review, see [55]). Recently, the application of novel inhibitors targeting mammalian GSK-3β or LRRK2, which are implicated in BR signaling, resulted in a significant increase in the number of embryos in barley [104, 105]. However, few BR-related genes have been directly linked to early stages of microspore culture. In barley, a BR signaling gene homologous to rice *CYCD3-2* has been identified [26]. In wheat, *TaBZR4* is upregulated during stress induction, while *TaBZR5*, also stress-responsive, reaches peak expression during early culture. Notably, *TaBZR5* expression is further enhanced when the highly efficient mannitol plus TSA treatment is used [29].

Although ABA is traditionally associated with stress responses, accumulative evidences highlight its central role in regulating developmental transitions [106]. In *Brassica*, exogenous ABA applied at the pre-globular or globular embryo stages significantly enhances embryogenesis frequency [107, 108]. Gene expression studies further confirm the ABA involvement in the early culture phase. In wheat, *EcMt*, a metallothionein gene linked to ABA signaling, is one of the earliest upregulated [14]. Similarly, in barley, a *VP1* homolog shows specific expression during early embryogenesis [19]. Moreover, expression profiling of ABA-related genes reveals that those involved in ABA biosynthesis (*ZEP2*, *AO5a*), catabolism (*ABA8′OH*, *BG3*), and signaling (*ABI1b*, *ABI5*) are activated during the initial stress treatment, subsequently downregulated, and reactivated during embryo development. In contrast, transcriptional regulators, such as *FUS3*, *VP1*, and *FCA*, exhibit stage-specific expression around the time of exine rupture [19].

Cell identity and fate are regulated by complex networks of TFs displaying tightly controlled spatial and temporal expression patterns. During the early hours of culture in barley, a limited number of TF-encoding genes are upregulated, including members of the bHLH (*HvICE1*) and C3H zinc finger (*HvZNF*) families, which are associated with stress response and developmental regulation [27]. Transcriptomic analyses in ELS reveal the expression dynamics of TFs previously identified as key markers of embryonic development in *Brassica* ME [22], supporting their conserved role in ME across species. In barley, *HAP3/LEC1* is activated at 2 days of culture, followed by *FUS3* and *VP1/ABI3-like* at 4 days. While *FUS3* expression sharply declines at exine rupture (8 days of culture), *LEC1*, *ABI3*, and *ABI5* reach maximum expression between 12 and 16 days of culture [19]. *TaNF-YA7*, likely part of the same complex as *LEC*, has been identified as a marker of mid-stage embryogenic development, showing genotype-specific expression in wheat and triticale [19, 101]. Members of the AP2/ERF TF family are also upregulated in both barley and wheat [25–27,29], including orthologs of well-characterized regulators of cell identity, proliferation, and morphogenesis such as *BBM* and its homolog *ALI5* [109]. Furthermore, *WUS4* and *WOX5* have been identified in barley and wheat, respectively ([26], Vallés, unpublished data). In addition, *SPL*, *GRF*, and *REV* genes have been reported as miRNA targets during early culture stages in barley [103].

TF families involved in hormone signaling, stress-response regulation, organ polarity establishment, primordium development, and nodule formation, such as NAC, WRKY, Whirly, SRS, YABBY, ARF, and Nin-like families, are also enriched [27, 29]. Moreover, TFs associated with broader regulatory functions (e.g., *HvBTB/POZ*) or specific developmental pathways (e.g., *RKD1*, a female reproductive cell marker) are upregulated as well [25, 27].

Comparative analyses between cultivars or treatments have identified TFs as key regulators of ME during culture. In barley, TF families such as B3, ERF, FAR1, HD-ZIP, MYB, NAC, NF-Y, and WOX, exhibit cultivar-specific expression patterns [27]. In wheat, a *BARD1* gene, implicated in the regulation of WUS, along with four *AP2/ERF* genes and one *BZR* gene have been associated with the most effective induction treatments [28, 34]. Moreover, WRKY TFs, linked to pathogen defense, have been identified both in a high-responding barley cultivar and under the most effective treatment in wheat [27, 28]. However, the precise roles of all these TFs in the initiation of embryogenesis during ME remain to be elucidated.

This review synthesizes findings from current studies on the transcriptional dynamics of ME in barley and wheat, providing an integrated overview of the molecular mechanisms that underlie the process and introducing new perspectives on their significance. To further advance our understanding, it will be essential to expand the number of transcriptomic studies, coupled with robust comparative approaches. Such efforts will help unravel the complex regulatory networks that appear to govern ME and elucidate their relationship with other biotechnologically relevant systems, including somatic embryogenesis. Moreover, the incorporation of high-resolution approaches—such as single-cell and spatio-temporal transcriptomics—together with comparative analyses using lines that, for example, overexpress key embryogenic regulators, is expected to shed new light on molecular players and interactions that remain elusive in this singular system of cellular reprogramming.

Acknowledgments

This work was supported by the grant PID2020-119540RB-100 funded by MICIU/AEI/10.13039/501100011033, the grant PID2023-150342OB-I00 funded by MICIU/AEI/10.13039/501100011033 and by ERDF/UE, and a grant of Reference Research Groups recognized by Diputació n General de Aragón, Spain (Group A08-23R). I. V-R was supported by the grant BES-2017- 080970 funded by MICIU/ AEI/10.13039/501100011033.

References

1. Soriano M, Li H, Boutilier K (2013) Microspore embryogenesis: establishment of embryo identity and pattern in culture. Plant Rep 26(3):181–196. https://doi.org/10.1007/s00497-013-0226-7
2. Castillo AM, Cistué L, Vallés MP, Soriano M (2009) Chromosome doubling in monocots. In: Touraev A, Forster BP, Jain SM (Eds.)Eds. Advances in haploid production in higher plants. Springer, Dordrecht, 10.1007/978-1-4020-8854-4_27
3. Forster BP, Thomas WT (2005) Doubled haploids in genetics and plant breeding. Plant Breed Rev 25:57–88. https://doi.org/10.1002/9780470650301.ch3
4. Guha S, Maheshwari SC (1964) In vitro production of embryos from anthers of *Datura*. Nature 204(4957):497–497. https://doi.org/10.1038/204497a0
5. Maluszynski M, Kasha KJ, Szarejko I. (2003). Published doubled haploid protocols in plant species. In: Maluszynski M, Kasha KJ, Forster BP, Szarejko, I. (eds) Doubled Haploid Production in Crop Plants. Springer, Dordrecht. https://doi.org/10.1007/978-94-017-1293-4_46
6. Seguí-Simarro JM, Moreno JB, Fernández MG, Mir R (2021) Species with haploid or doubled haploid protocols. In: Segui-Simarro JM (ed) Doubled haploid technology. Methods in molecular biology, vol 2287. Humana, New York, NY, pp 41–103 https://doi.org/10.1007/978-1-0716-1315-3_3
7. Clapham D (1973) Haploid *Hordeum* plants from anthers in vitro. Z Pflanzenzüchtg 65:285–292
8. Ouyang J, Hu H, Chuang CC, Tseng CC (1973) Induction of pollen plants from anthers of *Triticum aestivum* L. cultured in vitro. Sci Sin 16:79–95
9. Forster BP, Heberle-Bors E, Kasha KJ, Touraev A (2007) The resurgence of haploids in higher plants. Trends Plant Sci 12(8):368–375. https://doi.org/10.1016/J.TPLANTS.2007.06.007
10. Weyen J (2009) Barley and wheat doubled haploids in breeding. In: Touraev A, Forster BP, Jain SM (Eds.)Eds. Advances in haploid production in higher plants. Springer, Dordrecht, https://doi.org/10.1007/978-1-4020-8854-4_15
11. Hansen NJP, Andersen SB (1998) In vitro chromosome doubling with colchicine during microspore culture in wheat (*Triticum aestivum* L.). Euphytica 102(1):101–108. https://doi.org/10.1023/A:1018348816205
12. Castillo AM, Sánchez-Díaz RA, Vallés MP (2015) Effect of ovary induction on bread wheat anther culture: ovary genotype and developmental stage, and candidate gene association. Front Plant Sci 6:402. https://doi.org/10.3389/fpls.2015.00402
13. Reynolds TL, Kitto SL (1992) Identification of embryoid-abundant genes that are temporally expressed during pollen embryogenesis in wheat anther cultures. Plant Physiol 100(4):1744–1750. https://doi.org/10.1104/pp.100.4.1744
14. Reynolds TL, Crawford RL (1996) Changes in abundance of an abscisic acid-responsive, early cysteine-labeled metallothionein transcript during pollen embryogenesis in bread wheat (*Triticum aestivum*). Plant Mol Biol 32(5):823–829. https://doi.org/10.1007/BF00020480
15. Vrinten PL, Nakamura T, Kasha KJ (1999) Characterization of cDNAs expressed in the early stages of microspore embryogenesis in barley (*Hordeum vulgare* L.). Plant Mol Biol 41:455–463. https://doi.org/10.1023/a:1006383724443
16. Maraschin SF, Caspers M, Potokina E et al. (2006) cDNA array analysis of stress-induced gene expression in barley androgenesis. Physiol Plant 127:535–50. https://doi.org/10.1111/j.1399-3054.2006.00673.x
17. Potokina E, Sreenivasuslu N, Altschmied L et al. (2002) Differential gene expression during seed germination in barley (*Hordeum vulgare* L.). Funct Integr Genomics 2:28–39. https://doi.org/10.1007/s10142-002-0050-x
18. Muñoz-Amatriaín M, Svensson JT, Castillo AM et al. (2006) Transcriptome analysis of barley anthers: effect of mannitol treatment on microspore embryogenesis. Physiol Plant 127(4):551–560. https://doi.org/10.1111/j.1399-3054.2006.00729.x
19. Sánchez-Díaz RA (2014) Estudio de los mecanismos moleculares implicados en la embriogénesis de la microspora en cebada y trigo. (Study of the molecular mechanisms involved in microspore embryogenesis in barley

and wheat). PhD Dissertation. University of Zaragoza, Zaragoza. Spain
20. Close TJ, Wanamaker S, Caldo RA et al. (2004) A new resource for cereal genomics: 22K barley GeneChip comes of age. Plant Physiol 134:960–968. https://doi.org/10.1104/pp.103.034462
21. Joosen R, Cordewener J, Supena EDJ et al. (2007) Combined transcriptome and proteome analysis identifies pathways and markers associated with the establishment of rapeseed microspore-derived embryo development. Plant Physiol 144(1):155–172. https://doi.org/10.1104/pp.107.098723
22. Malik MR, Wang F, Dirpaul JM et al. (2007) Transcript profiling and identification of molecular markers for early microspore embryogenesis in *Brassica napus*. Plant Physiol 144(1):134–154. https://doi.org/10.1104/pp.106.092932
23. Tsuwamoto R, Fukuoka H, Takahata Y (2007) Identification and characterization of genes expressed in early embryogenesis from microspores of *Brassica napus*. Planta 225(3):641–652. https://doi.org/10.1007/s00425-006-0388-8
24. Stasolla C, Belmonte MF, Tahir M et al. (2008) Buthionine sulfoximine (BSO)-mediated improvement in cultured embryo quality in vitro entails changes in ascorbate metabolism, meristem development and embryo maturation. Planta 228(2):255–272. https://doi.org/10.1007/s00425-008-0735-z
25. Seifert F, Bössow S, Kumlehn J et al. (2016) Analysis of wheat microspore embryogenesis induction by transcriptome and small RNA sequencing using the highly responsive cultivar "Svilena". BMC Plant Biol 16:1–16. https://doi.org/10.1186/s12870-016-0782-8
26. Bélanger S, Marchand S, Jacques PÉ et al. (2018) Differential expression profiling of microspores during the early stages of isolated microspore culture using the responsive barley cultivar gobernadora. G3: Bethesda 8(5):1603–1614. https://doi.org/10.1534/g3.118.200208
27. Nowicka A, Kovacik M, Maksylewicz A et al. (2024) The transcriptional landscape of the developmental switch from regular pollen maturation towards microspore-derived plant regeneration in barley. Crop J 12(4):1064–1080. https://doi.org/10.1016/j.cj.2024.07.003
28. Jensen T (2021) The impact of cold and Trichostatin A treatment on the transcriptome of wheat microspores. MS Dissertation. University of Lethbridge, Lethbridge, Canada
29. Valero-Rubira I (2023) Microspore embryogenesis induction in bread wheat: application of epigenetic modifiers and characterization of associated mechanisms. PhD Dissertation. University of Zaragoza, Zaragoza, Spain
30. Jiang F, Ryabova D, Diedhiou J et al. (2017) Trichostatin A increases embryo and green plant regeneration in wheat. Plant Cell Rep 36:1701–1706. https://doi.org/10.1007/s00299-017-2183-3
31. Castillo AM, Valero-Rubira I, Burrell MÁ et al. (2020) Trichostatin A affects developmental reprogramming of bread wheat microspores towards an embryogenic route. Plants 9(11):1442. https://doi.org/10.3390/plants9111442
32. Nelms B, Walbot V (2022) Gametophyte genome activation occurs at pollen mitosis I in maize. Science 375:424–429. https://doi.org/10.1126/science.abl7392
33. Touraev A, Vicente O, Heberle-Bors E (1997) Initiation of microspore embryogenesis by stress. Trends Plant Sci 2(8):297–302. https://doi.org/10.1016/S1360-1385(97)89951-7
34. Valero-Rubira I, Castillo AM, Burrell MÁ, Vallés MP (2023) Microspore embryogenesis induction by mannitol and TSA results in a complex regulation of epigenetic dynamics and gene expression in bread wheat. Front Plant Sci 13:1058421. https://doi.org/10.3389/fpls.2022.1058421
35. Muñoz-Amatriaín M, Svensson JT, Castillo AM et al. (2009) Expression profiles in barley microspore embryogenesis. In: Touraev A, Forster BP, Jain SM (Eds.)Eds. Advances in haploid production in higher plants. Springer, Dordrecht, 127–134. 10.1007/978-1-4020-8854-4_9pp
36. Jacquard C, Mazeyrat-Gourbeyre F, Devaux P, Boutilier K et al. (2009) Microspore embryogenesis in barley: anther pre-treatment stimulates plant defence gene expression. Planta 229:393–402. https://doi.org/10.1007/s00425-008-0838-6
37. Soriano M (2008) Embriogénesis de la microspora en trigo panadero: optimización de la producción de plantas doblehaploides y estudio de genes de estrés durante el pretratamiento. (Microspore embryogenesis in bread wheat: optimization of doubled haploid plant production and analysis of stress-related genes during pretreatment). PhD Dissertation. University of Zaragoza, Zaragoza, Spain
38. Mittler R, Zandalinas SI, Fichman Y, Van Breusegem F (2022) Reactive oxygen species signalling in plant stress responses. Nat Rev Mol Cell Biol 23(10):663–679. https://doi.org/10.1038/s41580-022-00499-2
39. Żur I, Dubas E, Krzewska M et al. (2014) Antioxidant activity and ROS tolerance in triticale (× *Triticosecale* Wittm.) anthers affect the efficiency of microspore embryogenesis. Plant Cell Tiss Organ Cult 119:79–94

40. Żur I, Dubas E, Krzewska M et al. (2021) Triticale and barley microspore embryogenesis induction requires both reactive oxygen species generation and efficient system of antioxidative defence. Plant Cell Tiss Organ Cult 145:347–366. https://doi.org/10.1007/s11240-021-02012-7
41. Dubas E, Krzewska M, Surówka E et al. (2024) New prospects for improving microspore embryogenesis induction in highly recalcitrant winter wheat lines. Plants (Basel) 13(3):363. https://doi.org/10.3390/plants13030363
42. Żur I, Kopeć P, Surówka E et al. (2021) Impact of ascorbate—glutathione cycle components on the effectiveness of embryogenesis induction in isolated microspore cultures of barley and triticale. Antioxidants 10:1254. https://doi.org/10.3390/antiox10081254
43. Moons A (2005) Regulatory and functional interactions of plant growth regulators and plant glutathione S-transferases (GSTs). Vitam Horm 72:155–202. https://doi.org/10.1016/S0083-6729(05)72005-7
44. Muñoz-Amatriaín M, Svensson JT, Castillo AM et al. (2009) Microspore embryogenesis: assignment of genes to embryo formation and green vs. albino plant production. Funct Integr Genomics 9:311–23. https://doi.org/10.1007/s10142-009-0113-3
45. Hoekstra S, Van Bergen S, Van Brouwershaven IR et al. (1997) Androgenesis in *Hordeum vulgare* L.: effects of mannitol, calcium and abscisic acid on anther pretreatment. Plant Sci 126(2):211–218. https://doi.org/10.1016/S0168-9452(97)00096-4
46. Cistué L, Vallés MP, Echávarri B et al. (2004) Production of barley doubled haploids by anther and microspore culture. In: Touraev A, Forster BP, Jain SM (Eds.)Eds. In vitro application in crop improvement. CRC Press, Boca Raton, 1–17
47. Soriano M, Cistué L, Castillo AM (2008) Enhanced induction of microspore embryogenesis after n-butanol treatment in wheat (*Triticum aestivum* L.) anther culture. Plant Cell Rep 27(5):805–811. https://doi.org/10.1007/s00299-007-0500-y
48. Broughton S (2011) The application of n-butanol improves embryo and green plant production in anther culture of Australian wheat (*Triticum aestivum* L.) genotypes. Crop Pasture Sci 62(10):813–822. https://doi.org/10.1071/CP11204
49. Ahmadi B, Ahmadi M, Teixeira da Silva JA (2018) Microspore embryogenesis in Brassica: calcium signaling, epigenetic modification, and programmed cell death. Planta 248(6):1339–1350. https://doi.org/10.1007/s00425-018-2996-5
50. Zandalinas SI, Balfagón D, Gómez-Cadenas A, Mittler R (2022) Plant responses to climate change: metabolic changes under combined abiotic stresses. J Exp Bot 73(11):3339–3354
51. Uváčková Ľ, Takáč T, Boehm N et al. (2012) Proteomic and biochemical analysis of maize anthers after cold pretreatment and induction of androgenesis reveals an important role of anti-oxidative enzymes. J Prot 75(6):1886–1894. https://doi.org/10.1016/j.jprot.2011.12.033
52. Krzewska M, Gołębiowska-Pikania G, Dubas E et al. (2017) Identification of proteins related to microspore embryogenesis responsiveness in anther cultures of winter triticale (× *Triticosecale* Wittm.). Euphytica 213:1–17. https://doi.org/10.1007/s10681-017-1978-1
53. Krzewska M, Dubas E, Gołębiowska G et al. (2021) Comparative proteomic analysis provides new insights into regulation of microspore embryogenesis induction in winter triticale (× *Triticosecale* Wittm.) after 5-azacytidine treatment. Sci Rep 11(1):22215. https://doi.org/10.1038/s41598-021-01671-y
54. Perez-Piñar T, Hartmann A, Bössow S et al. (2024) Metabolic changes during wheat microspore embryogenesis induction using the highly responsive cultivar Svilena. J Plant Physiol 294:154193. https://doi.org/10.1016/j.jplph.2024.154193
55. Żur I, Dubas E, Krzewska M, Janowiak F (2015) Current insights into hormonal regulation of microspore embryogenesis. Front Plant Sci 6:424. https://doi.org/10.3389/fpls.2015.00424
56. Żur I, Dubas E, Krzewska M et al. (2015) Hormonal requirements for effective induction of microspore embryogenesis in triticale (× *Triticosecale* Wittm.) anther cultures. Plant Cell Rep 34:47–62. https://doi.org/10.1007/s00299-014-1686-4
57. Springer A, Krzewska M, Dubas E et al. (2025) Induction of microspore embryogenesis in bread wheat by mannitol pre-treatment is associated with the disruption of endogenous hormone balance and substantial accumulation of auxins. BMC Plant Biol 25(1):370. https://doi.org/10.1186/s12870-025-06389-x
58. Juzoń-Sikora K, Nowicka A, Plačková L et al. (2023) Hormonal homeostasis associated with effective induction of triticale microspore embryogenesis. Plant Cell Tiss Organ Cult (PCTOC) 152(3):583–604. https://doi.org/10.1007/s11240-022-02433-y
59. Eckardt NA, Avin-Wittenberg T, Bassham DC et al. (2024) The lowdown on breakdown: open questions in plant proteolysis. Plant Cell 36(9):2931–2975. https://doi.org/10.1093/plcell/koae193

60. Kacprzyk J, Burke R, Armengot L et al. (2024) Roadmap for the next decade of plant programmed cell death research. New Phytol 242(5):1865–1875. https://doi.org/10.1111/nph.19709
61. Rodríguez-Serrano M, Bárány I, Prem D et al. (2012) NO, ROS, and cell death associated with caspase-like activity increase in stress-induced microspore embryogenesis of barley. J Exp Bot 63(5):2007–2024. https://doi.org/10.1093/jxb/err400
62. Sinha RK, Eudes F (2015) Dimethyl tyrosine conjugated peptide prevents oxidative damage and death of triticale and wheat microspores. Plant Cell Tiss Organ Cult 122:227–237. https://doi.org/10.1007/s11240-015-0763-x
63. Sinha RK, Pospíšil P, Maheshwari P, Eudes F (2016) Bcl-2△ 21 and Ac-DEVD-CHO inhibit death of wheat microspores. Front Plant Scie 7:1931. https://doi.org/10.3389/fpls.2016.01931
64. Bárány I, Berenguer E, Solís MT et al. (2018) Autophagy is activated and involved in cell death with participation of cathepsins during stress-induced microspore embryogenesis in barley. J Exp Bot 69(6):1387–1402. https://doi.org/10.1093/jxb/erx455
65. Pérez-Pérez Y, El-Tantawy AA, Solís MT et al. (2019) Stress-induced microspore embryogenesis requires endogenous auxin synthesis and polar transport in barley. Front Plant Sci 10:1200. https://doi.org/10.3389/fpls.2019.01200
66. Rueda-Varela C, Carneros E, Caro E et al. (2025) Enhancing microspore embryogenesis initiation by reducing ROS, autophagy, and cell death with novel small molecules in rapeseed and barley. J Plant Physiol 311:154546. https://doi.org/10.1016/j.jplph.2025.154546
67. Dubas E, Żur I, Moravčiková J et al. (2021) Proteins, small peptides and other signaling molecules identified as inconspicuous but possibly important players in microspores reprogramming toward embryogenesis. Front Sustain Food Syst 5:745865. https://doi.org/10.3389/fsufs.2021.745865
68. Borderies G, Le Béchec M, Rossignol M et al. (2004) Characterization of proteins secreted during maize microspore culture: arabinogalactan proteins (AGPs) stimulate embryo development. Eur J Cell Biol 83(5):205–212. https://doi.org/10.1078/0171-9335-00378
69. Makowska K, Kałużniak M, Oleszczuk S et al. (2017) Arabinogalactan proteins improve plant regeneration in barley (*Hordeum vulgare* L.) anther culture. Plant Cell Tiss Organ Cult 131:247–257. https://doi.org/10.1007/s11240-017-1280-x
70. Shim YS, Kasha KJ (2003) The influence of pretreatment on cell stage progression and the time of DNA synthesis in barley (*Hordeum vulgare* L.) uninucleate microspores. Plant Cell Rep 21(11):1065–1071. https://doi.org/10.1007/s00299-003-0635-4
71. Li H, Soriano M, Cordewener J, Muiño JM, Riksen T, Fukuoka H, Angenent GC, Boutilier K (2013) The histone deacetylase inhibitor Trichostatin A promotes totipotency in the male gametophyte. Plant Cell 26:195–209. https://doi.org/10.1105/tpc.113.116491
72. Pandey P, Daghma DS, Houben A et al. (2017) Dynamics of post-translationally modified histones during barley pollen embryogenesis in the presence or absence of the epi-drug trichostatin A. Plant Reprod 30:95–105. https://doi.org/10.1007/s00497-017-0302-5
73. Wang HM, Enns JL, Nelson KL et al. (2019) Improving the efficiency of wheat microspore culture methodology: evaluation of pretreatments, gradients, and epigenetic chemicals. Plant Cell Tiss Organ Cult 139:589–599. https://doi.org/10.1007/s11240-019-01704-5
74. Sai CB, Chidambaranathan P, Samantaray S (2022) Role of histone deacetylase inhibitors in androgenic callus induction of *Oryza sativa* sub indica, in sight into evolution and mode of action of histone deacetylase genes. Mol Biol Rep 49:2169–2183. https://doi.org/10.1007/s11033-021-07036-y
75. Maheshwari P, Laurie JD (2021) Triticale isolated microspore culture for doubled haploid production. In: Segui-Simarro, J.M. (eds) Doubled haploid technology. Methods in molecular biology, vol 2287. Humana, New York, NY, pp 295–312. https://doi.org/10.1007/978-1-0716-1315-3_16
76. Berenguer E, Bárány I, Solís MT, Pérez-Pérez Y, Risueño MC, Testillano PS (2017) Inhibition of histone H3K9 methylation by BIX-01294 promotes stress-induced microspore totipotency and enhances embryogenesis initiation. Front Plant Sci 8:1161
77. Solís MT, El-Tantawy AA, Cano V et al. (2015) 5-azacytidine promotes microspore embryogenesis initiation by decreasing global DNA methylation, but prevents subsequent embryo development in rapeseed and barley. Front Plant Sci 6:472. https://doi.org/10.3389/fpls.2015.00472
78. Nowicka A, Juzoń K, Krzewska M et al. (2019) Chemically-induced DNA de-methylation alters the effectiveness of microspore embryogenesis in triticale. Plant Sci

287:110189. https://doi.org/10.1016/j.plantsci.2019.110189
79. Valero-Rubira I, Vallés MP, Echávarri B et al. (2024) New epigenetic modifier inhibitors enhance microspore embryogenesis in bread wheat. Plants 13(6):772. https://doi.org/10.3390/plants13060772
80. Bednarek PT, Orłowska R, Mańkowski DR et al. (2021) Structural equation modeling (SEM) analysis of sequence variation and green plant regeneration via anther culture in barley. Cells 10(10):2774. https://doi.org/10.3390/cells10102774
81. Maraschin S, Vennik M, Lamers GE et al. (2005) Time-lapse tracking of barley androgenesis reveals position-determined cell death within pro-embryos. Planta 220(4):531–540. https://doi.org/10.1007/s00425-004-1371-x
82. Dubas E, Castillo AM, Żur I et al. (2021) Microtubule organization changes severely after mannitol and n-butanol treatments inducing microspore embryogenesis in bread wheat. BMC Plant Biol 21:1–19. https://doi.org/10.1186/s12870-021-03345-3
83. Daghma DES, Hensel G, Rutten T et al. (2014) Cellular dynamics during early barley pollen embryogenesis revealed by time-lapse imaging. Front Plant Sci 5:675. https://doi.org/10.3389/fpls.2014.00675
84. Indrianto A, Barinova I, Touraev A, Heberle-Bors E (2001) Tracking individual wheat microspores in vitro: identification of embryogenic microspores and body axis formation in the embryo. Planta 212(2):163–174. https://doi.org/10.1007/s004250000375
85. Sablowski R, Gutierrez C (2022) Cycling in a crowd: coordination of plant cell division, growth, and cell fate. Plant Cell 34(1):193–208. https://doi.org/10.1093/plcell/koab222
86. El-Tantawy AA, Solís MT, Risueño MC, Testillano PS (2014) Changes in DNA methylation levels and nuclear distribution patterns after microspore reprogramming to embryogenesis in barley. Cytogenet Genome Res 143(1-3):200–208. https://doi.org/10.1159/000365232
87. Markulin L, Škiljaica A, Tokić M et al. (2021) Taking the wheel–de novo DNA methylation as a driving force of plant embryonic development. Front Plant Sci 12:764999. https://doi.org/10.3389/fpls.2021.764999
88. Testillano PS (2019) Microspore embryogenesis: targeting the determinant factors of stress-induced cell reprogramming for crop improvement. J Exp Bot 70(11):2965–2978. https://doi.org/10.1093/jxb/ery464
89. Yang F, Liu X, Qiao Y et al. (2024) Microspore embryogenesis: in vitro cultivation induced cell reprogramming for plant breeding. Seed Biol 3(1):e021. https://doi.org/10.48130/seedbio-0024-0019
90. Paire A, Devaux P, Lafitte C et al. (2003) Proteins produced by barley microspores and their derived androgenic structures promote in vitro zygotic maize embryo formation. Plant Cell Tiss Organ Cult 73(2):167–176. https://doi.org/10.1023/A:1022805623167
91. Letarte J, Simion E, Miner M, Kasha KJ (2006) Arabinogalactans and arabinogalactan-proteins induce embryogenesis in wheat (*Triticum aestivum* L.) microspore culture. Plant Cell Rep 24(12):691–698. https://doi.org/10.1007/s00299-005-0013-5
92. Klajmon A, Makowska K, Zimny J et al. (2023) Gum Arabic influences the activity of antioxidant enzymes during androgenesis in barley anthers. Plant Cell Tiss Organ Cult 153(1):145–157. https://doi.org/10.1007/s11240-023-02451-4
93. Konieczny R, Pawłowski TA, Surman M et al. (2025) Identification of proteins associated with gum Arabic-enhancement of androgenesis from barley anthers. Plant Cell Tiss Organ Cult 160(3):77. https://doi.org/10.1007/s11240-025-03015-4
94. Sánchez-Díaz RA, Castillo AM, Vallés MP (2013) Microspore embryogenesis in wheat: new marker genes for early, middle and late stages of embryo development. Plant Reprod 26:287–296. https://doi.org/10.1007/s00497-013-0225-8
95. Camacho-Fernández C, Corral-Martínez P, Calabuig-Serna A et al. (2024) The different response of *Brassica napus* genotypes to microspore embryogenesis induced by heat shock and trichostatin A is not determined by changes in cell wall structure and composition but by different stress tolerance. Physiol Plant 176(3):e14405. https://doi.org/10.1111/ppl.14405
96. Erb M, Kliebenstein DJ (2020) Plant secondary metabolites as defenses, regulators, and primary metabolites: the blurred functional trichotomy. Plant Physiol 184(1):39–52. https://doi.org/10.1104/pp.20.00433
97. Cistué L, Batlle RF, Echávarri B (2009) Improvements in the production of doubled haploids in durum wheat (*Triticum turgidum* L.) through isolated microspore culture. Plant Cell Rep 28:727–735. https://doi.org/10.1007/S11240-007-9281-9
98. Asif M, Eudes F, Goyal A et al. (2013) Organelle antioxidants improve microspore embryogenesis in wheat and triticale. InVitro Cell Dev Biol Plant 49:489–497. https://doi.org/10.1007/s11627-013-9514-z
99. Sprunck S, Baumann U, Edwards K, Langridge P, Dresselhaus T (2005)). The transcript

composition of egg cells changes significantly following fertilization in wheat (*Triticum aestivum* L.). Plant J 41(5):660–672
100. Leljak-Levanić D, Juranić M, Sprunck S (2013) De novo zygotic transcription in wheat (*Triticum aestivum* L.) includes genes encoding small putative secreted peptides and a protein involved in proteasomal degradation. Plant reproduction 26(3):267–285
101. Żur I, Dubas E, Krzewska M et al. (2014) Changes in gene expression patterns associated with microspore embryogenesis in hexaploid triticale (× *Triticosecale* Wittm.). Plant Cell Tiss Organ Cult 116:261–267. https://doi.org/10.1007/s11240-013-0399-7
102. Bao D, Chang S, Li X, Qi Y (2024) Advances in the study of auxin early response genes: Aux/IAA, GH3, and SAUR. Crop J 12(4):964–978. https://doi.org/10.1016/j.cj.2024.06.011
103. Bélanger S, Baldrich P, Lemay MA et al. (2020) The commitment of barley microspores into embryogenesis correlates with miRNA-directed regulation of members of the SPL, GRF and HD-ZIPIII transcription factor families. Plant Direct 4(12):e00289. https://doi.org/10.1002/pld3.289
104. Berenguer E, Carneros E, Pérez-Pérez Y et al. (2021) Small molecule inhibitors of mammalian GSK-3β promote in vitro plant cell reprogramming and somatic embryogenesis in crop and forest species. J Exp Bot 72(22):7808–7825. https://doi.org/10.1093/jxb/erab365
105. Carneros E, Berenguer E, Pérez-Pérez Y et al. (2024) Small molecule inhibitors of human LRRK2 enhance in vitro embryogenesis and microcallus formation for plant regeneration of crop and model species. J Plant Physiol 303:154334. https://doi.org/10.1016/j.jplph.2024.154334
106. Jiang Y, Jiang S, Liu L (2025) Understanding the multifaceted role of ABA signaling in orchestrating plant developmental transition. Stress Biol 5(1):5. https://doi.org/10.1007/s44154-024-00203-8
107. Ramesar-Fortner N, Yeung EC (2006) Physiological influences in the development and function of the shoot apical meristem of microspore-derived embryos of *Brassica napus* 'Topas'. Can J Bot 84(3):371–383. https://doi.org/10.1139/b06-007
108. Belmonte MF, Ambrose SJ, Ross ARS et al. (2006) Improved development of microspore-derived embryo cultures of *Brassica napus* cv Topaz following changes in glutathione metabolism. Phisiol Plant 127:690–700. https://doi.org/10.1111/j.1399-3054.2006.00707.x
109. Boutilier K, Offringa R, Sharma VK et al. (2002) Ectopic expression of BABY BOOM triggers a conversion from vegetative to embryonic growth. Plant Cell 14:1737–1749. https://doi.org/10.1105/tpc.001941

Chapter 3

Albinism in Cereal Androgenesis—Shedding Light on Molecular Mechanisms and Possible Solutions

Monika Gajecka-Lesik and Iwona Szarejko

Abstract

Albinism is a major limitation in androgenesis-based doubled haploid (DH) production in cereals, significantly limiting the efficiency of this powerful breeding tool. Albino plantlets, which do not develop functional chloroplasts, are unable to photosynthesize and thus cannot survive outside of in vitro culture. Despite numerous attempts to optimize androgenesis protocols, albinism remains difficult to eliminate, largely due to its strong genotype dependence. Recent studies have shed light on the molecular and cellular mechanisms underlying this phenomenon, linking it to early plastid differentiation during in vivo pollen development, plastid DNA instability, and disrupted chloroplast biogenesis. Premature activation of starch biosynthesis and amyloplast formation in microspores has been correlated with high rates of albino regeneration. This chapter summarizes key advances in understanding the developmental and genetic basis of albinism in androgenesis, focusing on plastid biology, regulation of transcription, and developmental checkpoints during chloroplast differentiation. Promising strategies, including utilization of early-mid uninucleate microspores to initiate androgenesis are discussed as potential solutions to overcome genotype-dependent albinism in cereals.

Keywords Albinism, Androgenesis, Doubled haploids, Microspore embryogenesis, Cereals

1 Introduction

Androgenesis is the most effective method for producing doubled haploids (DH); however, in all major cereals, its application is severely constrained by the frequent regeneration of albino plants. Albinos lack chlorophyll in tissues that are typically green and are, therefore, incapable of autotrophic growth, making them unable to survive outside in vitro culture conditions. Chlorophyll is essential for the absorption of light energy required for the photolysis of water; therefore, its absence inhibits the process of photosynthesis [1]. Albinism has been identified as a major limitation in the androgenesis of barley [2, 3], wheat [4, 5], rice [6, 7], and triticale [8]. Moreover, the occurrence of albinism is highly genotype-dependent, with some genotypes producing nearly 100% albino plantlets (Fig. 1). This

Ana María Castillo et al. (eds.), *Doubled Haploid Methods in Cereals*, Methods in Molecular Biology, vol. 3029,
https://doi.org/10.1007/978-1-0716-5245-9_3,

Fig. 1 Influence of genotype on albino plant production in spring barley microspore cultures

phenomenon severely limits the practical application of androgenesis in breeding programs, particularly for agronomically valuable cultivars/genotypes.

Many previous studies on androgenesis have focused on increasing the frequency of green plant regeneration by modifying in vitro culture conditions, such as pretreatments [9–14] or medium composition [15–17]. However, none of these external factor modifications has successfully eliminated the regeneration of albino plants. The main effect of such modifications has typically been an overall increase in regeneration capacity, which consequently results in a higher number of green plants, but the frequency of green plant regeneration remains unchanged.

Green androgenic plantlets contain well-developed chloroplasts with fully formed grana and several starch grains. Green plants regenerated from androgenic embryos have actively dividing plastids with a high thylakoid content and abundant plastid DNA. In contrast, plastids in albino plantlets are arrested at an early developmental stage, showing limited division activity, rudimentary thylakoid structures, high accumulation of starch grains, and a low plastid DNA content [18–20]. In barley, plastid development in albino plantlets is arrested at different stages depending on the genotype's overall ability to regenerate green versus albino plants. In cultivars producing over 90% albino plantlets, plastids exhibit prolamellar bodies along with disorganized prothylakoid/thylakoid structures. In contrast, cultivars producing less than 10% albino plantlets contain plastids with only a prolamellar body, with no prothylakoids [20].

Understanding the molecular mechanisms underlying chloroplasts development during androgenesis has enabled the identification of key biological factors and potential strategies for effectively reducing albinism in barley androgenesis. Here, we present a detailed characterization of chloroplast biogenesis in microspore-derived embryos and during the regeneration of androgenic plantlets, along with a proposed strategy to minimize the occurrence of albino plants.

1.1 Microspore and Plastid Ontogeny: Natural Versus Androgenic Development

Microspores are haploid cells formed during microsporogenesis within the pollen sac of the stamen. This process begins with the meiotic division of a microspore mother cell resulting in four haploid daughter cells arranged in a tetrad and enclosed by a callose wall [21]. Within the tapetum (the tissue lining the pollen sac), the enzyme callase is secreted and degrades the callose wall, thereby releasing individual microspores from the tetrad [22]. Early uninucleate microspores are characterized by a centrally located nucleus and numerous small, dispersed vacuoles. As development progresses, a large vacuole forms and expands to occupy most of the cell volume, pushing the nucleus to an eccentric position. At this late stage of development, the nucleus has already undergone DNA replication and is situated peripherally, while the cytoplasm becomes compressed between the vacuole and the cell wall [23].

Microspore development concludes with an asymmetric mitotic division resulting in two distinct daughter cells: a large vegetative cell and a smaller generative cell, together forming an immature pollen grain [24]. The generative cell later undergoes a second mitotic division, giving rise to two sperm cells, which are enclosed within the cytoplasm of the vegetative cell and surrounded by the exine. These cells serve distinct functions: the sperm cells contribute genetic material during double fertilization, while the vegetative cell functions as a supportive structure, accumulating starch grains as energy reserves to facilitate pollen tube germination and growth [21].

Cereal microspores contain proplastids, which are small, undifferentiated, and pigment-free plastids that do not perform any metabolic functions but are capable of differentiating into all plastid types [25]. In plants, proplastids are usually inherited maternally and are typically eliminated from the generative cell during the first mitotic division of the pollen grain or during its subsequent development. However, in cereals, proplastids are eliminated only during fertilization, as the generative cell occupies a much smaller volume than the egg cell. Due to this size difference, plastids present in the sperm cell do not contribute to fertilization [26]. In contrast, proplastids located in the vegetative cell differentiate into amyloplast during in vivo pollen grain development. It has been demonstrated that amyloplasts already appear in vacuolated microspores [27–29]. Starch is a polysaccharide made of two glucose polymers: linear amylose and branched amylopectin, and functions as the primary storage carbohydrate in plants, being accumulated in plastids across various tissues and mobilized as needed for carbon and energy supply. Synthesis of starch begins with the formation of ADP-glucose, catalyzed by glucose-1-phosphate adenylytransferase (AGPase). Chain elongation is carried out by starch synthases: Granule-bound starch synthase (GBSS) for amylose, and starch synthase (SS) for amylopectin, while branching is introduced by the starch branching enzyme (SBE). The final structure is modified by the disproportioning enzyme (DPE) [30].

A precise understanding of microspore development and plastid differentiation in vivo is crucial not only for basic biological research but also for applied plant breeding. Under exposure of vacuolated microspores to stress factors such as high or low temperature or nutrient deficiency, it is possible to reprogram the developmental pathway of the microspore from the gametophytic to the sporophytic route and induce androgenesis in vitro [31]. The stress required to induce androgenesis must be applied at the appropriate developmental stage, due to the narrow window in which gametophytic development can be arrested and redirected toward the sporophytic pathway [32]. Studies have shown that, in most cereals, the mid-to-late uninucleate (ML) microspore is the most responsive and effective stage for the induction of androgenesis [33]. This developmental stage is utilized in many androgenesis protocols for barley [34–36], oat [17], rye [37], triticale [38], and wheat [39, 40]. Moreover, the application of stress treatment prior to in vitro culture suppresses the synthesis of storage starch during the induction of androgenesis, which in turn facilitates chloroplast differentiation by enabling plastids to follow the photosynthetic developmental pathway [20].

The differentiation status of proplastids at the time of microspores sampling has been identified as a critical factor for successful green plant regeneration [27]. This has been demonstrated in studies with two spring barley cultivars "Jersey" and "Mercada," which both possess high regeneration potential but exhibit markedly different capacities for green plant production. Despite similar overall regeneration rates, "Mercada" predominantly produces albino regenerants (>90%), while "Jersey" yields mostly green plants. Cytological analysis reveals that in "Mercada," an early transition of proplastid into amyloplasts occurs early in pollen development. Thus, at the ML stage, routinely used for stress treatment, nearly 50% of plastids have already differentiated into amyloplasts, with the remainder consisting largely of proplastids and a minor fraction of undifferentiated plastids. Conversely, microspores of "Jersey" at the same developmental stage contain predominantly undifferentiated proplastids, with amyloplasts observed only in binucleate pollen [27]. Cultivars "Jersey" and "Mercada" also exhibit distinct expression patterns of starch biosynthesis genes (*SbeI*, *GBSS*, *SSII*, *Dpe2*) during pollen development. At early stages, expression levels are low and similar in both genotypes. However, in "Mercada," a sharp increase is already observed at the early-to-mid (EM) uninucleate stage, while in "Jersey," expression remains low until the ML stage. These results indicate a faster differentiation of amyloplast from proplastids in cv. "Mercada," a cultivar that predominantly produces albino plants.

These findings suggest that early plastid differentiation into nonphotosynthetic forms, such as amyloplasts, may compromise the potential for chloroplast development and, consequently, green plant regeneration. At the cellular level, plastids in albino plants remain arrested at an early developmental stage, reflecting the absence and/

or dysfunction of key factors necessary for plastid differentiation and proper chloroplast biogenesis.

1.2 Mechanisms Underlying Albino Plant Regeneration During Androgenesis

One of the earliest hypotheses proposed to explain defects in chloroplast development in albino plants during androgenesis is the occurrence of deletions within the plastid genome. The plastome exists as multiple copies of circular DNA located in the stroma. The number of plastome copies per plastid varies widely, ranging from 10 to 500 copies [41, 42]. The circular DNA consists of two single-copy regions: the long single copy (LSC) region and the short single copy (SSC) region, which are separated by two inverted repeat (IR) regions (Fig. 2). Each IR region contains the same set of genes, including all rRNAs required for plastid translation; therefore, these genes are present in two copies. The plastid genome encodes approximately 140 genes, including those for RNA polymerase (PEP; plastid-encoded

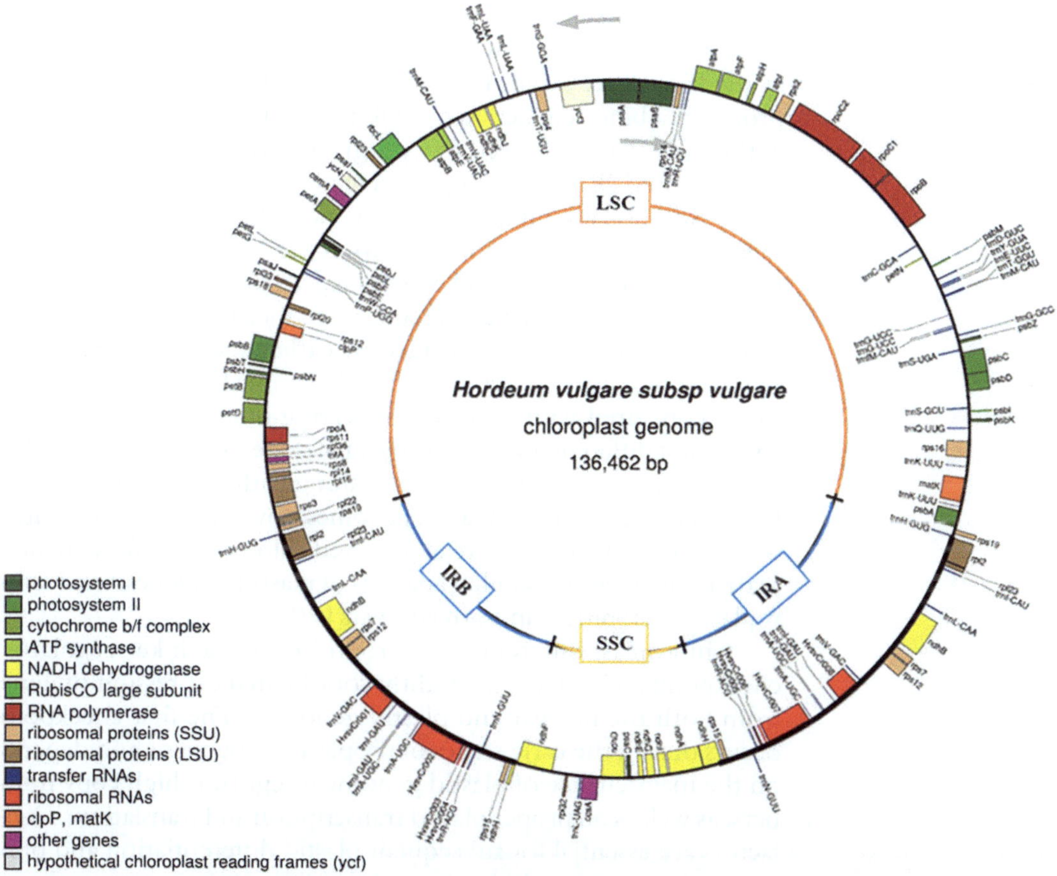

Fig. 2 Structure of the chloroplast genome of barley. LSC, long single copy region; SSC, short single copy region; IR, inverted repeat region. Arrows indicate the direction of transcription. Diagram generated using OGDraw version 1.3.1 and modified from Greiner et al. [62]

polymerase), tRNAs, rRNAs, several ribosomal subunits, and proteins that are components of the photosystems [43].

Deletions within the plastid genome have been reported in albino plantlets in several cereal species, including wheat [44–46], rice [47, 48], barley [45, 49, 50], and triticale [46]. Most of these deletions are localized to the LSC region, considered the most unstable part of the plastid genome. The LSC region primarily contains genes involved in photosynthesis, including those encoding components of photosystem I and II (Fig. 2). In addition to LSC deletions, albino plantlets derived from callus tissue have also exhibited deletions in the SSC and IR regions. These structural rearrangements are accompanied by reduced expression of key plastid genes, including *rbcL*, *psbD-psbC*, *16S rRNA*, and *23 rRNA* in albino compared to green plants, as demonstrated by Northern blot hybridization [49].

Comparative analysis of plastid genome copy number during in vivo pollen development revealed clear cultivar-specific differences associated with the overall capability of the genotype to regenerate green/albino plants in androgenesis. In cv. "Jersey," which produces mostly green plants, the circular DNA copy number remains stable from early to late uninucleate stage, with a decline observed only at the binucleate stage. In contrast, cv. "Mercada," which mainly produces albino regenerants, exhibits a higher initial copy number at the early stage, followed by a progressive reduction starting from the early-mid (EM)) microspore stage throughout pollen development. Additionally, high variability in individual gene copy numbers in early and EM microspores of "Mercada" indicates that in some microspores the onset of plastome degradation takes place before reaching the optimal developmental stage for stress application. Plastome degradation is accompanied by earlier and higher expression of starch biosynthesis genes at the EM uninucleate stage of "Mercada" compared with "Jersey," indicating premature activation of amyloplast differentiation at the optimal stage for androgenesis [27].

The occurrence of amyloplasts, along with early plastome instability, and a reduction in plastid genome copy number in "Mercada" microspores, may contribute to impaired chloroplasts differentiation, as demonstrated by detailed analyses of plastid biogenesis and chloroplast development in both cultivars [20].

Chloroplast differentiation progresses through key regulatory checkpoints relying on the tightly coordinated expression of genes from both the nuclear and plastid genomes. The first checkpoint occurs during the early stages of proplastid biogenesis and depends on the maintenance of plastid genome integrity at high copy numbers, as well as on proper plastid transcription and translation. These factors are essential for subsequent plastid differentiation and photomorphogenesis, a light-dependent differentiation stage that is fully controlled by the nuclear genome [51–53]. When proplastid development proceeds correctly, a retrograde signal is transmitted from the plastid to the nucleus [53, 54]. One of earliest signals is the transcription of tRNA^{Glu} (tRNA^{E}), which depends on the SIG2 factor and serves as a substrate for chlorophyll biosynthesis [54, 55].

Plastid gene transcription is carried out by two RNA polymerases: the nuclear-encoded NEP and the plastid-encoded PEP (Fig. 3). NEP is a single-subunit phage-type polymerase, whereas PEP is a multisubunit bacterial-type polymerase composed of subunits encoded by *rpoA*, *rpoB*, *rpoC1*, and *rpoC2*. PEP requires nuclear-encoded sigma (SIG) factors for transcription initiation, and upon light exposure, PEP forms a larger complex with PEP-associated proteins (PAPs) [56–58]. Plastid genes are often organized in operons and can be differentially expressed through alternative promoters. NEP and PEP recognize different promoter motifs, resulting in three classes of plastid genes: PEP-dependent (mainly photosynthesis-related genes), NEP/PEP-shared (housekeeping genes), and NEP-only (e.g., *rpoB*) [59]. During the early stages of plastid development, NEP is primarily active and ensures the accumulation of PEP subunits. As chloroplasts mature, PEP activity increases while NEP activity is downregulated. This transition is mediated by SIG2-dependent transcription of tRNA^{Glu}, which functions both as a translation component and as a precursor of chlorophyll biosynthesis. Accumulation of tRNA^{Glu} transcripts inhibits NEP activity, thereby promoting the switch to PEP-dependent transcription [60].

The expression levels of key transcription-related genes have been analyzed in barley cultivars during embryo formation and regeneration of green (cv. "Jersey") and albino (cv. "Mercada") plants. Expression profiles of PEP subunits differ markedly between the two cultivars. The *rpoA* gene (PEP-dependent) shows a gradual upregulation starting from the embryo-to-plant conversion stage in "Jersey," whereas in "Mercada," its expression increases only at the young plantlet stage. In contrast, *rpoB* (NEP-dependent) expression

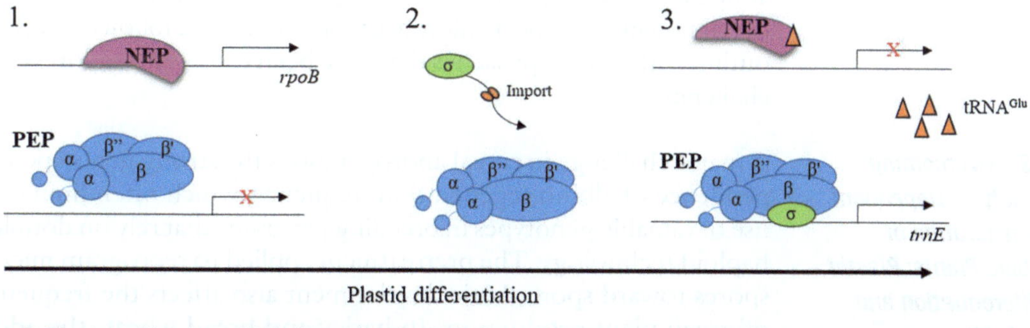

Fig. 3 The transition from NEP-dependent to PEP-dependent transcription. (**1**) In the early stage of proplastid development, the transcription of plastid genes is dependent on the NEP polymerase. Although the PEP polymerase is present, it is inactive due to the absence of sigma factors. (**2**) The SIG2 factor, essential for the transcription of plastid biogenesis-related genes, is imported into the proplastid. (**3**) As a result of the PEP complex activity, the *trnE* gene is expressed, encoding the tRNA^{Glu} molecule, which binds to the NEP and suppresses NEP-dependent gene expression

remains stable in "Jersey" during the embryo-to-plant conversion stage but increases sharply in "Mercada," indicating sustained NEP activity. To investigate the NEP-to-PEP transcriptional switch, the expression of *SIG2* and $tRNA^{Glu}$ has been analyzed. Both genes are more strongly upregulated at the embryo-to-plant conversion stage in Jersey, suggesting efficient activation of PEP and repression of NEP. In "Mercada" their expression is delayed, which may reflect impaired transcriptional reprogramming [20].

Due to the failed transition from NEP- to PEP-dependent transcription, no significant increase in plastid rRNA levels is observed in differentiated embryos or regenerated plants of "Mercada." In contrast, cv. "Jersey" exhibits a 20–30-fold increase in plastidial *16S* and *23S* rRNA genes during the embryo-to-plant conversion stage (Fig. 4). As a result of disrupted checkpoint progression, no activation of *GLK* transcription factors (*GLK1* and *GLK2*), which are key regulators of photosynthesis-associated nuclear genes, is detected in the embryos or albino regenerants of cv. "Mercada" [20]. A similar mechanism was observed in wheat. Canonge et al. [61] have investigated the occurrence of albino plantlets during wheat androgenesis by comparing two cultivars: "Pavon," which exhibits a high capacity for green plant regeneration, and "Paledor," which predominantly produces albino plants. They have analyzed the expression of key genes involved in chloroplast biogenesis at several developmental stages, from isolated microspores through multicellular structures to early proembryos (days 7 and 21 of culture). The cultivar "Paledor" exhibits the same key features as the barley cultivar "Mercada," including failure of the NEP- to PEP-transcriptional switch, low expression of plastid rRNAs, and delayed activation of *GLK1*, already detectable in multicellular structures and proembryos [61]. These findings underscore the conserved role of plastid biogenesis disruptions in albino plant regeneration across cereals. Based on this developmental insight, efforts have been made to develop strategies aimed at mitigating albinism by targeting critical stages of plastid development to support chloroplast formation. The following section outlines effective approaches that have been explored to address this challenge.

1.3 Overcoming Genotype-Dependent Regeneration of Albino Plants: Plastid Differentiation and Early Microspore Culture Initiation in Barley

A major challenge in cereal androgenesis is the strong genotype-dependence of albino regeneration frequency, which often limits the use of valuable genotypes in breeding programs that rely on doubled haploid technology. The pretreatment applied to reprogram microspores toward sporophytic development also affects the frequency of green plant production. In barley and bread wheat, the addition of dimethyl sulfoxide (DMSO) to the mannitol pretreatment medium increases the number of green regenerants [10]. In rye, osmotic stress induced by mannitol facilitates the regeneration of green plants [14]. The composition of the regeneration medium is also critical for achieving a high yield of green plants. In both barley

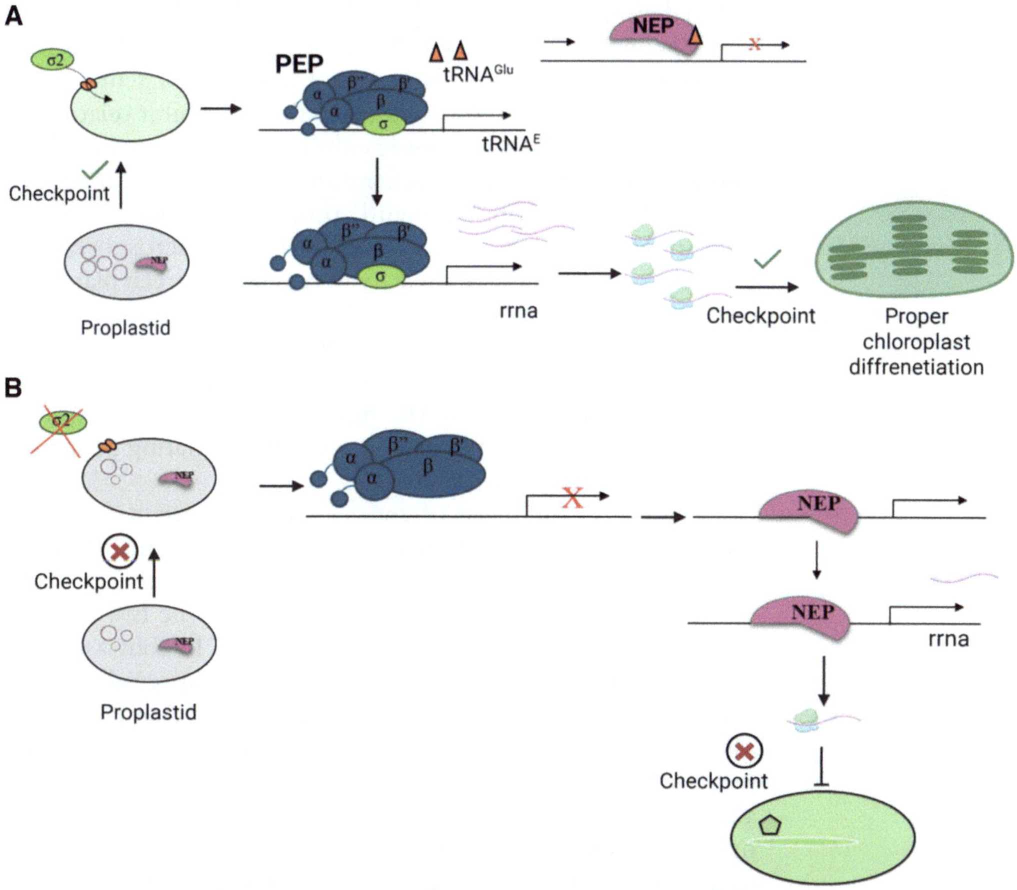

Fig. 4 Early stages of plastid development during the regeneration of androgenic plants. (**A**) Schematic representation of proper chloroplast development from proplastids in green plants regenerated from microspore cultures. A stable plastome (ptDNA) and correct transcriptional activity allow progression through the checkpoint and the import of sigma factor 2 (σ2). This enables a functional switch from NEP to PEP driven transcription. Efficient transcription of plastid rRNA genes (e.g., *rrna*) by the PEP complex ensures effective translation. Successful early plastid development permits the activation of photomorphogenesis, ultimately leading to the formation of fully differentiated chloroplast. (**B**) Schematic representation of disrupted plastid development in regenerating albino plants. Instability of the plastid genome, indicated by variable gene copy numbers, prevents checkpoint progression and blocks the import of sigma factor 2 (σ2). As a result, transcription remains NEP-dependent with no activation of PEP polymerase. Consequently, plastid rRNA gene expression remains low, leading to poor translational efficiency, failure to trigger photomorphogenesis, and developmental arrest of chloroplasts at an early stage

and triticale, the use of a regeneration medium supplemented with a copper source and Gamborg B-5 vitamins significantly increases the ability to produce green plantlets [8, 38].

Comparative studies on the barley cultivars "Jersey" and "Mercada" have shown that the genotype-dependent albino regeneration is closely linked to plastid differentiation [27]. To further investigate the relationship between the rate of amyloplast differentiation during pollen development and the frequency of albino plantlets, Gajecka et al. [27] conducted an extended analysis of amyloplast differentiation alongside the expression of starch synthesis-related genes across ten spring barley cultivars. Although all cultivars exhibit relatively high regeneration rates in microspore culture, they differ markedly in the frequency of green plant regeneration, which ranged from 10% to 98%. Cultivars that predominantly produce albino plantlets show a substantial accumulation of amyloplasts in ML microspores. In contrast, cultivars with a high capacity to regenerate green plants mostly contain undifferentiated proplastids in ML microspores, with only a few proplastids showing early signs of differentiation to amyloplasts. Consequently, a strong positive correlation is observed between the number of amyloplasts in ML microspores and the frequency of albino plants, whereas the presence of differentiating proplastids is negatively correlated with albinism.

The expression levels of genes associated with starch synthesis during pollen grain development confirm these results. In cultivars with green plant regeneration below 60%, a significant upregulation of *Dpe2* and *GBSS* genes is already observed at the EM microspore developmental stage, similar to cv. "Mercada." In contrast, cultivars with over 60% green regenerants exhibit no upregulation at the EM stage, with *Dpe2* and *GBSS* expression increasing only at ML microspore stage, similar to cv. "Jersey." Notably, the high increase in *GBSS* gene expression at the EM stage is strongly correlated with albino plant regeneration, indicating that it may serve as a reliable marker of albinism in barley androgenesis [27]. Identifying genotypes incapable of producing green plants prior to androgenesis allows for the provision of a sufficient number of donor plants and the establishment of cultures with a high number of replicates to produce the desired number of DH lines. It also facilitates the exclusion of genotypes that predominantly produce albino plantlets from breeding programs where the efficient generation of large numbers of DH lines is a primary goal. However, when such genotypes possess valuable agronomic traits and cannot be excluded from breeding programs, it is essential to develop strategies to overcome the genotype-dependent tendency for albino plant regeneration. Based on the observed correlation between a genotype ability to produce green plant at high frequency and the presence of undifferentiated proplastids in microspores, Gajecka et al. [27] have proposed to use EM microspore stage rather than the routinely used ML stage for androgenesis. In their studies, cultivars that predominantly produce green plantlets in cultures initiated from ML microspores show a 30% reduction in total regeneration rate when cultures are initiated at the EM stage; nevertheless, the proportion of green to albino regenerants remains unchanged.

In contrast, cultivars prone to producing mostly albino plantlets when using ML microspores respond very differently. No significant difference in total regeneration is observed between cultures

Fig. 5 Regeneration of androgenic plants in isolated microspore cultures initiated using early-mid (EM) and mid-to-late (ML) microspores of barley cultivars "Bordo," "Justina," and "Mercada"

initiated at the EM and ML stages. However, a significant three- to four-fold increase in the frequency of green plant regeneration is recorded when EM microspores are used for androgenesis. Two cultivars, "Bordo" and "Justina," produce over 80% green plants when cultures are initiated using EM microspores, compared to less than 20% from ML microspores. Similarly, in the cultivar "Mercada," green plantlets account for about 50% of the regenerants from EM microspores cultures, in contrast to only 10% from ML microspores (Fig. 5).

Although the use of early-mid-stage microspores is technically more demanding, due to their smaller size and the careful handling required during mechanical release and centrifugation, it significantly enhances green plant regeneration in cultivars that predominantly produce albino plants under standard protocols. Combining the use of *GBSS* expression as a marker to identify genotypes prone to albino regeneration together with the initiation of cultures from early-mid microspores provides an effective strategy to increase DH line production independently of genotype, which is an important advancement for cereal breeding programs.

References

1. Kumari M, Clarke HJ, Small I, Siddique KHM (2009) Albinism in plants: a major bottleneck in wide hybridization, androgenesis and doubled haploid culture. CRC Crit Rev Plant Sci 28:393–409. https://doi.org/10.1080/07352680903133252
2. Lantos C, Markó F, Mihály R, Pauk J (2024) Comparative analyses of green plantlet regeneration

in barley (*Hordeum vulgare* L.) anther culture. Agriculture (Switzerland) 14:1440. https://doi.org/10.3390/agriculture14091440
3. Makowska K, Oleszczuk S, Zimny A et al (2015) Androgenic capability among genotypes of winter and spring barley. Plant Breed 134:668–674. https://doi.org/10.1111/pbr.12312
4. Lantos C, Pauk J (2016) Anther culture as an effective tool in winter wheat (*Triticum aestivum* L.) breeding. Russ J Genet 52:794–801. https://doi.org/10.1134/S102279541608007X
5. Sharafat AR, Ari E, Izmirli A (2023) The effect of dimethyl sulfoxide on embryogenesis and green plant regeneration in wheat (*Triticum aestivum* L.) anther culture. Biol Life Sci Forum 27:46. https://doi.org/10.3390/IECAG2023-1667
6. He T, Yang Y, Tu SB et al (2006) Selection of interspecific hybrids for anther culture of indica rice. Plant Cell Tissue Organ Cult 86:271–277. https://doi.org/10.1007/s11240-006-9117-z
7. Sakina A, Mir S, Najeeb S et al (2020) Improved protocol for efficacious in vitro androgenesis and development of doubled haploids in temperate japonica rice. PLOS One 15:1–18. https://doi.org/10.1371/journal.pone.0241292
8. Kruppa J, Kanbar OZ, Tóth-Lencsés KA et al (2023) Induction of triticale (×Triticosecale Wittmack) in vitro androgenesis in anther cultures of F1 hybrid combinations, varieties and homogeneity testing of offspring generation. Life 13:1970. https://doi.org/10.3390/life13101970
9. Castillo AM, Nielsen NH, Jensen A, Vallés MP (2014) Effects of n-butanol on barley microspore embryogenesis. Plant Cell Tissue Organ Cult 117:411–418. https://doi.org/10.1007/s11240-014-0451-2
10. Echávarri B, Cistué L (2016) Enhancement in androgenesis efficiency in barley (*Hordeum vulgare* L.) and bread wheat (*Triticum aestivum* L.) by the addition of dimethyl sulfoxide to the mannitol pretreatment medium. Plant Cell Tissue Organ Cult (PCTOC) 125:11–22. https://doi.org/10.1007/S11240-015-0923-Z
11. Esteves P, Belzile F (2014) Improving the efficiency of isolated microspore culture in six-row spring barley: I-optimization of key physical factors. Plant Cell Rep 33:993–1001. https://doi.org/10.1007/s00299-014-1583-x
12. Khatun R, Shahinul Islam SM, Ara I et al (2012) Effect of cold pretreatment and different media in improving anther culture response in rice (*Oryza sativa* L.) in Bangladesh. Indian J Biotechnol 11:458–463
13. Würschum T, Tucker MR, Maurer HP (2014) Stress treatments influence efficiency of microspore embryogenesis and green plant regeneration in hexaploid triticale (×Triticosecale Wittmack L.). In Vitro Cell Dev Biol Plant 50:143–148. https://doi.org/10.1007/s11627-013-9539-3
14. Zieliński K, Krzewska M, Żur I et al (2020) The effect of glutathione and mannitol on androgenesis in anther and isolated microspore cultures of rye (*Secale cereale* L.). Plant Cell Tissue Organ Cult 140:577–592. https://doi.org/10.1007/s11240-019-01754-9
15. Asif M, Eudes F, Goyal A et al (2013) Organelle antioxidants improve microspore embryogenesis in wheat and triticale. In Vitro Cell Dev Biol Plant 49:489–497. https://doi.org/10.1007/s11627-013-9514-z
16. Esteves P, Clermont I, Marchand S, Belzile F (2014) Improving the efficiency of isolated microspore culture in six-row spring barley: II-exploring novel growth regulators to maximize embryogenesis and reduce albinism. Plant Cell Rep 33:871–879. https://doi.org/10.1007/s00299-014-1563-1
17. Ferrie AMR, Irmen KI, Beattie AD, Rossnagel BG (2014) Isolated microspore culture of oat (*Avena sativa* L.) for the production of doubled haploids: effect of pre-culture and post-culture conditions. Plant Cell Tissue Organ Cult 116:89–96. https://doi.org/10.1007/s11240-013-0385-0
18. Caredda S, Devaux P, Sangwan RS, Clement C (1999) Differential development of plastids during microspore embryogenesis in barley. Protoplasma 208:248–256. https://doi.org/10.1007/BF01279096
19. Caredda S, Devaux P, Sangwan RS et al (2004) Plastid ultrastructure and DNA related to albinism in androgenetic embryos of various barley (*Hordeum vulgare*) cultivars. Plant Cell Tissue Organ Cult 76:35–43. https://doi.org/10.1023/A:1025812621775
20. Gajecka M, Marzec M, Chmielewska B et al (2021) Changes in plastid biogenesis leading to the formation of albino regenerants in barley microspore culture. BMC Plant Biol 21:1–24. https://doi.org/10.21203/rs.3.rs-42246/v1
21. Sharma A, Singh MB, Bhalla PL (2015) Ultrastructure of microsporogenesis and microgametogenesis in *Brachypodium distachyon*. Protoplasma 252:1575–86. https://doi.org/10.1007/s00709-015-0793-6
22. Pacini E (2010) Relationships between tapetum, loculus, and pollen during development. Int J Plant Sci 171:1–11. https://doi.org/10.1086/647923
23. Szarejko I (2003) Anther culture for doubled haploid production in barley (*Hordeum vulgare*

L.). In: Maluszynski M, Kasha KJ, Forster BP, Szarejko I (Eds.) Doubled haploid production in crop plants. Kluwer Academic Publishers, Dordrecht, Netherlands, 35–42

24. Carrizo García C, Nepi M, Pacini E (2017) It is a matter of timing: asynchrony during pollen development and its consequences on pollen performance in angiosperms—a review. Protoplasma 254:57–73. https://doi.org/10.1007/s00709-016-0950-6
25. Jarvis P, López-Juez E (2013) Biogenesis and homeostasis of chloroplasts and other plastids. Nat Rev Mol Cell Biol 20:787–802. https://doi.org/10.1038/nrm3702
26. Greiner S, Sobanski J, Bock R (2015) Why are most organelle genomes transmitted maternally. Bioessays 37:80–94. https://doi.org/10.1002/bies.201400110
27. Gajecka M, Marzec M, Chmielewska B et al (2020) Plastid differentiation during microgametogenesis determines green plant regeneration in barley microspore culture. Plant Sci 291:110321. https://doi.org/10.1016/j.plantsci.2019.110321
28. Koonjul PK, Minhas JS, Nunes C et al (2005) Selective transcriptional down-regulation of anther invertases precedes the failure of pollen development in water-stressed wheat. J Exp Bot 56:179–190. https://doi.org/10.1093/jxb/eri018
29. Pacini E, Jacquard C, Clément C (2011) Pollen vacuoles and their significance. Planta 234:217–227. https://doi.org/10.1007/s00425-011-1462-4
30. Nakamura Y (2015) Biosynthesis of reserve starch. In: Nakamura Y (ed) Starch: metabolism and structure. Springer, Tokyo, pp. 161–209. https://doi.org/10.1007/978-4-431-55495-0
31. Islam SMS, Tuteja N (2012) Enhancement of androgenesis by abiotic stress and other pretreatments in major crop species. Plant Sci 182:134–144. https://doi.org/10.1016/j.plantsci.2011.10.001
32. Dunwell JM (2010) Haploids in flowering plants: origins and exploitation. Plant Biotechnol J 8:377–424. https://doi.org/10.1111/j.1467-7652.2009.00498.x
33. Maluszynski, M, Kasha, KJ, Forster, BP, Szarejko, I, 2003. Doubled haploid production in crop plants A manual. Kluwer Academic Publishers, Dordrecht, Netherlands
34. Hoekstra S, Van Bergen S, Van Brouwershaven IR et al (1997) Androgenesis in *Hordeum vulgare* L.: effects of mannitol, calcium and abscisic acid on anther pretreatment. Plant Sci 126:211–218. https://doi.org/10.1016/S0168-9452(97)00096-4
35. Kruczkowska H, Pawłowska H, Skucińska B (2002) Influence of anther pretreatment on the efficiency of androgenesis in barley. J Appl Genet 43:287–296
36. Li H, Devaux P (2005) Isolated microspore culture overperforms anther culture for green plant regeneration in barley (*Hordeum vulgare* L.). Acta Physiol Plant 27:611–619. https://doi.org/10.1007/s11738-005-0065-8
37. Guo Y-D, Pulli S (2000) Isolated microspore culture and plant regeneration in rye (*Secale cereale* L.). Plant Cell Rep 19:875–880. https://doi.org/10.1007/s002990000194
38. Lantos C, Bóna L, Boda K, Pauk J (2014) Comparative analysis of in vitro anther- and isolated microspore culture in hexaploid Triticale (X Triticosecale Wittmack) for androgenic parameters. Euphytica 197:27–37. https://doi.org/10.1007/s10681-013-1031-y
39. Castillo AM, Sanchez-Diaz RA, Valles MP (2015) Effect of ovary induction on bread wheat anther culture: ovary genotype and developmental stage, and candidate gene association. Front Plant Sci 6:1–12. https://doi.org/10.3389/fpls.2015.00402
40. Zheng MY, Liu W, Weng Y et al (2001) Culture of freshly isolated wheat (*Triticum aestivum* L.) microspores treated with inducer chemicals. Plant Cell Rep 20:685–690. https://doi.org/10.1007/s00299-001-0393-0
41. Sakamoto W, Miyagishima S, Jarvis P (2008) Chloroplast biogenesis: control of plastid development, protein import, division and inheritance. Arabidopsis Book 6:e0110. https://doi.org/10.1199/tab.0110
42. Pogson BJ, Ganguly D, Albrecht-Borth V (2015) Insights into chloroplast biogenesis and development. Biochim Biophys Acta Bioenerg 1847:1017–1024. https://doi.org/10.1016/j.bbabio.2015.02.003
43. Bock R (2007) Structure, function, and inheritance of plastid genomes. In: Bock R (Ed.)Eds. Cell and molecular biology of plastids. Springer, Berlin, Heidelberg, 29–63
44. Day A, Ellis THN (1984) Chloroplast DNA deletions associated with wheat plants regenerated from pollen: possible basis for maternal inheritance of chloroplasts. Cell 39:359–368. https://doi.org/10.1016/0092-8674(84)90014-X
45. Day A, Ellis THN (1985) Deleted forms of plastid DNA in albino plants from cereal anther culture. Curr Genet 9:671–678. https://doi.org/10.1007/BF00449820
46. Mozgova GV, Zaitseva OI, Lemesh VA (2012) Structural changes in chloroplast genome accompanying albinism in anther culture of wheat and triticale. Cereal Res Commun 40:467–475. https://doi.org/10.1556/CRC.2012.0007
47. Harada T, Sato T, Asaka D, Matsukawa I (1991) Large-scale deletions of rice plastid DNA in

anther culture. Theor Appl Genet 81:157–161. https://doi.org/10.1007/BF00215717
48. Yamagishi M (2002) Heterogeneous plastid genomes in anther culture-derived albino rice plants. Euphytica 123:67–74. https://doi.org/10.1023/A:1014493316433
49. Dunford R, Walden RM (1991) Plastid genome structure and plastid-related transcript levels in albino barley plants derived from anther culture. Curr Genet 20:339–347. https://doi.org/10.1007/BF00318524
50. Mouritzen P, Holm PB (1994) Chloroplast genome breakdown in microspore cultures of barley (*Hordeum vulgare* L.) occurs primarily during regeneration. J Plant Physiol 144:586–593. https://doi.org/10.1016/S0176-1617(11)82141-3
51. Lepage E, Zampini E, Brisson N (2013) Plastid genome instability leads to reactive oxygen species production and plastid-to-nucleus retrograde signaling in arabidopsis. Plant Physiol 163:867–881. https://doi.org/10.1104/pp.113.223560
52. Maréchal A, Brisson N (2010) Recombination and the maintenance of plant organelle genome stability. New Phytol 186:299–317. https://doi.org/10.1111/j.1469-8137.2010.03195.x
53. Pfannschmidt T (2010) Plastidial retrograde signalling—a true "plastid factor" or just metabolite signatures?. Trends Plant Sci 15:427–435. https://doi.org/10.1016/j.tplants.2010.05.009
54. Woodson JD, Perez-Ruiz JM, Schmitz RJ et al (2013) Sigma factor-mediated plastid retrograde signals control nuclear gene expression. Plant J 73:1–13. https://doi.org/10.1111/tpj.12011
55. Kanamaru K, Nagashima A, Fujiwara M et al (2001) An arabidopsis sigma factor (SIG2)-dependent expression of plastid-encoded tRNAs in chloroplasts. Plant Cell Physiol 42:1034–1043. https://doi.org/10.1093/pcp/pce155
56. Chi W, He B, Mao J et al (2015) Plastid sigma factors: their individual functions and regulation in transcription. Biochim Biophys Acta Bioenerg 1847:770–778. https://doi.org/10.1016/j.bbabio.2015.01.001
57. Pfannschmidt T, Blanvillain R, Merendino L et al (2015) Plastid RNA polymerases: orchestration of enzymes with different evolutionary origins controls chloroplast biogenesis during the plant life cycle. J Exp Bot 66:6957–6973. https://doi.org/10.1093/jxb/erv415
58. Yagi Y, Shiina T (2014) Recent advances in the study of chloroplast gene expression and its evolution. Front Plant Sci 5:61. https://doi.org/10.3389/fpls.2014.00061
59. Liebers M, Grübler B, Chevalier F et al (2017) Regulatory shifts in plastid transcription play a key role in morphological conversions of plastids during plant development. Front Plant Sci 8:23. https://doi.org/10.3389/fpls.2017.00023
60. Börner T, Aleynikova A, Zubo YO, Kusnetsov VV (2015) Chloroplast RNA polymerases: role in chloroplast biogenesis. Biochimi Biophys Acta Bioenerg 1847:761–769. https://doi.org/10.1016/j.bbabio.2015.02.004
61. Canonge J, Roby C, Hamon C et al (2021) Occurrence of albinism during wheat androgenesis is correlated with repression of the key genes required for proper chloroplast biogenesis. Planta 254:123. https://doi.org/10.1007/s00425-021-03773-3
62. Greiner S, Lehwark P, Bock R (2019) OrganellarGenomeDRAW (OGDRAW) version 1.3.1: expanded toolkit for the graphical visualization of organellar genomes. Nucleic Acids Res 47:W59–W64

Part II

Protocol Chapters

Check for updates

Chapter 4

Production of Wheat Doubled Haploids through Intergeneric Hybridization with Maize

Xizhen Guan, Junhua Peng and Daolin Fu

Abstract

Chromosome elimination resulting in haploids is achieved by the rapid loss of chromosomes from one parent during the zygote stage and is an important procedure to produce doubled haploid (DH) lines in plants. Particularly in distant hybridization systems like "wheat × maize," the complete elimination of maize chromosomes can result in wheat haploid embryos. Through embryo rescue and chromosome doubling treatments, pure lines with stable traits can be rapidly obtained. The main operational steps for maize pollen-induced wheat DH technology include planting wheat and maize, emasculating wheat, collecting and pollinating with maize pollen, hormone treatment, embryo rescue, doubling treatment, and harvesting individual plants. The integration of modern biotechnology tools, such as molecular marker-assisted selection, gene editing, induced mutagenesis, transgenics, and genomic selection with this technology will further shorten breeding cycles, enhance selection accuracy, and significantly improve the efficiency of genetic improvement in wheat.

Keywords Wheat, Chromosome elimination, Maize pollen induction, Speed breeding, Doubled haploid

1 Introduction

In many countries, distinctness, uniformity, and stability (DUS) are prerequisites for a new plant variety to obtain protection and registration [1]. For wheat varieties, these attributes largely depend on the ratio of the homozygous DNA loci and seed purity. The higher the ratio of homozygous DNA loci, the better the uniformity and stability of the variety. For instance, varieties with a ratio of homozygous DNA loci of less than 90% have poor uniformity and stability [2]. Seed purity is an important indicator used to evaluate the quality of wheat seeds. The China National Standard (GB 4404.1-2008) stipulates that the purity of wheat foundation seeds should not be less than 99.9%, and the purity of wheat variety seeds must be at least 99.0% [3]. Fulfillment of these requirements needs a long breeding cycle. To accelerate the breeding process, haploid

Ana María Castillo et al. (eds.), *Doubled Haploid Methods in Cereals*, Methods in Molecular Biology, vol. 3029,
https://doi.org/10.1007/978-1-0716-5245-9_4,

induction and chromosome doubling techniques have become crucial procedures to produce doubled haploids (DH) plants [4]. This technology can produce pure lines with 100% homozygosity in a shorter time frame, overcoming the slow process of achieving homozygosity in conventional breeding methods [5]. However, DH plants rarely occur spontaneously in nature and usually need to be produced using in vitro or in vivo induction methods [6]. In wheat, the production of DH plant by anther or microspore culture is limited due to its genotype dependence, time-consuming regeneration, unstable ploidy, and albino seedlings [7, 8]. In contrast, intergeneric hybridization between wheat and maize is considered one of the most efficient methods for DH production. It achieves 2–3 times higher efficiency than anther culture [9] and has been successfully applied across diverse wheat genotypes [10–15]. Furthermore, this method is less dependent on the wheat genotype and does not need haploid identification. This methodology has been continuously optimized by previous researchers and has become the most practical in common wheat [16–18]. After 3 years of research, we have identified and optimized the factors affecting the efficiency of wheat DH line production in crosses with maize. These factors include wheat and maize genotypes, wheat growth conditions, pollination and hormone treatment methods, embryo rescue, and chromosome doubling treatments. Based on our findings, we propose a scheme suitable for the efficient production of wheat DH lines.

2 Materials

2.1 Plant Material

Each year, approximately 400 new wheat hybrid combinations (F_1) are generated and planted in the field in autumn (a single row of 1 m length for each combination). In the following year, about 200 F_1 are selected after phenotypic evaluation. In the same autumn, F_1 seeds are planted in the field (plot area of 7.5 m^2 and a seeding rate of 22.5 kg/ha). After field screening, about 20 F_2 are selected and further advanced to the F_3 generation. Induction of DH lines is conducted on the selected F_2 and F_3 combinations (using the remaining seeds from the previous generation).

2.2 Equipment and Instruments

1. Vernalization Chamber (Fig. 1A): Temperature 4°C, light intensity 2880 lux at 15 cm below the LED, provided by PHILIPS 13W/68×0.2W/LED Module, with 19 h/5 h (light/dark).
2. Maize Greenhouse (Fig. 1B): Temperature 25 ± 1/20 ± 1°C (day/night), light intensity 20,000 lux at 15 cm below the lamp, provided by XUBANG 150W UFO (full spectrum plant growth light lamp), with 12 h/12 h (light/dark).

Fig. 1 Wheat vernalization chamber and maize greenhouse. (**A**) Wheat vernalization chamber. (**B**) Maize greenhouse

3. Wheat Greenhouse: Temperature 23±1/18±1°C (day/night), light intensity 20,000 lux at 15 cm below the lamp, provided by XUBANG 150W UFO (full spectrum plant growth light lamp), with 20 h/4 h (light/dark).
4. Tissue Culture Room: Temperature 20–24°C, light intensity of 4800 lux, provided by WEGA 30W/T8 (full spectrum plant growth light) with 16 h/8 h (light/dark).
5. -20°C Freezer.
6. 4°C Refrigerator.
7. Laminar Flow Cabinet.
8. Fume Hood.
9. Stereomicroscope (1–6.5×).
10. Glass Bead Sterilizer.
11. Autoclave.
12. Oven (37°C).
13. Distilled Water Unit.
14. pH Meter.
15. Magnetic Stirrers.
16. Shaker.
17. Digital Balance (0.0001 accuracy).
18. Pipette (1 mL).

2.3 Plant Growth and Laboratory Materials

1. Seed germination and seedling cultivation substrate composition is shown in Table 1.
2. Seedling growth substrate is composed of peat soil, coconut coir, perlite, and vermiculite in a ratio of 4:2:1:1.

Table 1
Composition of substrate for the seed germination and seedling cultivation

Substrate component	Concentration (g/L)
Nitrate-N	2527.2
Ammonium-N	1814.4
Phosphorus (P)	2138.4
Phosphorus (P_2O_5)	4924.8
Potassium (K)	7128.0
Potassium (K_2O)	8553.6
Magnesium (Mg)	518.4
Magnesium (MgO)	907.2
Boron (B)	19.4
Molybdenum (Mo)	71.3
Copper (Cu)	77.8
Manganese (Mn)	136.1
Zinc (Zn)	45.4
Iron (Fe)	51.8
Wetting agent	0.1 mL/L

3. Smart-Release Fertilizer: Osmocote 15-9-12 (N:P:K)+2 MgO+TE (trace elements).
4. Plastic pots (Ø 17.0 cm × H 16.0 cm) for wheat.
5. Plastic pots (Ø 39.0 cm × H 48.5 cm) for maize.
6. Paper cups (Top Ø 5.0 cm × Bottom Ø 3.5 cm × H 5.0 cm) for wheat and maize.
7. 32-cell plastic seeding tray (Tray: L 54.0 cm × W28.0 cm; Single cell: Top Ø 6.0 cm × H 5.3 cm × Bottom Ø 2.8 cm).
8. 72-cell plastic seeding tray (Tray: L 53.0 cm × W 27.0 cm, Single cell: Top Ø 4.0 cm × H 4.2 cm × Bottom Ø 2.0 cm).
9. Surgical scissors.
10. Forceps (pointed tip).
11. Glassine bags (L 16.0 cm × W 6.0 cm).
12. Stapler.
13. Pencils.
14. Brushes.
15. Spray bottles.
16. Screw-cap 50 mL centrifuge tubes.

17. Centrifuge tubes (2 mL).
18. Scalpels.
19. Medical gauze.
20. Alcohol swabs.
21. Glass Petri dishes (Ø 90.0 mm).
22. Filter paper (Ø 90.0 mm).
23. Culture bottles (650 mL).
24. Waste liquid bucket (Ø 15.0 cm × H 18.0 cm).
25. Disposable Petri dishes (Ø 90.0 mm).
26. Culture bottles (Ø 5.8 cm × H 9.0 cm) (200 mL).
27. Sterilin jars (240 mL).
28. Paper labels (L 29 mm × W 19 mm).
29. Plastic wire.
30. Wide-mouth amber bottles (Ø 9.1 cm × H 10.0 cm) (500 mL).
31. Safety goggles.
32. Gas masks.
33. Rubber gloves.
34. Disposable gloves.
35. Brown paper bags (L 14.5 cm × W 9.0 cm).
36. Mulch film (width 3.0 m).
37. Ground labels.
38. Beakers.
39. Graduated cylinders (1 L).
40. Mesh bags.

2.4 Solutions and Culture Medium

The weighing of chemicals and preparation of solutions require the use of a digital balance (0.0001 g accuracy), pipette (1 mL), and magnetic stirrers.

1. 0.4% KH_2PO_4 Solution: 100 g KH_2PO_4 in 25 L water.
2. Insecticide (Aphids): 20 mL of 2.5% lambda-cyhalothrin (Syngenta) + 3 g of imidacloprid (Syngenta) + 20 L water.
3. Insecticide (Thrips): 20 mL of lambda-cyhalothrin (Syngenta) + 4.5 mL of 5.7% abamectin (Biaoji) + 7.5 mL of 20% dinotefuran (Saijing) in 20 L water.
4. Fungicide: 60 mL of tebuconazole (Syngenta) in 20 L water.
5. 2,4-Dichlorophenoxyacetic acid (2,4-D) stock solution (0.5%): Dissolve 1 g of 2,4-D in 200 mL of 95% ethanol. Aliquot into 2 mL centrifuge tubes and store at −20°C.
6. 2,4-D working solution (0.01%): Dilute 2 mL of the stock solution with distilled water to 100 mL.

7. Seed Disinfection Solutions: 500 mL 70% ethanol + 250 μL Tween 20 and 500 mL 10% NaClO + 250 μL Tween 20.
8. MS medium with vitamins powder (Zkorigin ZK-PYJ-008).
9. 1/2 MS Medium: Dissolve 2.47 g of MS medium with vitamin (Table 2), 20 g of sucrose, 2.4 g of Phytagel (Sigma-Aldrich P8169) in 1 L distilled water. Adjust pH to 5.8, autoclave at 123°C for 15–20 min, allow to cool, and then dispense into 90 mm Petri dishes (25 mL per dish) or culture bottles (40 mL per bottle) for later use.
10. Colchicine Solution (0.05%): Dissolve 0.5 g of colchicine in 15 mL of dimethyl sulfoxide (DMSO), then dilute to a final volume of 1 L with distilled water.
11. Hydrogen Peroxide Solution (H_2O_2) (1%): Dilute 50 mL of 30% H_2O_2 with distilled water to a final volume of 1.5 L.
12. Distilled water.

Table 2
Composition of MS medium (1/2)

Medium component	Concentration (mg/L)
KNO_3	950
NH_4NO_3	825
KH_2PO_4	85
$MgSO_4 \cdot 7H_2O$	185
$CaCl_2 \cdot 2H_2O$	220
KI	0.83
H_3BO_3	6.2
$MnSO_4 \cdot 4H_2O$	22.3
$ZnSO_4 \cdot 7H_2O$	8.6
$Na_2MoO_4 \cdot 2H_2O$	0.25
$CuSO_4 \cdot 5H_2O$	0.025
$CoCl_2 \cdot 6H_2O$	0.025
$Na_2 \cdot EDTA$	37.3
$FeSO_4 \cdot 7H_2O$	27.8
Myo-inositol	100
Glycine	2
Thiamine HCl	0.1
Pyridoxine HCl	0.5
Nicotinic acid	0.5

3 Methods

3.1 Wheat and Maize Seed Planting

To optimize this process, screen a large collection of maize genotypes against different wheat genotypes to identify those with high embryo formation frequency and haploid regeneration frequency (*see* **Note 1**).

1. Select wheat seeds based on field phenotyping, focusing on early-generation breeding materials (F_2 and F_3) or materials requiring purification for research.
2. Select maize seeds for high haploid embryo induction rates and abundant pollen production (*see* **Note 2**). After screening numerous commercial sweet or waxy maize hybrid varieties, choose four maize hybrids.
3. Plant the selected wheat seeds and the four maize hybrids continuously, with sowing conducted once a week for maize (at 7-day intervals).
4. After germination, sow wheat seeds into 72-cell plastic seeding trays and vernalize at 4°C in the vernalization chamber (Fig. 1A) for 40 days (*see* **Note 3**).
5. After vernalization, transplant the wheat seedlings into plastic pots (Ø 17 cm × H 16 cm) containing seedling substrate and 1.5 g of Osmocote fertilizer, with 3 plants per pot, and place them in the wheat greenhouse (23 ± 1/18 ± 1°C day/night, 20 h/4 h light/dark) (*see* **Note 4**).
6. Plant maize seeds in paper cups (Top Ø 5 cm × Bottom Ø 3.5 cm × H 5 cm), with two seedlings per cup. At the three-leaf stage, transplant maize seedlings into plastic pots (Ø 39 cm × H 48.5 cm) containing seedling substrate and 8 g of Osmocote fertilizer, with 2 plants per pot, and place them in the maize greenhouse (25 ± 1/20 ± 1°C day/night, 12 h/12 h light/dark) (Fig. 1B) (*see* **Note 5**).

3.2 Wheat Emasculation

1. Manually emasculate wheat flowers after heading (Fig. 2A). There is an effect of spikelet position on wheat × maize compatibility, with more success observed in the middle spikelets of a spike [19]. Use surgical scissors to cut off the underdeveloped spikelets at the top and base of the spike, leaving 6–8 spikelets in the middle (Fig. 2B). Remove the small flowers in the middle of each spikelet and emasculate the flowers on both sides with forceps. Make a horizontal cut in the middle of each flower to expose the pistil and anthers (Fig. 2B), then remove the anthers with forceps (Fig. 2C).
2. Cover the emasculated spikes with glassine bags, fold and seal the bottom of the bags with a stapler, and label the bags with the wheat material name or ID and the emasculation date using a pencil (Fig. 2D).

Fig. 2 Manual emasculation of wheat spikes. (**A**) Select suitable wheat spikes for emasculation. (**B**) Cut the glume in the middle of each spikelet. (**C**) Remove the three anthers from each flower. (**D**) Cover the emasculated spike with glassine bags and label each bag with the wheat material ID and the date of emasculation

3.3 Maize Pollen Collection and Pollination

1. Collect fresh maize pollen around 10 a.m. from maize tassels in the flowering stage. Gently shake the maize male tassels of different hybrids varieties over A4 paper (Fig. 3A) (*see* **Note 6**).
2. Transfer the pollen into a Petri dish (Ø 90 mm) and gently mix it with a brush (Fig. 3B).
3. Select wheat spikes that were emasculated 1–2 days ago based on the date labeled on the glassine bags. Open the glassine bags and pollinate the wheat spikes with pistils in the feathery stigma stage [11] using a brush (Fig. 3C, 3D) (*see* **Note 7**).
4. Label the pollination time on the glassine bags. For spikes not yet in the feathery stigma stage, fold and reseal the bags and wait until this stage is reached.

3.4 Hormone Treatment

The application of 2,4-D (0.01%) is highly effective for inducing embryo formation in hexaploid wheat haploids [20–22].

1. 24 h after pollination, treat the pollinated spikes with the 0.01% 2,4-D solution (Fig. 4A). After opening the sealed glassine bag, use a small spray bottle with the 2,4-D solution to spray the pollinated spikes, ensuring full coverage so that the wheat spikes are in a water-soaked state (Fig. 4B).

Fig. 3 Maize pollen collection and pollination of wheat spikes. (**A**) Select flowering maize tassels for pollen collection. (**B**) Mix the collected pollen from different maize hybrid varieties. (**C**) Choose suitable wheat spikes for pollination. (**D**) Perform the pollination

Fig. 4 Hormone treatment. (**A**) Application of 2,4-D solution. (**B**) State of wheat spike after hormone spraying

3.5 Embryo Rescue and Haploid Seedling Production

1. Based on the pollination date labeled on the glassine bags, collect wheat spikes 15 days post-pollination and carefully separate the grains (*see* **Note 8**).
2. Disinfect the isolated wheat grains by placing them in screw-cap centrifuge tubes (50 mL) according to wheat material type. Add 70% ethanol and place the tubes on a shaker for surface sterilization for 30 s. After pouring off the ethanol, add sodium

hypochlorite with a final concentration of 10% and sterilize on a shaker for 15 min.

3. After rinsing the grains five times with sterile water in a laminar flow hood, transfer the grains into a Petri dish with filter paper. Dissect the disinfected wheat grains under a stereomicroscope using a scalpel (Fig. 5A) and extract the haploid embryos (Fig. 5B). After dissecting a batch of grains, wipe the scalpel with alcohol swabs and place it in a glass bead sterilizer for sterilization (*see* **Note 9**).

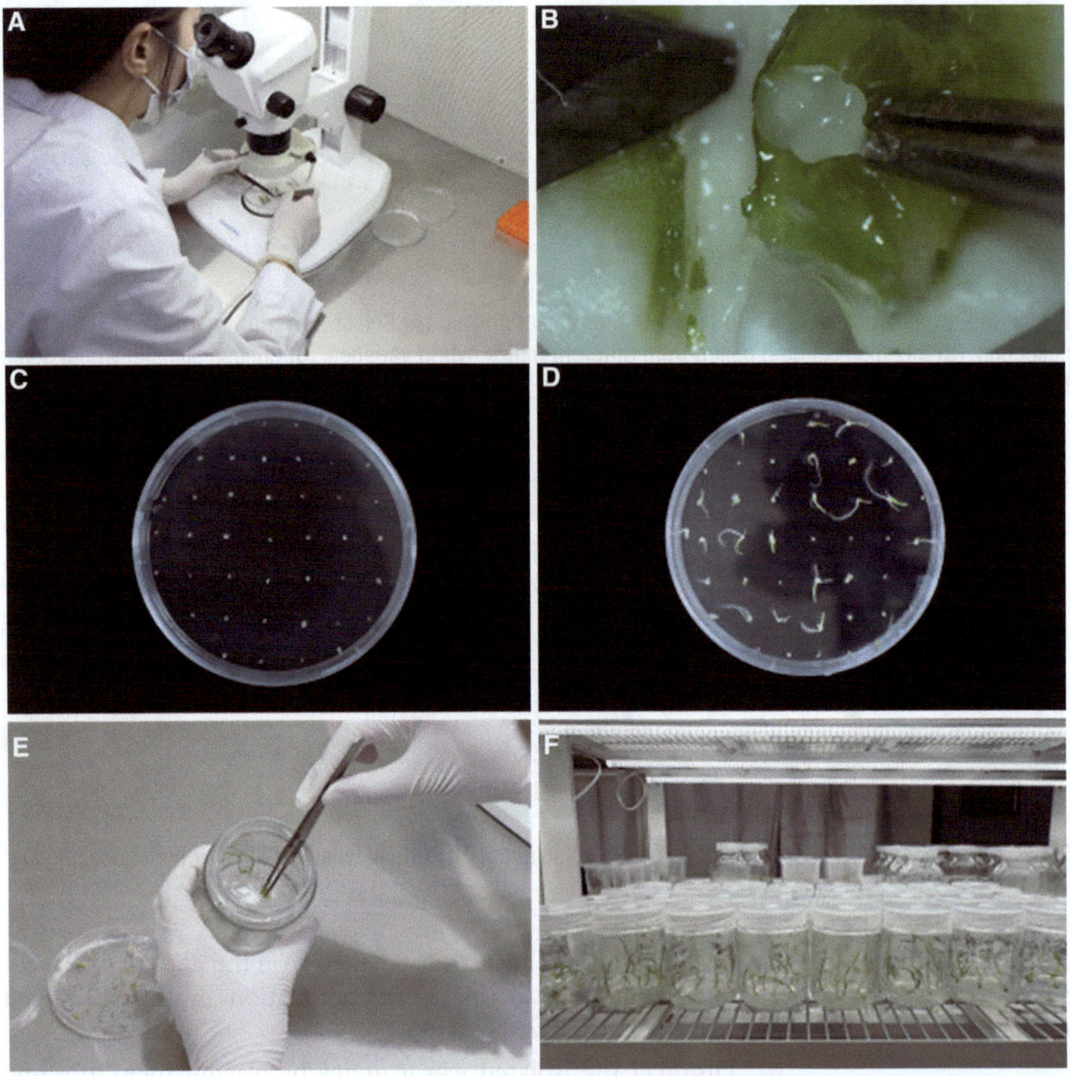

Fig. 5 Wheat haploid embryo rescue process. (**A**) Isolation of embryos. (**B**) Haploid embryos. (**C**) Inoculation of embryos onto culture medium. (**D**) Germination of embryos. (**E**) Transfer of seedlings to jars. (**F**) Seedling growth

4. Inoculate the embryos onto 1/2 MS medium [1/2 MS with 20 g/L sucrose and 2.4 g/L Phytagel (Sigma-Aldrich P8169), pH 5.8].
5. Place the Petri dishes (Ø 90 mm) with embryos (Fig. 5C) in an incubator at 20–24°C with a 16 h/8 h (light/dark) cycle and light intensity of 4800 lux until emergence of roots and shoots (Fig. 5D).
6. Under sterile conditions, transfer the seedlings with roots and shoots to Sterilin jars (240 mL) containing 1/2 MS medium (Fig. 5E) and incubate at 20–24°C with a 16 h/8 h (light/dark) cycle (Fig. 5F) until the seedlings reach the three-leaf stage.

3.6 Haploid Seedling Hardening

1. When the haploid seedlings reach the three-leaf stage, remove the lids from the jars and leave the jars open for 24 h (Fig. 6A).
2. Transplant the haploid seedlings into 32-cell plastic seeding trays filled with seedling substrate (Fig. 6B).
3. Place the seedlings in the wheat greenhouse (23 ± 1/18 ± 1°C day/night, 20 h/4 h light/dark) for hardening.
4. At this stage, collect leaf samples for DNA extraction and genomic analysis.

3.7 Chromosome Doubling Treatment

1. After growing for 15 days, pull out the haploid seedlings (Fig. 6C), clean the roots, and trim them to a length of 2–3 cm (Fig. 6D).
2. Bundle the seedlings with plastic wire according to their wheat genotype or ID, with the binding position set 1 cm above the wheat tiller node (*see* **Note 10**). The other end of the wire should be positioned 1–2 cm above the leaves. Attach a label to this elevated position, indicating the wheat genotype or ID (Fig. 6E).
3. Place the haploid seedlings in a 500 mL wide-mouth amber bottle containing a 0.05% colchicine solution and soak for 16 h in a fume hood. The colchicine solution level should be above the tiller nodes of the haploid seedlings. The treatment conditions should be in the dark, at 25°C (Fig. 6F) (*see* **Note 11**).
4. After completing the doubling treatment, take out the seedlings and wash them under running water for 30 min.

3.8 Hardening, Vernalization, and Transplanting of the Treated Haploid Seedlings and Seed Production

1. Transplant the treated seedlings into 32-cell plastic seeding trays filled with seedling substrate and place them in the wheat greenhouse (23 ± 1/18 ± 1°C day/night, 20 h/4 h light/dark) in a low-light area (Fig. 7A). Ensure the seedlings do not dry out until they produce new tillers (Fig. 7B) or new leaves (Fig. 7C).

Fig. 6 Hardening, transplanting, and chromosome doubling treatment of haploid seedlings. (**A**) Acclimatization of the haploid seedlings. (**B**) Transplanting the haploid seedlings into soil. (**C**) Pulling out the seedlings suitable for chromosome doubling. (**D**) Trimming the roots of haploid seedlings. (**E**) Bundling haploid seedlings by genotype. (**F**) Soaking treatment in colchicine solution

2. Place the seedlings that have developed new tillers or leaves in a vernalization chamber at 4°C for 40 days.
3. After the vernalization treatment, transplant the seedlings into plastic pots (Ø 17 × H 16 cm) filled with the seedling substrate (Fig. 7D), planting two seedlings per pot.
4. Place the pots in the wheat greenhouse (23 ± 1/18 ± 1°C day/night, 20 h/4 h light/dark) (Fig. 7E) until harvest.

Fig. 7 Post-chromosome doubling treatment, transplanting after vernalization of treated seedlings, and harvesting of DH lines. (**A**) Hardening the haploid seedlings following chromosome doubling. (**B**) Emergence of new tillers after doubling. (**C**) New leaves emerging after doubling. (**D**) Transplanting seedlings after vernalization. (**E**) Seedlings in the greenhouse after transplanting. (**F**) Grain development of the individual DH plants. (**G**) Harvesting individual DH plants

5. Harvest the plants when the grains are fully developed and mature (Fig. 7F). Harvest each plant individually (Fig. 7G). Cut the mature spikes and place them into small mesh bags, labeling the bags with the wheat material genotype or ID.
6. Place the harvested wheat materials into an oven (37°C) for drying (48 h), then thresh the dried spikes to obtain the clean seeds and put them into brown paper bags.

Fig. 8 Winter propagation and field performance of DH lines. (**A**) Seeds in paper cups subjected to dormancy-breaking treatment. (**B**) Winter field planting covered with plastic film. (**C–D**) DH lines in the field. (**E–G**) Plot performance of DH lines

3.9 Field Planting and Evaluation

1. Plant the seeds harvested from DH lines in two batches each year. Sow one batch in early October, using seeds harvested before August, which are sown directly into the field. Plant the other batch in late December, where newly harvested

seeds from the last month need dormancy-breaking treatment (Fig. 8A) (*see* **Note 12**). Directly sow seeds harvested more than a month ago.

2. Cover the field with plastic film after sowing (Fig. 8B).
3. Sow the seeds into plots (L 5 m×W 1.5 m) (Fig. 8B–D) in early autumn (early October) and select outstanding DH line plots based on a comprehensive evaluation of traits throughout the growing season, including quality and yield (Fig. 8E–G), for variety registration.
4. Harvest the plants when matured in early summer of the following year (early June).

4 Notes

1. The genotype of wheat and maize, as well as the interaction between wheat and maize genotypes, significantly affect the induction of wheat DH by maize pollen. By leveraging the effects of maize genotypes and the interaction between wheat and maize genotypes, the efficiency of wheat DH production through intergeneric hybridization with maize can be improved [23].
2. Previous attempts using maize inbred lines were less successful due to the poor growth and low pollen production of these lines under greenhouse conditions, which is not conducive to large-scale doubled haploid production.
3. To ensure that all wheat materials or genotypes can spike normally and grow healthily, regardless of winter and spring habits, the vernalization duration is standardized to 40 days.
4. During greenhouse growth, water the plants every 2 days to prevent drought stress on wheat plants. After the jointing stage, apply fungicides and insecticides weekly to control aphids, thrips, and fungal diseases. Spray 0.4 KH_2PO_4 once a week from the jointing stage until harvest of the young embryos.
5. Maize is watered weekly during growth and receives 0.4% KH_2PO_4 weekly after six leaves until tasseling. The primary pest control for maize is the management of thrips. These measures ensure the healthy growth of wheat and maize in the greenhouse, resulting in improved quality of wheat pistils and maize pollen, as well as enhanced recovery and robustness of haploid embryos in the wheat×maize system.
6. Researchers have confirmed that maize genotypes significantly impact the efficiency of wheat×maize system [24–26]. Verma et al. [27] demonstrated that wheat genotypes significantly influence embryo formation frequency, though

their effect is not as pronounced as that of maize genotypes. Avcı and Kutlu [26] conducted wide hybridization experiments with three sweet corn varieties and hexaploid wheat, and found that a pollen mixture from multiple sweet corn genotypes significantly enhanced the regeneration of haploid plants compared to pollen from a single maize genotype. After testing many maize genotypes, we found that a pollen mixture of four sweet corn hybrids produced a relatively high and stable rate of wheat DH plants in our studies [23].

7. To perform pollination efficiently and save maize pollen, open the seal of the glassine bag and shake a brush with pollen at the sealing place (Fig. 3D). Pollination using this method only needs to be done once and does not require repetition.
8. If there are many wheat spikes reaching the state of embryo harvest on the same day, some can be cut from the plant base and placed in water, then stored in the vernalization chamber (4°C). Complete the embryo harvest work within 2 days.
9. Before using the laminar flow hood for embryo rescue, wipe the bench surface with alcohol-soaked medical gauze to ensure sterility.
10. When bundling the haploid seedlings, do not tie too tightly. Leave space between the bases of each wheat haploid seedling to allow the colchicine solution to penetrate the tissues (Fig. 6E).
11. Throughout the entire process of preparing the colchicine solution and conducting the colchicine doubling treatment, the personnel must implement proper safety measures, including wearing lab coats, rubber gloves, safety goggles, a gas mask, and safety shoes, and the use of a fume hood. All toxic solutions should be collected into dedicated containers (waste liquid buckets) and sent to a waste disposal company for processing.
12. Place newly harvested seeds from different wheat materials/genotypes into small paper cups (Fig. 8A), label each cup with an identification code, prepare a 1% H_2O_2 solution, and pour it into the cups to submerge the seeds. After 24 h, rinse the seeds with clean water and sow them in the field.

Acknowledgments

This work was supported by the National Key Research and Development Program of China (2022YFF1002300), the Quancheng '5150' Talent Program (No. 07962021047), the Key R&D Program of Shandong Province, China (2024LZGCQY005), and the Jinan Agricultural Science and Technology Research Project 2025.

References

1. Wang L, Zheng Y, Duan L et al (2022) Artificial selection trend of wheat varieties released in Huang-Huai-Hai region in China evaluated using DUS testing characteristics. Front Plant Sci 13:898102. https://doi.org/10.3389/fpls.2022.898102
2. Wang L, Chang L, Li H et al. (2009) Method of wheat seeds purity testing by molecular markers. J Triticeae Crops 29:1–8
3. General Administration of Quality Supervision, Inspection and Quarantine of the People's Republic of China, Standardization Administration of the People's Republic of China (2008) Seed of Food Crops-Part 1: Cereals, GB 4404.1-2008. Standards Press of China, Beijing
4. Dunwell JM (2010) Haploids in flowering plants: origins and exploitation. Plant Biotechnol J 8:377–424. https://doi.org/10.1111/j.1467-7652.2009.00498.x
5. Prigge V, Xu X, Li L et al. (2012) New insights into the genetics of *in vivo* induction of maternal haploids, the backbone of doubled haploid technology in maize. Genetics 190:781–793. https://doi.org/10.1534/genetics.111.133066
6. Dwivedi SL, Britt AB, Tripathi L et al. (2015) Haploids: constraints and opportunities in plant breeding. Biotechnol Adv 33:812–829. https://doi.org/10.1016/j.biotechadv.2015.07.001
7. Liu W, Zheng MY, Polle EA, Konzak CF (2002) Highly efficient doubled-haploid production in wheat (*Triticum aestivum L.*) via induced microspore embryogenesis. Crop Sci 42:686–692. https://doi.org/10.2135/cropsci2002.0686
8. Tawkaz S (2011) Response of some wheat genotypes to anther culture technique for doubled haploid production. Master's Thesis, Sudan Academy of Science, Khartoum
9. Niroula RK, Thapa DB (2009) Response of wheat genotypes to maize mediated polyhaploid production. Am-Eurasian J Agron 2:156–161. https://www.researchgate.net/publication/230642921
10. Laurie DA, Bennett MD (1986) Wheat × maize hybridization. Can J Genet Cytol 28:313–316. https://doi.org/10.1139/g86-046
11. Laurie DA (1989) Factors affecting fertilization frequency in crosses of *Triticum aestivum* cv. 'Highbury' × *Zea mays* cv. 'Seneca 60'. Plant Breed 103:133–140. https://doi.org/10.1111/j.1439-0523.1989.tb00361.x
12. Laurie DA, Snape JW (1990) The agronomic performance of wheat doubled haploid lines derived from wheat × maize crosses. Theor Appl Genet 79:813–816. https://doi.org/10.1007/BF00224250
13. Laurie DA, Reymondie S (1991) High frequencies of fertilization and haploid seedling production in crosses between commercial hexaploid wheat varieties and maize. Plant Breed 106:182–189. https://doi.org/10.1111/j.1439-0523.1991.tb00499.x
14. Campbell AW, Griffin WB, Burritt DJ, Conner AJ (2000) Production of wheat doubled haploids via wide crosses in New Zealand wheat. N Z J Crop Hortic Sci 28:185–194. https://doi.org/10.1080/01140671.2000.9514138
15. Jauhar PP, Xu SS, Baenziger PS (2009) Haploidy in cultivated wheats: induction and utility in basic and applied research. Crop Sci 49:737–755. https://doi.org/10.2135/cropsci2008.08.0462
16. Niroula RK, Bimb HP, Thapa DB et al. (2007) Production of haploid wheat plants from wheat (*Triticum aestivum L.*) x maize (*Zea mays L.*) cross system. Himal J Sci 4:65–69
17. Moradi P, Haghnazari A, Bozorgipour R, Sharma B (2009) Development of yellow rust resistant doubled haploid lines of wheat through wheat × maize crosses. Int J Plant Prod 3:77–88. https://doi.org/10.22069/IJPP.2012.654
18. Li H, Li S, Abdelkhalik S et al. (2020) Development of thermo-photo sensitive genic male sterile lines in wheat using doubled haploid breeding. BMC Plant Biol 20: 246. https://doi.org/10.1186/s12870-020-02458-5
19. Martins-Lopes PF, Guedes-Pinto H, Pinto-Carnide O, Snape J (2001) The effect of spikelet position on the success frequencies of wheat haploid production using the maize cross system. Euphytica 121:265–271. https://doi.org/10.1023/A:1012091610195
20. Suenaga K, Nakajima K (1989) Efficient production of haploid wheat (*Triticum aestivum*) through crosses between Japanese wheat and maize (*Zea mays*). Plant Cell Rep 8:263–266. https://doi.org/10.1007/bf00274125
21. Kaushik N, Sirohi M, Khanna VK (2004) Influence of age of the embryo and method of hormone application on haploid embryo formation in wheat × maize crosses. Proceedings of the 4th International Crop Science Congress, Brisbane, Australia, 26 September–1 October, 2004, p 771
22. Usha P, Khanna VK (2017) Effect of hormonal treatments on haploid formation and *in vitro* haploid regeneration in wheat × maize system. Int J Plant Sci 12:234–239
23. Guan X, Peng JH, Fu D (2024) Technology for production of wheat doubled haploid

via maize pollen induction—updated review. Agronomy 14:375. https://doi.org/10.3390/agronomy14020375
24. Kalinowska K, Chamas S, Unkel K et al. (2019) State-of-the-art and novel developments of *in vivo* haploid technologies. Theor Appl Genet 132:593–605. https://doi.org/10.1007/s00122-018-3261-9
25. Kapoor C, Chaudhary HK, Sharma P et al. (2021) *In vivo* haploid induction potential of Himalayan maize (*Zea mays*) and cogon grass (*Imperata cylindrica*) gene pools in different segregational cycles of intra and inter-generic crosses of wheat. Plant Genet Resour Charact Util 19:522–529. https://doi.org/10.1017/S1479262121000642
26. Avcı S, Kutluİ (2020) Comparison of orchard-grass and sweet maize for doubled haploid plant production via wide hybridization in bread wheat. Turk J Agric-Food Sci Technol 8:1548–1552. https://doi.org/10.24925/turjaf.v8i7.1548-1552.3436
27. Verma V, Bains NS, Mangat GS et al. (1999) Maize genotypes show striking differences for induction and regeneration of haploid wheat embryos in the wheat × maize system. Crop Sci 39:1722–1727. https://doi.org/10.2135/cropsci1999.3961722x

Check for updates

Chapter 5

Wheat Doubled Haploid Production Through Maize Crossing

Sadiye Hayta, Meltem Bayraktar and Viktor Korzun

Abstract

Doubled haploid (DH) technology is a powerful tool in cereal breeding, enabling the rapid production of homozygous lines from heterozygous plants. This technique accelerates the development of new cereal varieties by producing genetically uniform plants in just one generation, a process that would otherwise require several generations of self-pollination. DH methods are particularly valuable in breeding programs for wheat, barley, maize, and rice, where they enhance selection efficiency and shorten the breeding cycle. The process typically involves inducing haploidy through interspecies hybridization, using inducer genotypes, anther or microspore culture, followed by chromosome doubling to restore fertility. The resulting DHs are genetically stable and carry fixed traits, making them ideal for genetic studies, trait selection, and commercial cultivation. In cereals, commonly used methods include wide hybridization, androgenesis, and gynogenesis. Wide hybridization, such as wheat-maize crosses, is one of the most frequently used approaches to producing haploids, while treatments like colchicine are used for chromosome doubling to produce fertile, DH plants. By combining these techniques, breeders can rapidly develop high-yielding, disease-resistant, and stress-tolerant cereal varieties tailored to the demands of modern agriculture.

Keywords Chromosome elimination, Double haploid, Wheat breeding, Haploid, Homozygous lines, Interspecific hybridization, Maize, *Triticum aestivum* L., Wheat, *Zea mays* L.

1 Introduction

A haploid plant (n) is described as a plant that contains the gametic number of chromosomes in its somatic cells [1]. Haploid plants can be converted into doubled haploid (DH) either spontaneously or by induced chromosome duplication of a haploid.

Haploid technology has proven to be a fast and very efficient method for developing new varieties. Since each DH plant is derived from different gametophytic cells, a population of DH plants derived from the same cross represents a collection of meiotically recombined but genetically fixed individuals. In this sense, DH populations represent both an ideal source of genetic variation, from which superior genotypes can be selected by phenotypic and/or genomic selection and a useful resource for genetic mapping and gene cloning.

Ana María Castillo et al. (eds.), *Doubled Haploid Methods in Cereals*, Methods in Molecular Biology, vol. 3029,
https://doi.org/10.1007/978-1-0716-5245-9_5,

DH technology offers several advantages that make it a powerful tool in modern plant breeding: (i) DH allows breeders to produce fully homozygous lines in just one generation, significantly reducing the time required to develop new cultivars [2, 3]; (ii) Because DH lines are completely homozygous, they provide a uniform genetic background, making it easier to identify and select for desirable traits [4]; (iii) The ability to rapidly produce homozygous lines enables breeders to test and select from a smaller number of lines, optimizing the use of resources such as land, labor, and time; (iv) DH lines are invaluable in genetic studies, including QTL mapping and the development of molecular markers, due to their genetic uniformity and stability [5]; (v) Since DH lines are completely homozygous, they offer a great opportunity to apply genomic selection in the early phase of the selection process [6].

DHs are genetically uniform and stable, providing a consistent genetic base for breeding programs [7]. They can be used in the development of hybrid wheat, where the genetic uniformity of the parents is critical for hybrid vigor. DH lines are also ideal for genetic studies such as QTL mapping, association studies, marker-assisted selection, and the application of genomic selection [8–10].

Various methods for producing DH in crops have been documented, but in wheat, the most effective approach to date has been intergeneric hybridization with maize [11]. Haploid technology in wheat involves inducing haploidy through interspecies hybridization, anther or microspore culture, followed by either spontaneously or induced chromosome doubling to restore fertility [8]. Haploid production through chromosome elimination occurs when chromosomes from one parent are rapidly lost during the zygote stage, making it a crucial method for producing DH lines in plants. In crosses between emasculated wheat (*Triticum aestivum* L.) and maize (*Zea mays* L.) used as a pollen donor, the maize chromosomes are completely eliminated, leading to the formation of wheat haploid embryos. By employing embryo rescue and chromosome doubling, pure lines with stable traits can be produced efficiently. This method is referred to as the "Wheat × Maize System" [12] (Fig. 1).

In this chapter, we will describe wide hybridization, such as wheat–maize crosses, which is one of the most frequently used approaches to produce haploids in wheat. The intergeneric hybridization method provides a fast way to produce homozygous wheat lines in just one generation, saving time compared to conventional breeding methods and showing high frequency of haploid embryo production compared to traditional methods like wheat × wheat hybridization [12]. Embryo rescue and in vitro culture are relatively straightforward, with a high success rate when managed properly. Wheat × Maize System has some challenges: (i) Maize pollen needs to be viable and available at the correct time for successful pollination [13]; (ii) This system typically requires in vitro procedures

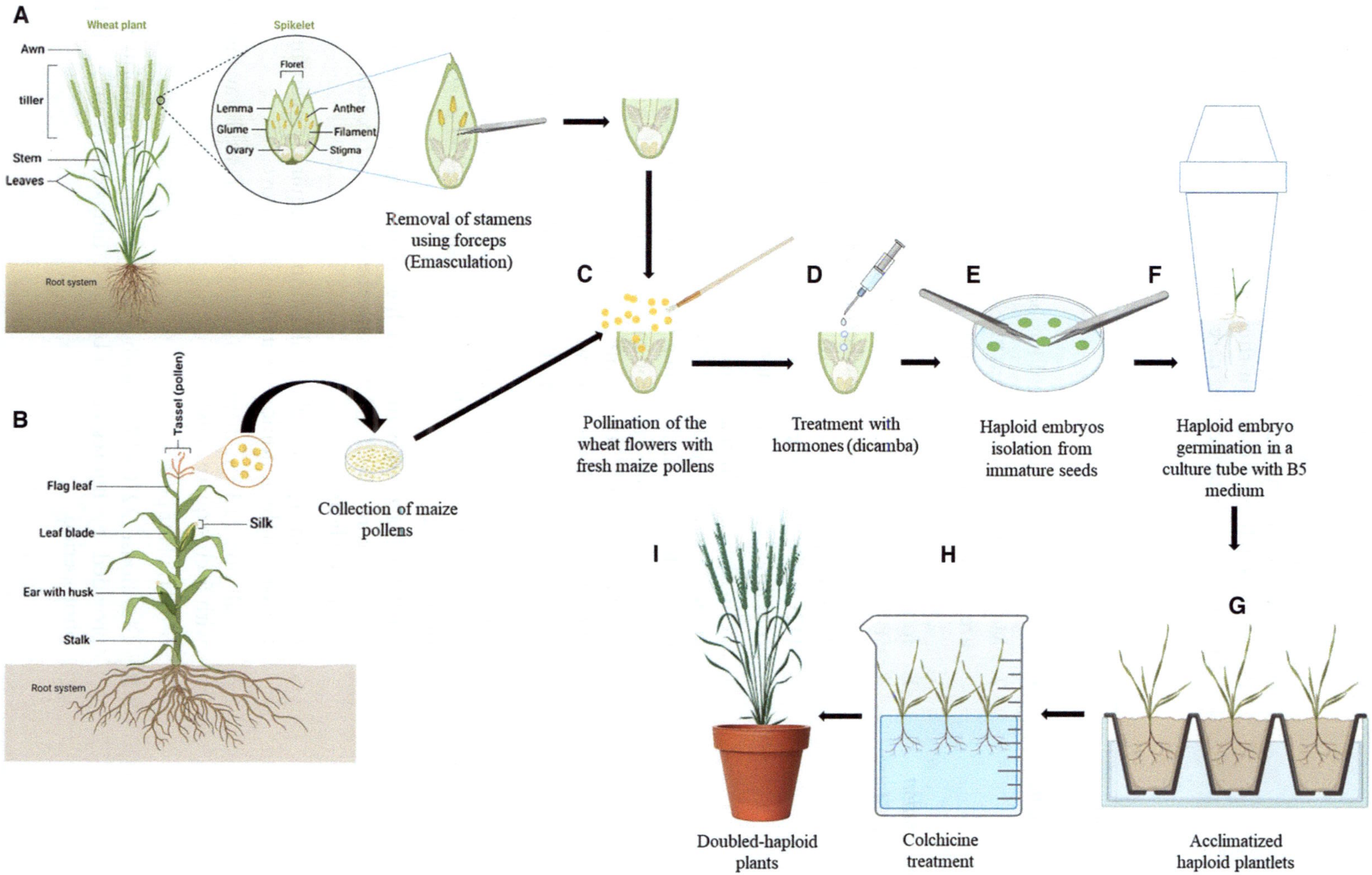

Fig. 1 Steps to produce wheat doubled haploid plants through maize crossing: (**A**) Wheat spike emasculation. (**B**) Maize pollen collection. (**C**) Pollination of the wheat flowers with maize pollen (**D**) Hormone treatment. (**E**) Excision of haploid embryo. (**F**) In vitro haploid plantlets before transfer to soil. (**G**) Acclimatized haploid plantlets. (**H**) Colchicine treatment of haploid plants for chromosome doubling. (**I**) Seed production from doubled haploid plants

to rescue immature embryos from seeds prone to abortion, often caused by insufficient development of the surrounding fertilization product, the endosperm. This embryo rescue technique requires skill in dissecting and handling small embryos; (iii) Chromosome doubling is achieved through the application of colchicine. It is a highly toxic substance, and care must be taken when handling colchicine [14].

2 Materials

2.1 Plant Material

1. Wheat seeds. This protocol has been applied to 5–10 F_1.
2. Maize seeds: Varieties such as "Kelvedon Glory" and "Applause" are recommended for their prolonged pollen production.

2.2 Preparation of Wheat and Maize Plants

1. John Innes "Cereal Mix": 40% (*v/v*) medium-grade peat, 40% (*v/v*) sterilized soil, 20% (*v/v*) horticultural grit, 1.3 kg/m^3 PG Mix 14-16-18 + Te Base Fertilizer, 1 kg/m^3 Osmocote Mini 16-8-11 2 mg + Te 0.02% B, Wetting agent (e.g., 500 g/m^3 H2Gro®, ICL Specialty Fertilizers, Ipswich, UK), kg/m^3 Maglime, 300 g/m^3 Exemptor.
2. Pesticides used in maize and wheat: SB Invigorator 1 mL for aphids/mildew, Hallmark Zeon 0.25 mL for aphids, Teppeki 0.7 mL for aphids, Dynamec 0.5 mL for thrips, Gazelle 0.5 g for aphids, Talius 1 mL for mildew, Flexity 1 mL for mildew, Eradicoat 20 mL for 20 mL, Sequoia 0.25 mL for apfids, Proline 275 1 mL for mildew.
3. Controlled Environment Room (CER) for wheat: Donor plants are grown at 20 ± 1°C day and 15 ± 1°C night, 70% humidity, with light levels of 600 μmol m^{-2} s^{-1} provided by fluorescent tubes and tungsten lighting and 16 h photoperiod.
4. CER for maize growth under higher temperature conditions ranging from 18–25 ± 1°C, with the same light conditions as used for wheat.
5. Cold room 5 ± 1°C (if vernalization is required) maintaining the same light conditions as in the CER above.
6. Small pots (13-cm-diameter) or trays (24CT seed tray inserts with 5-cm-square cells).
7. 5–10 L plant pots.
8. Glasshouse 22°C–17°C with 70% humidity, 600 μmol m^{-2} s^{-1} provided by fluorescent tubes and tungsten lighting, 16 h photoperiod.

2.3 Crossing Procedure (Wheat Spike Emasculation, Maize Pollen Preparation, Pollination)

1. 100% ethanol.
2. Pair of fine scissors.
3. Paintbrush No. 0 or 1.
4. Crossing tags.
5. Glassine crossing bags.
6. Wire crossing ties.
7. Plant labels.
8. Folded piece of A4 paper.
9. 90-mm-diameter Petri dishes.

2.4 Auxin Application

1. Dicamba.
2. 100% ethanol.
3. Distilled water.
4. Spray bottle.
5. 0.5 mL insulin syringe.
6. Eppendorf tube.
7. 250 mL bottle.
8. Petroleum jelly (Vaseline).

2.5 Embryo Excision

1. Dissection microscope (Wild M3B with 6× magnification).
2. Dissecting needles.
3. Fine forceps.
4. Scalpels.
5. Laminar flow hood.
6. Incubator at 20°C without light.
7. Culture room at 24°C ± 1°C with light levels of 100 μmol $m^{-2} s^{-1}$ provided by fluorescent tubes with a 16-h photoperiod.
8. 70% ethanol.
9. 10% (v/v) sodium hypochlorite (Fluka 71696).
10. Sterile water.
11. Distilled water.
12. Tween 20.
13. Culture media: Gamborg B5 medium including vitamins (Duchefa G0210) [15].
14. Agar (Formedium AGA03).
15. De Wit culture tubes (Duchefa W 1607).
16. De Wit tray (Duchefa T1608).

17. pH meter.
18. Analytical balance (precision 0.1 mg).
19. Autoclave.
20. Bunsen burner
21. Aluminium foil wrap.
22. Parafilm.
23. Protective gloves.
24. Protective glasses.
25. Pipettes (capable of delivering 10 μL to 5 mL) and pipette tips.
26. 90-mm-diameter single-vent Petri dishes.
27. Beakers.

2.6 Seedling Care

1. Seed and Modular Compost (Scotts Levington F2 Compost): 100% proprietary low nutrient peat without any fertilizer.
2. P24 plant trays 50.6 cm (length) × 37.8 cm (width) × 8 cm (height).
3. Propagator.
4. Plant labels.
5. 1 L plant pots.
6. John Innes "Cereal Mix."
7. Seed trays.

2.7 Colchicine Treatment and Plant After Care

1. Colchicine (SIGMA code C-9754).
2. Dimethyl sulfoxide (DMSO).
3. Tween 20.
4. Gloves.
5. 1 L beaker.
6. Plastic bags.
7. Fume hood.
8. 1 L plant pots.

2.8 Solutions and Culture Medium

1. Hypochlorite solution (10% *v/v*): Take 10 mL of commercial bleach and make up the volume to 100 mL.
2. Dicamba solution (1 mg/mL 10×): Weigh 100 mg of Dicamba and add 4 mL of 100% ethanol to dissolve it. Shake gently until fully dissolved. Transfer the dissolved solution to a Duran bottle and then make up the volume to 100 mL with distilled water. Store the solution at 4°C in the dark until use. Prepare fresh stock if kept for more than 1 month.

3. Gamborg's B5 medium (Table 1): Dissolve 3.164 g Gamborg B5 medium including vitamins (Duchefa G0210) in 900 mL water. Add 20 g sucrose and the remaining 100 mL of water. Adjust pH to 5.8 using 1 M KOH. Add 10 g of Formedium Agar, autoclave, and dispense 8 mL into De Wit culture tubes (Duchefa W 1607). Place the caps on the tubes and arrange them in a tray.
4. Colchicine Solution (0.05%): Dissolve 0.25 g of colchicine in 500 mL of distilled water, add 7.5 mL of DMSO and 7–8 drops of Tween 20.

Table 1
Gamborg's B5 medium (Gamborg et al. [15])

Medium components	Concentration (mg/L)
Macro nutrients	
NaH_2PO_4	130.44
KNO_3	2500
$(NH_4)_2SO_4$	134
$MgSO_4$	121.56
$CaCl_2$	113.23
Micro nutrients	
$MnSO_4{\cdot}H_2O$	10
H_3BO_3	3
$ZnSO_4{\cdot}7H_2O$	2
$Na_2MoO_4{\cdot}2H_2O$	0.25
$CuSO_4{\cdot}5H_2O$	0.025
$CoCl_2{\cdot}6H_2O$	0.025
KI	0.75
FeNa EDTA	36.70
Vitamins	
Nicotinic acid	1
Thiamine HCl	10
Pyridoxine HCl	1
Myo-inositol	100
Miscellaneous	
Agar	8000
Sucrose	20,000
pH	5.5

3 Methods

3.1 Growth of Wheat and Maize Plants

Using healthy plants and controlling diseases and insects inside the glasshouse and vernalization room have crucial importance for success. Make sure that the glasshouse, vernalization room and every material used are completely cleaned and disinfected.

To ensure synchronization of pollen maturation with the anthesis of the wheat recipient, maize should be planted approximately 6–9 weeks prior to the desired pollination period. Varieties such as "Kelvedon Glory" and "Applause" are recommended for their prolonged pollen production (*see* **Notes 1 and 2**).

1. Wheat (*Triticum aestivum* L.): Grow wheat plants until they reach anthesis (flowering stage), usually around 6–7 weeks after sowing.
   - Sowing seed: Place a 24CT insert inside a seed tray and fill with John Innes "Cereal Mix." Make a hole ~1.5–2 cm deep in the center of the "Cereal Mix." Place a single wheat seed in each hole (or 2–3 seed if concerned about poor germination. They can be thinned out later) and cover with a little more "Cereal Mix" (if needed). Tap the tray to level the "Cereal Mix." Water well if compost is dry. Label pots/trays with labels (*see* **Note 3**).
   - Move the plant material to the growing location: Transfer the sown seeds, not requiring vernalization, to a glasshouse. If vernalization is required, a short period (2–3 weeks) of pre vernalization growth in a CER at optimal temperature (18°C–20°C) is recommended, before plants are transferred to the vernalization room (5°C–6°C) for 5–6 weeks (*see* **Note 4**).
   - Potting on: Transplant the seedlings from the original pot or tray into 13-cm-diameter pots with John Innes "Cereal Mix" for supporting mature plants growth. If multiple seed have germinated prior to potting on, remove the weakest seedlings. Make a small hole in the center of the compost using your finger. The hole should be about the size of the 5-cm modules, in which your wheat seedlings germinated. Water the seedlings that are about to be transplanted. Loosen the seedling roots from the module's walls by gently squeezing and release the seedling. Keep the root ball intact as much as possible by supporting the seedling from underneath. Gently plant the seedling into the hole in the fresh compost. Firm the compost around the roots gently. Grow the plants in a glasshouse under the conditions listed above until they are used. The compost should be kept moist but not overwatered.

2. Maize (*Zea mays* L.):
   - Pregerminate six maize plants each week in "Cereal Mix" (three on Monday and three on Thursday) to provide sufficient pollen for up to 100 crosses per week.

- Maintain higher temperatures in a CER for maize growth (18°C–25°C) compared to wheat. Use larger pots (5–10 L) to support maize plants growing to maturity. The pot size depends on the plant's vigor. Ensure that the maize plants are shedding viable pollen during the wheat flowering stage, so they are ready for cross-pollination (*see* **Note 5**).

3.2 Wheat Spike Emasculation

1. Timing: Emasculate wheat spikes between growth stages 55–58 (mid heading to late heading), when 50%–80% of the spike is visible (Fig. 1A).
2. Use a pair of AA forceps sterilized in 100% ethanol. Carefully remove less mature florets from the center, tip, and base of each spikelet to ensure the remaining florets are at a similar developmental stage.
3. Use fine scissors to cut the emasculated spikes to half their length to facilitate pollen application.
4. Extract the three anthers from each floret. Ensure they are green, as yellow anthers indicate mature pollen that may have already self-pollinated the ovule.
5. Label the ear with a cross number, details of the plant, and the date of emasculation. Cover the emasculated ear with a Glassine crossing bag (*see* **Note 6**) and secure it with a wire tie to prevent desiccation and cross-pollination.

3.3 Maize Pollen Preparation and Pollination

1. Shake maize plants with flowering tassels at the end of each day to remove old anthers. Use maize pollen in the morning when it is fresh. Ensure that florets with visible anthers, which have a light-yellow color, have reached anthesis within the last 12 h.
2. Two days after emasculation, check the wheat florets for receptiveness. If the stigmas have fanned out and appear fluffy, they are ready for pollination. Continue checking daily until receptive.
3. Shake the maize tassel over a folded piece of paper to collect the pollen in the fold and transfer the pollen to a Petri plate (Fig. 1B).
4. Use a fine paint brush to apply pollen to the receptive stigmas (*see* **Note 7**). Repeat for all florets (*see* **Note 8**) (Fig. 1C).
5. Record the pollination date and replace the Glassine bag over the spike (*see* **Note 6**).

3.4 Auxin Application

1. 24 h after the maize pollination, treat the wheat plant with a 100 mg/L Dicamba (*see* **Note 9**). For injection, take 5 mL of the stock solution and dilute it to 50 mL with distilled water. Using a 0.5 mL insulin syringe, make two holes in the wheat stem: one above the last internode and one below the collar of the spike. Inject 0.5 mL of the Dicamba solution gently through the lower hole until a drop of auxin appears through

the top hole (Fig. 1D). Seal both holes with petroleum jelly (Vaseline) to prevent leakage.

2. Alternatively, spray a 20 mg/L Dicamba solution directly onto the florets. For this, take 1 mL of the stock solution and dilute it to 50 mL with distilled water (*see* **Note 9**) (Fig. 1D).
3. Replace the Glassine crossing bag over the spike. Allow embryos to grow and differentiate for 14–21 days. Adjust the development period based on glasshouse conditions: If day length exceeds 10 h or temperature exceeds 24°C, reduce the time allowed for embryo development.

3.5 Embryo Excision

1. Remove the embryos from the maternal plant at least 14 days after pollination. If left too long, as the spikes begin to ripen in preparation for harvest, the embryos will no longer be supported and will die (*see* **Note 10**).
2. Cut the spikes from the main plant and place them in a beaker of water until they are ready to be excised. Perform the excision of the embryos in a sterilized laminar flow hood. Clean the laminar flow hood with 70% ethanol before seed sterilization. Wash the seeds with sterile water. Wash with 70% ethanol for 2 min on a shaker. Rinse with sterile water twice. Immerse in 10% (*v/v*) sodium hypochlorite for 9 min on a shaker. Rinse four times with sterile water.
3. Perform in a sterilized Laminar Flow Hood. Sterilize tools and hands with 70% ethanol. Open seeds under a dissection microscope and remove embryos with a sterilized needle or pair of fine forceps (Fig. 1E). Place embryos scutellum down on the Gamborg's B5 medium in De Wit culture tubes (*see* **Note 11**).

3.6 Incubation and Seedling Care

1. Incubate embryos in a dark incubator at 20°C until germination. Transfer to the culture room set at 24°C and 16 h light after germination (Fig. 1F). If no germination occurs after 1 month, apply a cold shock at 4°C for 7 days to promote germination (*see* **Note 12**). Discard embryos that still fail to germinate after another month.
2. Provide vernalization to plantlets if required while still in vitro. Transplant plantlets into Seed and Modular compost once they grow to the top of the tube. Label, water, and place in a propagator to avoid desiccation. Remove the propagator lid once plants are established and grow until the 3–4 tiller stage (growth stage 22–23) (Fig. 1G).

3.7 Colchicine Treatment and Plant Aftercare

1. Use 12 mL of the colchicine solution per plants to be treated. Prepare the colchicine solution in a fume hood. Wear gloves (nitrile) during colchicine solution preparation and application.

2. Approximately 1 month after transplantation, the plants will be ready for colchicine treatment to double their chromosome complement. Remove plants from pots, discard excess soil, wash and trim roots to 2–3 cm, and leaves to 10–15 cm. Immerse roots and base of tillers in the colchicine solution in a beaker (Fig. 1H). Encase in a clear, labeled, sealed plastic bag, and place in a fume hood. Leave plants in light at room temperature for 5.5 h.

3.8 Post-treatment

1. After 5.5 h, wash the roots of the plants and repot plants (with corresponding labels) into 1 L pots of John Innes "Cereal Mix" compost (Fig. 1I).
2. Place pots on a seed tray to prevent colchicine residue from affecting other plants (*see* **Note 13**).
3. Bag all emerging ears. Expect a seed set of 50%+ compared to normal selfed plants.
4. One generation of seed multiplication is usually required before field testing.

4 Notes

1. Grow both maize and wheat plants in the same glasshouse to synchronize their development and ensure availability of fresh maize pollen.
2. Fresh seeds (a few weeks old) may be dormant. To break dormancy, place the sown seeds in pots/trays in a vernalization room for 2–3 days.
3. Alternatively, germinate seeds on damp filter paper in a Petri dish. Germinated seedlings are then transferred to John Innes "Cereal Mix" in pot/tray when roots have just emerged. Sow 1 seed (or 2–3 seeds if concerned about poor germination) per hole. They can be thinned out later. Label pots/trays with labels.
4. Usually, glasshouse plants must not be moved to a CER or vernalization room to avoid the spread of pests and disease. Occasionally, plants are pregrown in a clean glasshouse (rather than a CER) before being moved to vernalization. In this case, the plants must be inspected before moving. After vernalization, plants can be moved back to the CER or to a glasshouse.
5. Make sure maize plants are healthy and stress-free (neither too cold nor too hot), as stress can affect pollen viability.
6. Cover the wheat spikes with glassine waxy paper bags (also known as maize shoot bags) to protect them effectively.

7. The optimal pollination time is in the morning, preferably between 10 a.m. and 11 a.m.
8. The time between the collection of pollen from maize and the pollination of wheat spikes should not exceed 10–15 min. Thus, it is recommended to split the pollen collection to make sure efficient fertilization.
9. Remember to use protective gloves and glasses to avoid contact of auxin solution with fingers and eyes when the spike treatment is carried out. If you have many crosses, applying auxins with a small spray can help speed up the process.
10. Rescue the embryos after 14 days of pollination, though waiting a bit longer (~18 days) may improve the survival rate, depending on the genotype.
11. Continue with sterile techniques throughout the process to avoid contamination. Antibiotics can be added to the embryo rescue media to suppress contamination growth: 160 mg/L Timentin (Melford T0190.0002) or 250 mg/L Cefotaxime (Formedium Limited CEFO005).
12. Store the embryos in the dark at approximately 4°C until the coleoptile starts growing to about 1/2 cm, which is preferred.
13. Discard toxic waste safely.

References

1. Magoon ML, Khanna KR (1963) Haploids. Caryologia 16(1):191–235
2. Maqbool MA, Beshir A, Khokhar ES (2020) Doubled haploids in maize: development, deployment, and challenges. Crop Sci 60:2815–2840
3. Thakur P, Kumari N, Kumar A et al (2024) Recent advances in development and utilization of double haploids (DHs) in economically important vegetable crops. Plant Cell Tiss Organ Cult 156:15
4. Humphreys DG, Knox RE (2015) Doubled haploid breeding in cereals. In: Al-Khayri J, Jain S, Johnson D (eds) Advances in plant breeding strategies: breeding, biotechnology and molecular tools. Springer, Cham, pp 241–290
5. Weyen J (2021) Applications of doubled haploids in plant breeding and applied research. In: Segui-Simarro JM (ed) Doubled haploid technology: volume 1: general topics, alliaceae, cereals, methods in molecular biology, vol 2287. Humana, New York, NY, pp 23–39
6. Eliby S, Bekkuzhina S, Kishchenko O et al (2022) Developments and prospects for doubled haploid wheat. Biotechnol Adv 60:108007
7. Bocianowski J, Starosta E, Jamruszka T et al (2024) Quantifying genetic parameters for blackleg resistance in rapeseed: a comparative study. Plants 13:2710
8. Patial M, Pal D, Thakur A et al (2019) Doubled haploidy techniques in wheat (*Triticum aestivum* L.): an overview. Proc Natl Acad Sci India Sect B Biol Sci 89:27–41
9. Patial M, Chauhan R, Chaudhary HK et al (2023) Au-courant and novel technologies for efficient doubled haploid development in barley (*Hordeum vulgare* L.). Crit Rev Biotechnol 43(4):575–593
10. Gunundu R, Shimelis H, Mashilo J (2023) Genomic selection and enablers for agronomic traits in maize (*Zea mays*): a review. Plant Breed 42:573–593
11. Devaux P (2021) Production of wheat doubled haploids through intergeneric hybridization with maize. In: Seguí-Simarro JM (ed) Doubled haploid technology: volume 1: general topics, alliaceae, cereals, methods in molecular biology, vol 2287. Humana, New York, NY, pp 267–269

12. Niu Z, Jiang A, Abu Hammad W et al (2014) Review of doubled haploid production in durum and common wheat through wheat × maize hybridization. Plant Breed 133:313–320
13. Humphreys DG, Knox RE (2015) Doubled haploid breeding in cereals. In: Al-Khayri JM, Jain SM, Johnson DV (eds) Advances in plant breeding strategies: breeding, biotechnology and molecular tools, vol 1. Springer, pp 241–290
14. Jacquier NMA, Gilles LM, Pyott DE et al (2020) Puzzling out plant reproduction by haploid induction for innovations in plant breeding. Nat Plants 6:610–619
15. Gamborg OL, Miller RA, Olima K (1968) Nutrient requirements of suspension cultures of soybean root cell. Exp Cell Res 50:151–15

Chapter 6

Bread Wheat (*Triticum aestivum* L.) Doubled Haploid Production by Interspecific Crosses with *Imperata cylindrica*

Madhu Patial, Santosh Kumar Bishnoi, Dharam Pal and Kallol Kumar Pramanick

Abstract

The development of doubled haploid (DH) lines in bread wheat (*Triticum aestivum* L.) through interspecific hybridization with *Imperata cylindrica* represents a significant advancement in wheat breeding. This methodology exploits the intrinsic mechanism of chromosomal elimination during wide hybridization, wherein the chromosomes of *I. cylindrica* are selectively eliminated, resulting in the formation of haploid wheat embryos. These embryos are cultured, and haploid plants are subsequently subjected to colchicine treatment to induce chromosome doubling, thereby yielding fully homozygous DH plants within a single generational cycle. The *I. cylindrica*-mediated protocol confers several advantages over wheat×maize DH induction technique, including augmented efficiency, enhanced pollen viability, and the natural synchronization of *I. cylindrica* and wheat anthesis. This technique has demonstrated exceptional efficacy not only in bread wheat but also in triticale×wheat and wheat×rye derivatives, where alternative methodologies of DH induction have proven suboptimal. The resultant DH plants are indispensable for expediting genetic enhancement programs, facilitating the rapid development of novel wheat cultivars with superior traits such as increased yield, disease resistance, and abiotic stress tolerance. This chapter underscores the protocol of *I. cylindrica*-mediated doubled haploidy in wheat breeding, emphasizing its utility in the rapid generation of genetically uniform lines, which are essential for both basic genomic research and applied breeding programs.

Keywords Wheat, Doubled haploid, *Imperata cylindrica*, Chromosome elimination, Embryo rescue, Colchicine

1 Introduction

Wheat (*Triticum aestivum*), a globally significant cereal crop, has been the focus of extensive genetic enhancement efforts aimed at improving yield, quality, and resilience. Wheat breeding traditionally involves recurrent selfing from the F_1 generation to achieve homozygosity in loci-controlling desirable traits, a process that requires 8–9 generations. This iterative process ensures the uniformity of lines derived from a breeding population but is time-consuming and

Ana María Castillo et al. (eds.), *Doubled Haploid Methods in Cereals*, Methods in Molecular Biology, vol. 3029,
https://doi.org/10.1007/978-1-0716-5245-9_6,

labor-intensive. An outstanding methodology in this domain is the use of doubled haploids (DHs), which significantly speed up the breeding cycle [1, 2], and enable the production of 100% homozygosity at all loci in a single generation.

DH plants are rare in nature but can be induced by in-vitro or in-vivo approaches [3]. Androgenesis (anther or microspore culture) and gynogenesis (bud, ovary, or unfertilized ovule culture) are two methods of in-vitro haploid production. In-vivo methods include the chromosome elimination approach (intra- or interspecific and intergeneric) and parthenogenesis induction through inactive pollenpollination (using ionizing radiation, chemical agents, or high temperature). Use of in-vitro anther or microspore culture is limited in wheat DH breeding due to genotypic specificity, long regeneration time, differential rate of haploid and DH development, and albino seedling incidence [4]. These limitations led to the discovery of wide hybridization, which was pioneered by Barclay [5] in 1975, who produced wheat haploids by crossing the wheat variety "Chinese Spring" with *Hordeum bulbosum*. Although this technique was groundbreaking, it was genotype specific due to the presence of the dominant crossability inhibitor genes *Kr1* and *Kr2* located on the long arms of the *5B* and *5A* chromosomes, respectively [6]. Later, wheat × maize hybridization was discovered by Laurie and Bennett [7], which involves fertilization followed by maize chromosome elimination. Following this, Laurie and Bennett [8] found that maize pollen was not affected by these inhibitor genes, suggesting that the wheat × maize cross was a more effective way because it produces approximately two to three times more plants than anther culture, has low genotype specificity, and does not require haploid identification [9, 10]. However, the unsynchronized flowering of these two species hampered the process of hybridization.

Additional research conducted into other plant species, including sorghum, pearl millet, and Job's tears (*Coix lacryma-jobi*), reveal that crossing wheat with the wild grass *Imperata cylindrica*, which is distantly related to wheat, proved to be the most effective strategy [11, 12] and is currently in practical use [12–14]. *I. cylindrica* ($2n = 20$), or cogon grass, is a perennial wild grass commonly found around wheat fields that blooms synchronously with wheat, eliminating the need for a greenhouse. When compared to maize, *I. cylindrica* shows a noticeably higher frequency of haploid embryo development and regeneration [15]. Because of its genotype-nonspecificity and high pollen availability during the wheat hybridization period, it can hybridize with a variety of bread, durum wheat and also with triticale × wheat and wheat × rye derivates. This process employs uniparental chromosomeelimination, in which the haploid wheat embryos arise due to the specific removal of *I. cylindrica* chromosomes during the hybrid zygote's embryogenic development [16].

This method avoids common problems with other DH techniques, such as nonsynchronous flowering, albino plant emergence, somaclonal variation, etc.

The use of *I. cylindrica* to induce double haploidy in wheat and create DH lines from wheat–wheat, triticale–wheat and wheat–rye derivatives was first described by Chaudhary et al. [11]. The *I. cylindrica*-mediated DH method has been extensively utilized across various wheat species, including winter and spring bread wheat [17–19], durum wheat [20], triticale × wheat derivatives [21], and wheat × rye derivatives [17]. The Indian Council of Agricultural Research-Indian Agricultural Research Institute (ICAR-IARI), Regional Station situated at Shimla (Himachal Pradesh, India) refined, standardized, and used this method of breeding wheat for rust resistance [10, 22, 23], resulting in the development of "DH-1" (INGR21119), the first DH genetic stock resistant to wheat rust, registered at National Bureau of Plant Genetic Resources (NBPGR), New Delhi, India [2, 6, 24]. This avant-garde technique holds a significant promise for commercial exploitation, enabling the large-scale production of DHs in wheat. This chapter, therefore, covers an optimized protocol to produce wheat DHs via wide hybridization with *I. cylindrica* (Fig. 1).

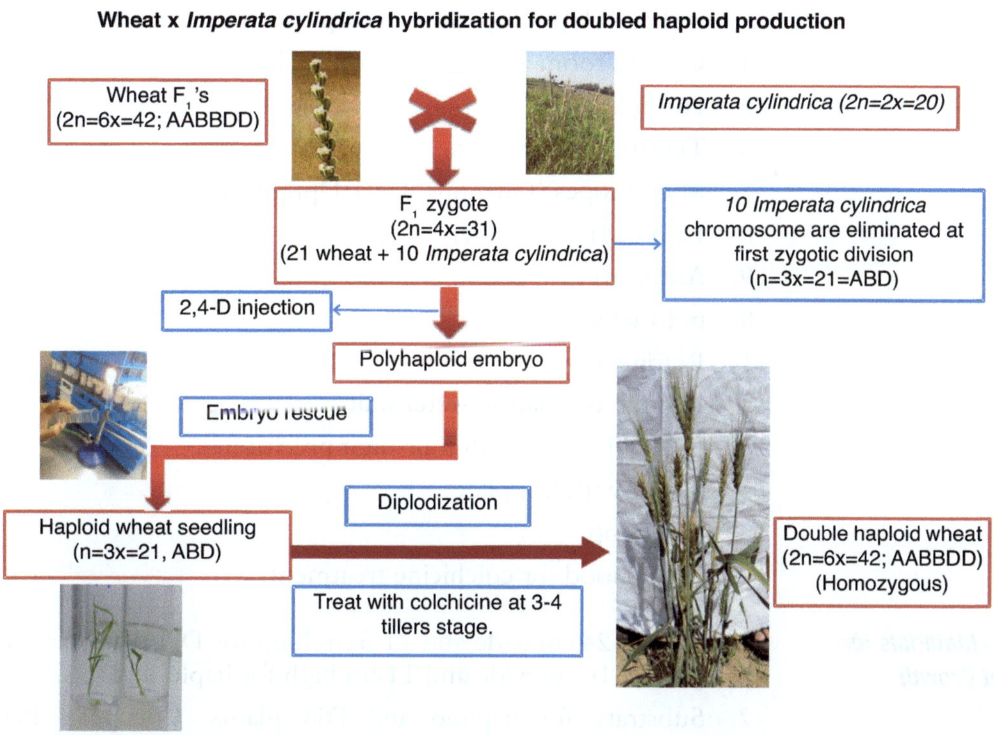

Fig. 1 Schematic presentation of wheat × *Imperata cylindrica*-mediated chromosome elimination technique for doubled haploid development in bread wheat

2 Materials

To develop wheat DH using the *I. cylindrica* mediated approach, the step wise requirement of different equipment, materials, reagents, and chemicals are as follows:

2.1 Plant Material

1. Wheat Varieties: Selected wheat parents are initially crossed to procure F_1 seeds. Predominantly, wheat F_1 seeds are used. However, F_2 or subsequent generational can be used for the development of wheat DHs. Generally, 500–600 wheat plants are required to produce a sufficient number of DH plants.
2. *I. cylindrica* grass as a pollen source: The "Ic-Pbr" genotype from Northwestern Himalayan, which is commonly found near the wheat fields.

2.2 Plant Growth and Laboratory Equipment

1. Vernalization growth chamber with temperature range of 0–4°C under 12:12 h dark:light regime, and 100 μmol $m^{-2}s^{-1}$.
2. Controlled glasshouse or polyhouse with air-conditioning and humidification, with day and night temperatures of 25°C and 15°C, respectively, and a 16 h photoperiod with a light intensity of 1000 μmol $m^{-2}s^{-1}$.
3. Controlled plant growth room or culture room with air-conditioning and humidification, and with 20/15°C (day/night) with a 16 h photoperiod, 1000 μmol $m^{-2}s^{-1}$, and 80% humidity.
4. Culture room at 20 ± 2°C without light.
5. Humidifier.
6. Thermo-hygrometer.
7. Microscopes (Olympus BX51 Upright model).
8. Laminar flow cabinet.
9. Autoclave.
10. pH meter.
11. Refrigerator.
12. Double-distillation water unit.
13. Digital balance with four-digit precision.
14. Stirrer with hot plate.
15. Air-pump.
16. Fume hood for colchicine treatment.

2.3 Materials for Plant Growth

1. Pots of 24 cm wide and 21.5 cm high for DH plants and small pots of 10 cm wide and 11 cm high for haploid plants.
2. Substrate for haploid and DH plants: Cocopeat: Perlite: Vermiculite in a 3:1:1 ratio.

3. Fertilizer: Well-decomposed organic farmyard manure is used for soil enrichment in polyhouse and pots. The organic fertilizers include Nitrogen (Urea), Potassium (Potassium Chloride), and Phosphorus (Monocalcium Phosphate Dihydrate) in a 120:60:30 ratio in the polyhouse and fields, whereas a ready-made N:P:K fertilizer mix (20:20:20) (2 g/L of water) was used for the pots.
4. Germination sheet (45 × 30 cm).
5. Bavistin (50% Carbendazim).

2.4 Pollination Tools

1. Fine-tipped forceps.
2. Fine brushes.
3. Glass Petri dishes (90 × 17 mm).
4. Butter paper bags.
5. Surgical scissors.
6. Crossing tags.
7. Stapler and pencils.

2.5 Growth Regulator Treatment

1. 2,4-Dichlorophenoxyacetic acid (2,4-D).
2. 3 6-Dichloro-2-methoxybenzoic acid (Dicamba).
3. Surgical syringe (5 mL).
4. Glass beaker (1 L) and magnetic stirrer.

2.6 Materials for Culture Medium Preparation

1. Full-strength Murashige and Skoog medium (MS) including vitamins (Sigma, M5519-10L): Use 4.4 g of powder per liter of medium.
2. Cotton plugs.
3. Sterile double-distilled water.
4. Glass Beakers (1 L).
5. Glass rod.
6. Magnetic stirrer.
7. Sucrose.
8. Agar-agar (Merck, JD5JF65105).
9. Activated charcoal.
10. Pipettes and pipette tips.
11. Aluminum foil.

2.7 Embryo Isolation and Culture

1. Glass vials (5 mL).
2. Fine-tipped forceps.
3. Table lamp.
4. Scalpels.

5. Glass Petri dishes (90 × 17 mm).
6. Whatman filter paper.
7. Alcohol burner.
8. Glass culture tubes (25 × 150 mm).
9. Culture tube racks.
10. Commercial liquid bleach (10% sodium hypochlorite).
11. 75% ethanol.
12. Sterile doubled-distilled water.

2.8 Colchicine Treatment

1. Colchicine.
2. Sterile double-distilled water.
3. Dimethylsulphoxide (DMSO).
4. Tween 20.
5. Amber glass bottle.
6. Glass tank (5 L).
7. Fume hood and exhaust.
8. Aeration pump.

2.9 Preparation of Solutions and Culture Media

1. Dicamba (50 mg/L) + 2,4-D (50 mg/L): Dissolve 50 mg/L 2,4-D in 15 mL of 75% ethanol and dissolve 50 mg/L dicamba in 15 mL of 75% ethanol separately. Stir and mix them, making the final volume up to 1 L. Adjust the pH to 10.36.
2. Hydrochloric acid (1 N) and sodium hydroxide (1 and 10 N) solutions to adjust the pH of the media.
3. Sodium Hypochlorite (0.6%): Pipette 60 mL of commercial liquid bleach (10% sodium hypochlorite) in 1 L glass bottle and add 940 mL of water. During sterilization, add 2–3 drops of Tween 20 as a surfactant.
4. Phenyl acetic acid (PAA) stock solution (1 mg/mL): Weigh 100 mg of PAA and put it in 100 mL glass bottle. Add 2–4 mL of 75% ethanol to dissolve the powder. Once completely dissolved, adjust the volume to 100 mL with double-distilled water while stirring continuously.
5. Kinetin stock solution (1 mg/mL): Weigh 100 mg of kinetin and put it in 100 mL glass bottle and add 2–4 mL of 75% ethanol to dissolve the powder. Once completely dissolved, adjust the volume to 100 mL with double-distilled water while stirring continuously.
6. Culture Media: Half strength Murashige and Skoog (1/2 MS) medium is used for embryo culture. This medium is prepared by diluting commercially available full-strength MS medium to half-strength (use 2.22 g/L of full-strength

MS medium) and supplementing with 100 mg/L myo-inositol, 30 g/L sucrose, 0.2 mL of Kinetin (0.2 mg/L Kinetin stock solution), and 0.1 mL of PAA (0.1 mg/L PAA stock solution). Adjust the pH to 5.8 and then add 8 g/L agar-agar. Make up the volume of the media to 1 L. Dispense 20 mL of medium in culture tubes and sterilize by autoclaving at 121°C for 15 min. The same medium (1/2 MS) supplemented with 2 g/L charcoal is used for haploid plant regeneration. The media are used the next day.

7. For 1/4 strength Murashige and Skoog (1/4 MS) medium preparation, dilute full-strength MS media to 1/4 strength, i.e., use 1.11 g/L of full-strength MS medium.
8. Colchicine (0.8% *w/v* + 1% DMSO): Prepare the colchicine solution by dissolving 80 mg of colchicine in 10 mL of DMSO solution and adding 1 mL of Tween 20 directly into a glass bottle. Make up the volume to 1 L with double-distilled water in an amber-colored bottle.

3 Methods

A step-by-step protocol for the *I. cylindrica*-mediated DH approach in wheat is detailed below:

3.1 Scientific Management of Wheat and *I. cylindrica*

1. Subject F_1 winter wheat seeds to treatment with Bavistin (50% Carbendazim) and vernalization by sowing and transferring them to a vernalization growth chamber at 4°C for 8 weeks with a 12:12 h dark/light regime.
2. After vernalization, transplant the wheat seedlings into pots to a controlled glasshouse environment, with day and night temperatures of 25°C and 15°C, respectively, and a 16 h photoperiod. Apply appropriate organic and inorganic fertilizers and water the plants on alternate days. Alternatively, the seedlings can be sown in a polyhouse or field if proper wheat growing conditions are available.
3. Implement staggered planting of wheat F_1s at intervals of 10–12 days to manage workload efficiently and minimize the demand for large quantities of *I. cylindrica* pollen on any single day.
4. Growing conditions of *I. cylindrica*: To encourage healthy growth and flowering, ensure regular watering and weeding of *I. cylindrica* grown near the bunds of wheat fields (*see* **Note 1**).

3.2 Emasculation of Wheat Florets

1. Identify and select robust, healthy, and vigorous F_1 wheat plants derived from crosses between parental lines. These plants, free from diseases and pests, will serve as female parents for wide hybridization.
2. Identify the wheat spikes at the booting stage and look for florets which are approximately two-thirds of the way up the spike with the help of a forceps (Fig. 2A). When the stigma appears fluffy and the anthers change color from dark green to light green, the spike is ready to be emasculated (Fig. 2B).
3. Carefully remove the central florets and leave the two lateral florets for emasculation using fine-tipped forceps (*see* **Note 2**).
4. Using forceps, remove the three anthers from each floret systematically, starting from the base and working upwards (Fig. 3A). The terminal or basal spikelets may be removed completely, if they are tiny and difficult to emasculate.
5. After emasculation, enclose the spike in a butter paper bag and make sure it is properly labeled in crossing tags with the cross designation and the emasculation date (*see* **Note 3**).

3.3 Pollen Collection from *I. cylindrica*

1. In the morning, select *I. cylindrica* plants at the appropriate stage for pollen collection (Fig. 3B). Make sure that the Petri dish used for pollen collection is dry and clean.
2. Gently tap the *I. cylindrica* inflorescences over a Petri dish. The pollen grains will dislodge and fall into the dish. Take care not to disturb other anthers to avoid losing pollen. For maximum viability,

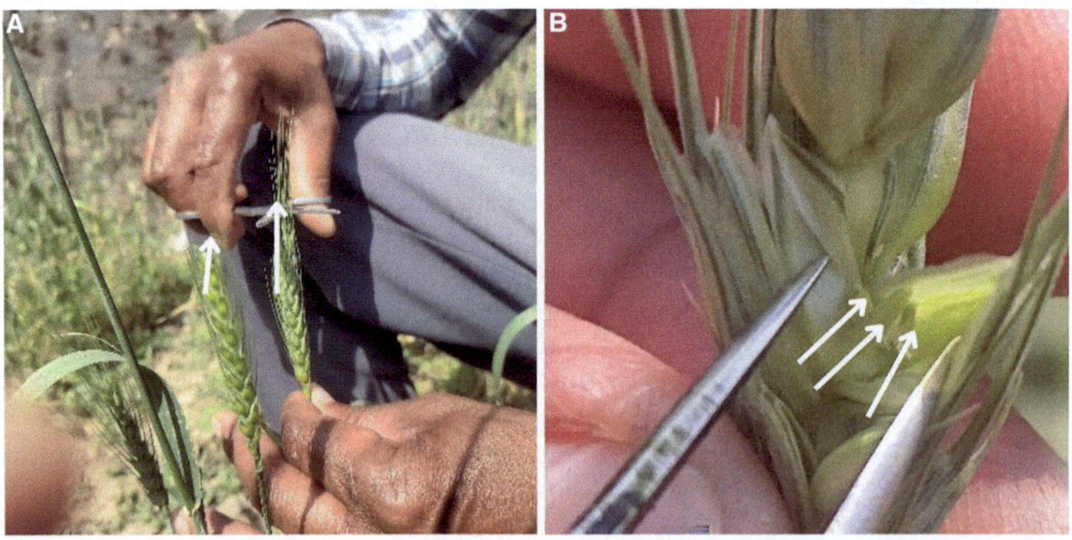

Fig. 2 Identification of the proper stage of wheat florets for emasculation. (**A**) Wheat spike ready for emasculation (left arrow) and clipping of spikelets prior to anther removal (right arrow). (**B**) Wheat anthers (arrows indicate the three anthers inside a flower)

Fig. 3 Emasculation of wheat florets and pollen collection from *Imperata cylindrica*. (**A**) Removal of anthers. (**B**) *I. cylindrica* donor pollen plant. (**C**) Pollen collected for *I. cylindrica*

the pollen should be light or pale yellow in color (Fig. 3C). Use pollen within 30–40 min, before it turns dark yellow (*see* **Note 4**).

3.4 Pollination Process

1. Remove the butter paper bags from emasculated wheat spikes 1 day after emasculation or when the stigma appears feathery (Fig. 4A).
2. Brush pollen onto each feathery stigma on the emasculated spikes using a small paintbrush. The pollen should flow easily from the brush to the emasculated florets (*see* **Note 5**, Fig. 4B).
3. Once the entire spike is pollinated, record the pollination date on the tag and replace the butter paper bag over the spike (*see* **Note 6**, Fig. 5A). Pollination is generally preferred during the midmorning hours.

3.5 Growth Regulators Application

1. Inject 2,4-D (50 mg/L) + dicamba (50 mg/L) solutions into the uppermost internodes of the pollinated wheat spikes 24 h after pollination (*see* **Note 7**, Fig. 5B).
2. Administer the injections for 2 consecutive days to promote pseudoseed formation. After growth regulator treatment, cover the spikes with butter paper bags and secure them with a stapler to ensure protection and containment.

3.6 Embryo Sorting and Excision

1. After pollination, allow the fertilized ovules of wheat to undergo embryogenesis for a period of approximately 21–24 days.
2. Harvest the spikes, including the stems, and cut them at approximately 20 cm in length. Submerge them in a container with double distilled water (Fig. 6A).

Fig. 4 Pollination process for haploid induction. (**A**) Feathery stigma ready for pollination (arrows indicate wheat pistil). (**B**) Pollination of wheat pistil with *Imperata cylindrica* pollen using a brush

Fig. 5 Tagging and growth regulator application of pollinated plants. (**A**) Tagging and bagging the pollinated wheat spikes with the butter paper bags. (**B**) Application of 2,4-dichlorophenoxyacetic acid (2,4-D) injections to the pollinated wheat spikes

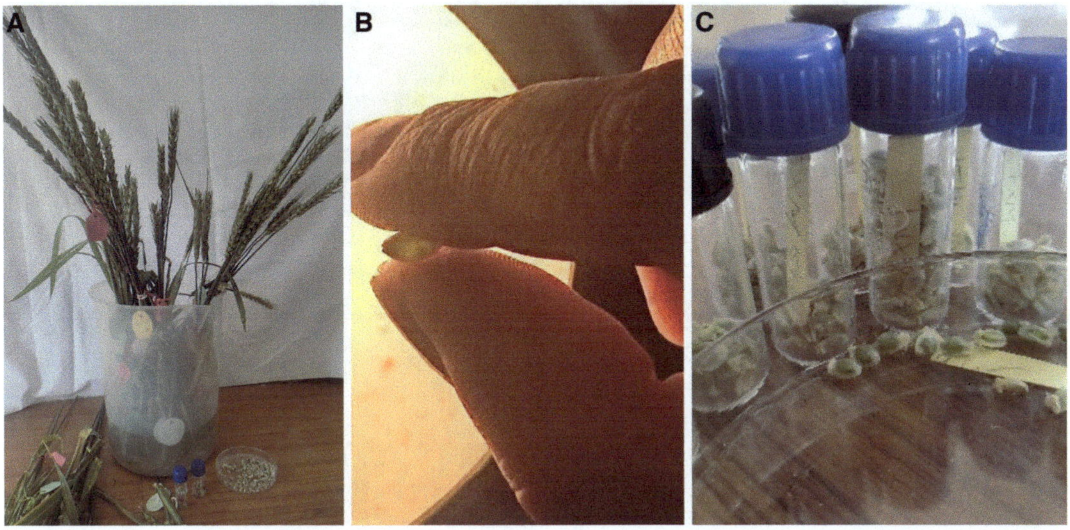

Fig. 6 Harvesting wheat spikes, embryo sorting, and storage. (**A**) Placing the wheat spikes in distilled water before embryo identification. (**B**) Sorting of pseudoseed containing embryo. (**C**) Storage of pseudoseeds with haploid embryo at 4°C before culturing

3. Carefully remove the pseudoseeds (hybrid seeds) from the spikes using fine forceps.
4. For the identification and sorting of haploid embryos, examine the caryopses under a light source to distinguish those containing embryos floating in the endosperm [25] (*see* **Note 8**, Fig. 6B).

3.7 Sterilization and Embryo Excision

1. After harvest, store the pseudoseeds containing embryos at 4°C for one day before rescuing the embryos (*see* **Note 9**, Fig. 6C).
2. Sterilize the pseudoseeds under sterile conditions within a laminar flow hood to prevent contamination. First, subject the pseudoseeds to surface sterilization by spraying them with 75% ethanol for 1 min. Following this, drain the ethanol. Then, treat the pseudoseeds with a 0.6% sodium hypochlorite solution for 10 min, incorporating 2–3 drops of Tween 20 as a surfactant. Finally, rinse the pseudoseeds three times with sterilized double-distilled water to remove any residual hypochlorite (*see* **Note 10**, Fig. 7A).
3. Transfer the sterilized pseudoseeds to sterile Petri dishes lined with Whatman filter paper (Fig. 7B). Perform dissection using a dissecting microscope and a dissecting kit, which includes pointed forceps and sharp scalpels. Secure the pseudoseeds at the embryo end with forceps. Use a scalpel to excise the brush end of the pseudoseed. Tilt the open end of the pseudoseeds towards the microscope. The embryo, seen as a small white

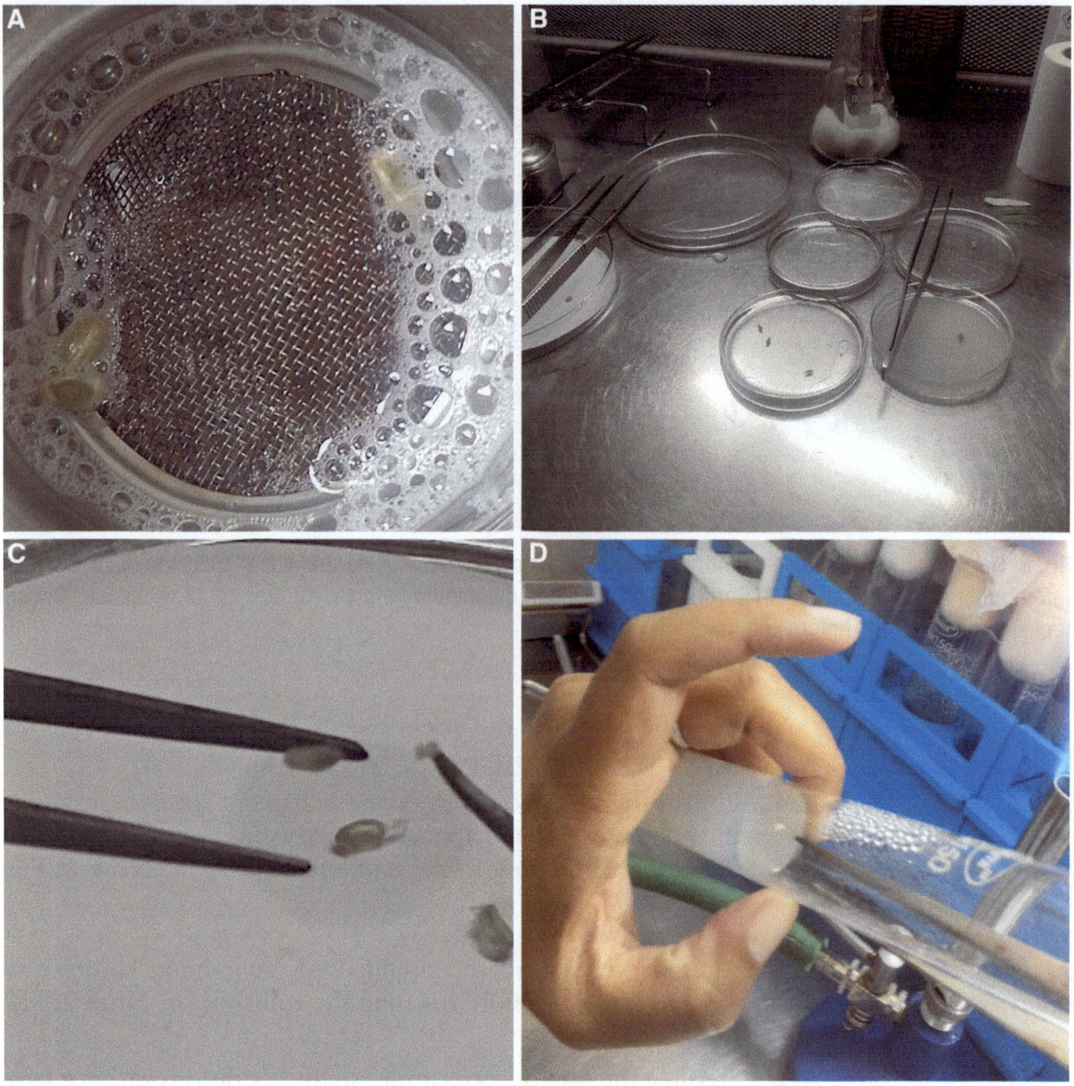

Fig. 7 Excision and culture of haploid embryos. (**A**) Sterilization of pseudoseeds. (**B**) Transfer of pseudoseeds to sterile Petri dish lined with Whatman filter paper. (**C**) Dissection of haploid embryo. (**D**) Culture of dissected haploid embryo

or translucent structure, will be visible within the pseudoseed (Fig. 7C). Due to the absence of solid endosperm, the embryo will be fairly easy to identify (*see* **Note 11**).

3.8 Culturing and Regeneration of Embryos

1. After excision, place the haploid embryos carefully in a 1/2 MS basal medium, supplemented with 0.1 mg/L PAA and 0.2 mg/L Kinetin to facilitate optimal embryo growth and development (*see* **Note 12**, Fig. 7D).
2. It is imperative to place the embryos correctly on the medium (with the embryo axis facing upwards) to promote germination rather than callus formation (*see* **Note 13**).

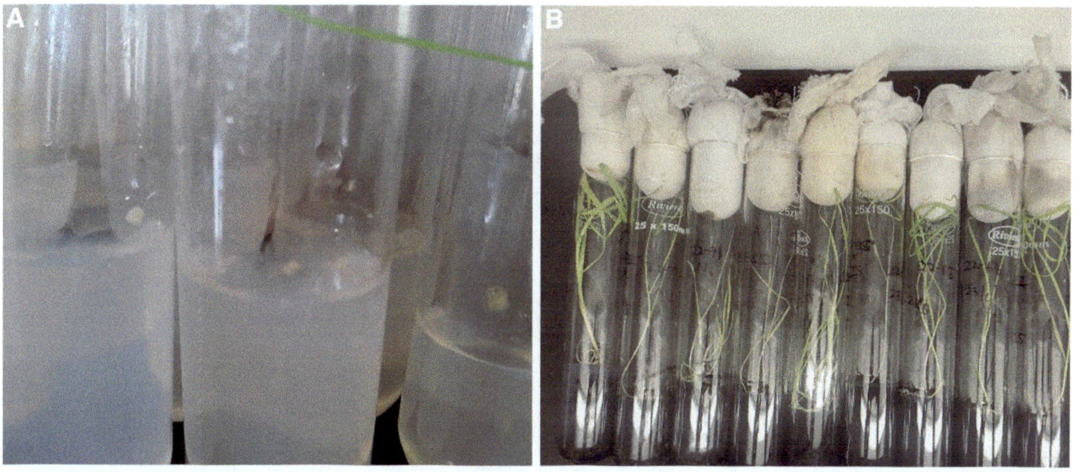

Fig. 8 Regeneration of haploid embryos. (**A**) Regeneration of the wheat embryos in controlled dark conditions. (**B**) Growth of haploid wheat plantlets in an incubator with light

3. Sterilize all tools and work surfaces repeatedly throughout the work session to maintain aseptic conditions.
4. Incubate the cultured embryos in a culture room at 20 ± 2°C for 1–2 weeks with lights off (Fig. 8A).
5. Once the embryos germinate, transfer them to a culture room with installed air conditioner and humidifiers with day and night temperatures of 20 ± 2 and 15°C, respectively, and a 16 h photoperiod, 1000 μmol/m^2/s PAR, and 80% humidity (Fig. 8B).

3.9 Vernalization of Haploid Plantlets Developed from Winter Wheat Crosses

1. Once the regenerated plantlets achieve vigorous root and shoot development within the culture tubes, place them in a vernalization growth chamber maintained at 4°C for 8 weeks with 16:8 h day/light regime (*see* **Note 14**).
2. After vernalization, transplant the haploid plantlets into a soil composed of cocopeat: perlite: vermiculite in a ratio of 3:1:1, and subsequently transfer them into pots of 10 cm wide and 11 cm high (Fig. 9A, B). Cover the plants with polybags having holes to maintain moisture for acclimatization.
3. After an 8-week growth period, or once the haploid plants exhibit healthy development and attain the 3–4 tiller stage, they are suitable for chromosome doubling.

3.10 Chromosome Doubling using Colchicine Treatment

The incidence of spontaneous chromosome doubling in haploid wheat plants is notably low. Therefore, colchicine treatment is applied for chromosome doubling (*see* **Note 15**).

1. The procedure involves uprooting the haploid plants from the soil and thoroughly washing the roots to eliminate any soil

Fig. 9 Hardening of haploid plants, seed production, and selection of DH1 lines. (**A**) Hardening of haploid wheat plants under controlled conditions. (**B**) Hardening of haploid plants by maintaining high humidity. (**C**) Matured plants after colchicine treatment. (**D**) Selection of doubled haploid (DH1) plants in station trials for different traits

residue. Then, trim the roots to approximately 3/4 of the crown before immersing them in the colchicine solution.

2. Perform colchicine treatment under a fume hood (*see* **Note 16**). Submerge the crown region of the plants in a solution containing 0.08% colchicine for 5 h, with enhanced aeration using an aeration pump.
3. Dispose the colchicine solution in a waste container and rinse the roots inside with tap water for 4–5 times. Subsequently, keep the rinsed roots under running tap water for 2 h to ensure the removal of any residual colchicine before replanting them into pots.

3.11 Hardening and Acclimatization of Plants

1. Transplant the plantlets into well-drained sandy-loam soil.
2. Transfer the plantlets to a greenhouse under rigorously controlled environmental parameters, with day and night temperatures of 25°C and 15°C, respectively, a 16 h photoperiod, and 80% humidity. During the initial posttransplantation phase, subject the plantlets to frequent misting with water to mitigate desiccation stress until they establish successfully (*see* **Note 17**).
3. Gradually increase the photonic exposure of the plantlets to natural sunlight during the hardening process. This can be done by placing the plants in direct light initially for 10 min for 15 days and subsequently increasing to 20, 30, 40, 50, 60 min for the next 15 days of each exposure. Routinely irrigate the plantlets with a 1/4 MS nutrient solution during this acclimatization phase ensuring continued hydration and sufficient nutrient absorption.
4. Given optimal nutritional and hydric conditions, the plantlets (DH0) typically attain phenological maturity within an approximate temporal frame of 4 months (Fig. 9C).

3.12 Selection and Evaluation

1. Thoroughly evaluate the DH1 plants (developed by selfing DH0 plants) phenotypically to identify superior agronomic traits, such as yield potential, pathogen resistance, and biotic/abiotic stress tolerance (Fig. 9D).
2. Verify the homozygosity and genetic stability of the DH1 lines through extensive molecular and cytogenetic methods.
3. Select the most promising DH1 lines for integration into advanced breeding programs and subsequent cultivar development initiatives based on their optimal performance metrics.

4 Notes

1. Make sure *I. cylindrica* and wheat F_1 plants are free of pests, diseases, and drought stress. Immediate preventive measures, along with routine monitoring are essential. Use effective irrigation systems and balanced fertilizers. For optimal plant health, implement real-time monitoring and integrated disease/pest management.
2. The stigma must not be damaged during the emasculation process as it is essential for proper fertilization and embryo development. Since the flag leaf and awns are important in providing vital nutrients to the growing embryos, they should not be cut or damaged.

3. In wheat breeding programs, this meticulous procedure is essential for avoiding autogamy and enabling controlled cross-pollination.
4. In wheat, DH production can be prolonged by at least 1 month by storing *I. cylindrica* pollen at −20°C and 65% relative humidity. Pollen viability is preserved under these conditions. Make sure that the Petri dish used for pollen collection is dry and clean.
5. Pollen should be sufficiently spread onto the stigma during its receptive time to increase the chances of successful fertilization. The probability of getting a greater number of haploid embryos is considerably raised by pollinating twice within 2 days. Several pollination attempts offset the effects of possible environmental factors that may affect the first pollination attempt, such as humidity, temperature swings, and pollen viability.
6. Crossing tags record important information such as the details of the cross, the emasculation and pollination dates, and the hormone spray applications. Different colored tags can be used to make tracking these activities easier. For example, we choose a different color tag for the first day of each hormone spray application. This color-coding method reduces errors and improves operational efficiency.
7. Growth regulator treatment should start 24–48 h after pollination. The use of protective equipment such as gloves, lab coats, and goggles is essential due to the toxicity of 2,4-D and Dicamba.
8. Identification of haploid embryos is easy as the caryopsis is filled with fluid instead of endosperm. The cross-pollination of wheat with *I. cylindrica* pollen leads to fertilization, but embryo development is slow, and the endosperm aborts, resulting in the development of a loose embryo in a fluid-filled state. The nonhaploid status of the embryo is confirmed by the presence of a solid endosperm, indicating self-pollination.
9. It is imperative to carry out embryo rescue within 5 days of spike harvest to optimize the success rate of embryo recovery and, ultimately, haploid plant development. The embryos are still relatively immature during this 5-day window, which makes them more receptive to in-vitro culture conditions. Postponing the rescue beyond this point may cause embryos overdevelopment, raising the risk of degeneration or desiccation, and decreasing the likelihood of successful recovery and subsequent growth.

10. Maintaining a sterile environment requires repeatedly sterilizing all tools and work surfaces during the work session. Microorganisms can be effectively eliminated, and contamination can be avoided by following strict sterilization protocols.
11. Sometimes the embryo may fall into the Petri dish after separation from the caryopsis. Using a moist scalpel to lift the embryo makes handling easier. It is important to avoid damaging the embryo axis or accidentally pushing the embryo into the medium during culturing.
12. Very small embryos may require specific growth conditions to develop. Hormone-free media do not support their growth. Instead, they should be immersed in a medium supplemented with 0.1 mg/L PAA and 0.2 mg/L kinetin. Whereas PAA facilitates root formation and cell elongation, kinetin encourages cell division. The hormone-rich medium promotes embryo development and raises the chances of survival.
13. The embryo should be carefully placed on the culture medium, with the embryo axis facing upwards. This orientation ensures that the medium will not obstruct shoot formation. Optimal nutrient absorption occurs when the scutellum side is positioned downward. Its precise placement is essential for the proper embryo growth and development.
14. Vernalization is not necessary for spring wheat, in contrast to winter wheat. After their strong roots and shoots emerge, spring haploid plants can be planted directly into the soil.
15. Colchicine, an alkaloid derived from *Colchicum* species, disrupts microtubule organization and impedes normal chromatid segregation during mitosis, thereby generating cells with a doubled chromosome complement. Colchicine treatment extends the growth period but greatly increases the chance of haploid plants survival.
16. Colchicine is an extremely toxic alkaloid that should be handled carefully because it can have serious negative effects. To avoid unintentional exposure, the use of protective gear such as safety glasses, lab coats, and gloves is crucial. Strict safety precautions must be followed because even tiny dosages can cause major health issues.
17. Frequent watering is essential to reduce the cytotoxic effects of colchicine and foster the recovery and tillering of plants with doubled chromosomes. This is achieved by applying a fine mist of water three to four times a day for a week using specialized misting nozzles. Simultaneously, loosen the soil periodically with a small garden shovel to improve root respiration and increase tillering.

Acknowledgments

This work is supported by in-house program of Indian Council of Agricultural Research-Indian Agricultural Research Institute (ICAR-IARI), New Delhi. The authors also acknowledge Dr Harinder Kumar Chaudhary and Chaudhary Sarwan Kumar Himachal Pradesh Agricultural University, Palampur, India for initial help in establishing the DH facility.

References

1. Patial M, Pal D, Thakur A et al (2019) Doubled haploidy techniques in wheat (*Triticum aestivum* L.): an overview. Proc Natl Acad Sci, India Sect B Biol Sci 89:27–41
2. Patial M, Chaudhary HK, Sharma N et al (2022) Developing genetic stock for yellow and brown rust resistance in *Triticum aestivum* L. via *Imperata cylindrica*-mediated doubled haploidy technique. Cereal Res Commun 50:439–448
3. Patial M, Pal D (2017) Wheat improvement via doubled haploidy breeding. In: Kumar A, Kumar A, Prasad B (eds) Wheat a premier food crop. Kalyani Publisher, pp 67–76
4. Islam MT, Arif MR, Hasan MT, Robin AHK (2023) Anther culture in crop plants: progress and perspectives. Plant Breed Biotechnol 11:69–96
5. Barclay IR (1975) High frequencies of haploid production in wheat (*Triticum aestivum*) by chromosome elimination. Nature 256:410–411
6. Patial M, Verma R (2023) Developing wheat lines for rust resistance through *Imperata cylindrica* mediated chromosome elimination doubled haploidy technique: a review. J Cereal Res 15(3):309–319
7. Laurie DA, Bennett MD (1986) Wheat × maize hybridization. Can J Genet Cytol 28:313–316
8. Laurie DA, Bennett MD (1987) The effect of the crossability loci Kr1 and Kr2 on fertilization frequency in hexaploid wheat × maize crosses. Theor Appl Genet 73:403–409
9. Kumar NR, Prasad BH (2009) Overview of wheat × maize system of crosses for dihaploid induction in wheat. World Appl Sci J 7:1037–1045
10. Gandhi S, Chaudhary HK, Pal D et al (2023) *Imperata cylindrica* mediated doubled haploidy in wheat (*Triticum aestivum* L.): quittances and breakthrough. J Cereal Res 15(1):31–40
11. Chaudhary HK, Sethi GS, Singh S et al (2005) Efficient haploid induction in wheat by using pollen of *Imperata cylindrica*. Plant Breed 124:96–98
12. Patial M, Pal D, Chaudhary HK et al (2016) Exploring the potential of *Imperata cylindrica*, a wild grass for development of doubled haploids in wheat. Int J Trop Agric 34:1211–1213
13. Patial M, Pal D, Kumar J, Chaudhary H (2015) Doubled haploids production in wheat (*Triticum aestivum* L.) *via Imperata cylindrica* mediated chromosome elimination approach. Int J Trop Agric 33:3333–3335
14. Patial M (2024) *Imperata cylindrica* mediated doubled haploidy breeding in wheat. Vigyan Varta 5(2):133–136
15. Chaudhary HK, Tayeng T, Kaila V, Rather SA (2013) Enhancing the efficiency of wide hybridization mediated chromosome engineering for high precision crop improvement with special reference to wheat × *Imperata cylindrica* system. The Nucleus 56(1):7–14
16. Komeda N, Chaudhary HK, Suzuki G, Mukai Y (2007) Cytological evidence for chromosome elimination in wheat × *Imperata cylindrica* hybrids. Genes Genet Syst 82:241–248
17. Kishore N, Chaudhary HK, Chahota RK et al (2011) Relative efficiency of the maize and *Imperata cylindrica* mediated chromosome elimination approaches for induction of haploids of wheat-rye derivatives. Plant Breed 130:192–194
18. Patial M, Pal D, Chaudhary HK et al (2017) Exploration of wild grass *Imperata cylindrica* for development of doubled haploids in winter × spring wheat (*Triticum aestivum* L.) hybrids accompanied with combining ability and hybrid potential estimation. Indian J Genet Plant Breed 77(2):316–320
19. Patial M, Chaudhary HK, Pramanick KK et al (2021) Doubled haploidy breeding in wheat and barley. Int J Trop Agric 39(4):403–405
20. Mahato A, Chaudhary HK (2015) Relative efficiency of maize and *Imperata cylindrica* for haploid induction in *Triticum* durum following chromosome elimination-mediated approach of doubled haploid breeding. Plant Breed 134:379–383

21. Pratap A, Sethi GS, Chaudhary HK (2005) Relative efficiency of different *Gramineae* genera for haploid induction in Triticale and Triticale × wheat hybrids through the chromosome elimination technique. Plant Breed 124:147–153
22. Verma T, Gandhi S, Bishnoi SK et al (2023) Doubled haploidy techniques for wheat improvement: an introduction. Krishisewa. https://www.krishisewa.com/resource-management/1508-regenerative-agriculture-a-step-towards-conservation-and-sustainability.html
23. Patial M, Chaudhary HK, Sharma N et al (2020) Effect of different *in vitro* and *in vivo* variables on the efficiency of doubled haploid in *T. aestivum* L. using *Imperata cylindrica*-mediated chromosome elimination technique. Cereal Res Commun 49:133–140
24. Patial M, Chaudhary HK, Pal D et al (2023) DH-1 (IC640653; INGR21119), a wheat (*Triticum aestivum* L.) germplasm with resistance to yellow and brown rust in seedling stage adult plat stage. Indian J Genet Resour 36(3):463–465
25. Bains NS, Mangat GS, Singh K, Nanda GS (1998) A simple technique for the identification of embryo-carrying seeds from wheat × maize crosses prior to dissection. Plant Breed 117:191–192

Chapter 7

Hexaploid Wheat × Barley (*Triticum aestivum* × *Hordeum vulgare*) Intergeneric Hybrids: Production and Analysis

Dávid Polgári, Adél Sepsi, Mohammad Ali, András Kis and László Sági

Abstract

Hybrids of taxonomically distant plant species have been important sources of genetic variation for introgression breeding in cereals and the generation of maternal haploid lines via paternal genome elimination followed by whole genome duplication. Despite a long history of crossing wheat and barley, the two most important temperate-zone cereals, efficient hybridization between the two species has not been achieved until recently. Here, we describe the crossing technique between suitable wheat and barley genotypes and demonstrate two procedures to confirm and characterize the resulting plants. First, multiplex PCR analysis allows the quick determination of chromosome composition in young plants. Further, advanced in situ hybridization tools such as FISH–GISH provide additional information about the fine structure of the hybrid chromosome complement.

Keywords Cereals, Chromosome, DNA marker, GISH, In situ hybridization, Molecular cytogenetics, Multiplex PCR

1 Introduction

Wheat (*Triticum aestivum* L. $2n = 6\times = 42$, AABBDD genomes) and barley (*Hordeum vulgare* L. $2n = 2\times = 14$, HH genome), two of the most important small grain cereals worldwide, constitute the only intergeneric cross-combination in monocotyledonous plants known to produce all possible outcomes [1]—from maternal haploids at a high frequency [2 and references therein] to partial hybrids (hypoploids) and full hybrids [3].

All these outcomes can be of practical significance and beneficial for crop genetic improvement. Maternal haploid or doubled haploid (DH) lines represent pure breeding material and models to study the genome elimination process. Partial or full hybrids, on the other hand, provide valuable prebreeding lines for the sexual, non-GM introgression of useful agronomic or disease-resistance traits.

The relatively low plant yield (1%–2% of pollinated florets) obtained from this cross-combination hampers the more widespread

Ana María Castillo et al. (eds.), *Doubled Haploid Methods in Cereals*, Methods in Molecular Biology, vol. 3029, https://doi.org/10.1007/978-1-0716-5245-9_7,

application of wheat × barley hybrids and derived haploids. Among the underlying reasons, technical shortcomings and dependency on a limited range of genotypes can be mentioned.

We have achieved over 10% plant yield in various wheat and barley genotype combinations with haploid frequencies ranging between 20.5% and 76% [2, 3 and unpublished] by optimizing the crossing procedure described here in detail. In addition, step-by-step protocols are provided to confirm and analyze the obtained plants via multiplex PCR and state-of-the-art molecular cytogenetic techniques. Flow cytometry can also be applied as a preliminary screening for the genome composition of the regenerated plants [2]. However, multiplex PCR (MPCR) allows for quick and large-scale analysis of DNA samples. Using bioinformatically tailored and pooled chromosome-specific markers for the four (A, B, D, and H) subgenomes, the presence or absence of each homoeologous chromosome can be precisely determined in an early developmental phase of the individual plants [4]. Finally, multiprobe and hyphenated in situ hybridizations (e.g., FISH–GISH, [5]) permit a more detailed characterization of chromosome preparations.

2 Materials

Prepare all solutions using ultrapure Milli-Q water (resistivity of ≥ 18.2 MΩ·cm) or double-distilled water (DDW) and analytical or molecular-grade reagents. When disposing of waste materials, follow the relevant guidelines and regulations.

Materials are listed in the order of their first use during the execution of the protocols.

2.1 Plant Material

1. *Triticum aestivum* L. "M1" seeds, a maternal doubled haploid line derived from the hexaploid spring wheat landrace "Sichuan" [2] (*see* **Note 1**).
2. *Hordeum vulgare* L. "Golden Promise" seeds, a two-row spring barley mutant [6] (*see* **Note 1**).
3. "M1" × "Golden Promise" F1 hybrid plants.

2.2 Generation of F1 Hybrids

1. Petri dishes (Ø 9 cm).
2. Round filter paper (Ø 7 cm).
3. Gibberellic acid (GA_3) (10 mg/L, 28.9 μM): Dissolve 10 mg in the minimal amount of ethanol, and take the volume to 1 L with double-distilled water, store at 4°C.
4. Peat pellet (Ø 33 mm).
5. Plant soil pots (1.5–2 L).

6. Growth cabinet for vernalization set at 2–4°C and 8-h photoperiod with low light intensity (50–100 μmol $m^{-2} s^{-1}$) provided by LED lamps (color temperature at 4000 K).
7. Growth cabinet for vegetative growth set at 15/12°C (day/night), 60% ± 10% relative humidity (RH), and 12-h photoperiod with 520 μmol $m^{-2} s^{-1}$ provided by LED lamps (4000 K).
8. Growth cabinet for flowering set at 18°C, 60% ± 10% RH, and 16-h photoperiod with 520 μmol $m^{-2} s^{-1}$ provided by LED lamps (4000 K).
9. Growth cabinet for embryo development in spikes set at 23°C, 60% ± 10% RH, and 16-h photoperiod with 520 μmol $m^{-2} s^{-1}$ provided by LED lamps (4000 K).
10. Growth cabinet for embryo development in vitro and plant regeneration set at 18°C, 60% ± 10% RH, and 12-h photoperiod with 520 μmol $m^{-2} s^{-1}$ provided by LED lamps (4000 K).
11. Growth cabinet for plant acclimatization set at 18°C, 100% RH, and 12-h photoperiod with 520 μmol $m^{-2} s^{-1}$ provided by LED lamps (4000 K).
12. Scissors and needle-tip tweezers.
13. Glass jar (1–1.5 L).
14. 2,4-Dichlorophenoxyacetic acid (2,4-D) (100 mg/L): Dissolve 10 mg of 2,4-D in the minimal amount of ethanol 96%, add water to take the volume to 100 mL, and store at 4°C.
15. 5-Azacytidine (50 mM): Gently suspend 122 mg of 5-azacytidine in 10 mL dimethyl sulfoxide (DMSO) with a pipette, distribute the solution into 100 μL aliquots, and store the stock at −80°C.
16. Seed Development Inducing (SDI) solution: Prepare freshly by mixing 1 mL of 100 mg/L 2,4-D solution (see point 14) with 10 μL of 5-azacytidine stock solution (see point 15) to get a final concentration of 0.45 mM of 2,4-D, 0.5 mM 5-azacytidine and 1% (*v/v*) DMSO).
17. Standard insulin syringe (1 mL).
18. Disposable nitrile gloves for the injection of SDI solution.
19. Automatic pipette set and sterile polypropylene pipette tips.
20. Cotton gauze sheets (5 × 5 cm).
21. Pins and small self-adhesive labels.
22. Ethanol: 70% (*v/v*) aqueous solution.
23. Sodium hypochlorite solution (1%–2%, *w/v*): Dilute household bleach (3%–6% *w/v* sodium hypochlorite) with two volumes of distilled water.

24. Pointed preparation needles.
25. Stereo microscope.
26. Solid N6D medium [7] in Petri dishes: Macroelements and microelements of N6 medium [8] supplemented and modified (Table 1), pH 5.8. Prepare stock solutions from macronutrients (10×), micronutrients (1000×), vitamins (500×), organic acids and sugars (100×), indolebutyric acid (1000×), and kinetin (1000×) growth regulators. Store in the fridge, except for the vitamins and plant growth regulators, which are frozen in aliquots. To prepare 1 L of medium, mix the indicated volumes (Table 1, column 4) of the macronutrients and micronutrients, and organic acids and sugars stock solutions with 500 mL of Milli-Q water, dissolve the other compounds directly in this solution, then add 5 g/L Phytagel (Sigma P8169) and take the volume to 990 mL. The medium is autoclaved at 121°C for 15 min, then the filter-sterilized (0.22 mm pore size) vitamins and plant growth regulators stock solutions are added. Pour ca. 30 mL of the medium into Ø 9 cm Petri dishes.

2.3 MPCR Analysis

1. Plant DNA isolation and direct PCR kit (Extract-N-Amp™ Plant PCR Kit (Merck, Sigma-Aldrich) (*see* **Note 2**).
2. Sterile microcentrifuge tubes (1.5 mL).
3. Sterile stainless-steel beads (Ø 3 mm).
4. Mixer mill (Bullet Blender Storm Pro).
5. Ice produced by a flake ice machine.
6. Sterile blades.
7. Thermoblock (95°C).
8. Vortex.
9. Microcentrifuge.
10. Sterile PCR tubes (0.2 mL).
11. 2× Phusion U Green Multiplex PCR Master Mix (Thermo Scientific) (*see* **Note 3**).
12. Chromosome-specific DNA primers at 100 μM concentration, stored at –20°C.
13. Genome-specific DNA primer mixes for the subgenomes A, B, D, and H (Table 2): Mix well equal volumes of the seven forward and reverse chromosome-specific primer pairs (see point 12) to make a primer mix for a subgenome.
14. Thermal cycler for PCR (*see* **Note 4**).
15. Agarose, molecular biology grade.

Table 1
Components of N6D medium

Stock solutions	Compound	Stock solution (g/L)	mL stock/L medium	Final cc. (mg/L)
Macroelements	$(NH_4)_2SO_4$	4.62	100	462
	KNO_3	28.3		2830
	KH_2PO_4	4		400
	$CaCl_2{\cdot}2H_2O$	1.66		166
	$MgSO_4{\cdot}7H_2O$	1.86		186
Microelements	$MnSO_4{\cdot}4H_2O$	4	1	4
	H_3BO_3	0.5		0.5
	$ZnSO_4{\cdot}7H_2O$	0.5		0.5
	$Na_2MoO_4{\cdot}2H_2O$	0.025		0.025
	$CuSO_4{\cdot}5H_2O$	0.025		0.025
	$CoCl_2{\cdot}6H_2O$	0.025		0.025
Vitamins	Retinol (vitamin A1)	0.005	2	0.01
	Thiamine (vitamin B1)	0.5		1
	Nicotinic acid (vitamin B3)	0.5		1
	Riboflavin (vitamin B2)	0.1		0.2
	Ca-pantothenate (vit. B5)	0.5		1
	Folic acid (vitamin B9)	0.2		0.4
	Pyridoxine (vitamin B6)	0.5		1
	Cobalamine (vitamin B12)	0.01		0.02
	Ascorbic acid (vitamin C)	1		2
	Calciferol (vitamin D)	0.005		0.01
	Biotin (vitamin B7)	0.005		0.01
	Cholin chloride	0.5		1
	p-Aminobenzoic acid	0.01		0.02
Organic acids and sugars	Malic acid	4	10	40
	Citric acid	4		40
	Fumaric acid	4		40
	Na-pyruvate	2		20
	myo-Inositol	10		100
	Xylose	15		150

(continued)

Table 1
(continued)

Stock solutions	Compound	Stock solution (g/L)	mL stock/L medium	Final cc. (mg/L)
Others	NaFeEDTA		Directly dissolved	25
	Glutamine			750
	Casein hydrolisate			250
	Maltose			54000
Plant growth regulators	Indolebutyric acid	0.5	1	0.5
	Kinetin	0.5	1	0.5

16. 10× TBE buffer [9]: 108 g/L Tris base, 55 g/L boric acid (0.89 M), and 9.3 g/L (0.025 M) disodium EDTA·2 H_2O (or 40 mL of 0.5 M EDTA, pH 8.0), pH 8.3.
17. Microwave oven.
18. UltraPure™ Ethidium Bromide, 10 mg/mL (Invitrogen) aqueous stock solution, stored in the dark.
19. DNA molecular mass marker (GeneRuler™ 100 bp Plus DNA Ladder, Thermo Scientific).
20. Electrophoresis equipment (chamber, tray, and comb) and power supply.
21. Gel documentation system (ChemiDoc Go Imaging System, Bio-Rad).

2.4 Molecular Cytogenetic Analysis

1. Cold room.
2. Fume hood.
3. Glass vials (10 mL).
4. Thermocool (styrofoam) box.
5. Centrifuge tubes (50 mL).
6. Clarke's fixative [10]: one volume of glacial acetic acid with three volumes of absolute ethanol, freshly prepared.
7. Acetocarmine (1% *w/v*): dissolve 5 g of carmine powder in 500 mL of 45% acetic acid in a 1000 mL glass beaker under a fume hood. Slowly bring the liquid to boiling using an electric heater (carefully increase the heat progressively as the liquid is prone to run out with heating). Boil the mixture at low heat for 60 min. Allow the liquid to cool down and filter it through filter paper placed into a glass funnel. Store the acetocarmine stain solution in a dark bottle at 4°C.
8. Acetic acid [45% (*v/v*) and 60% (*v/v*)] aqueous solution.

Table 2

Characterization of 28 chromosome-specific primer pairs used for four subgenomic (A, B, D, and H) pooling in MPCR analysis

Primer name	Sequence (5′-3′)	Length (nt)	Tm (°C)	Genomic location (on reference genome)	Product size (bp)
100bp_1A_F	GTGTTCCGGTAGTGGGTGGA	20	59.1	1A:538121084-538121103	100
100bp_1A_R	GAGAGGCAATCGGCTAGCAG	20	57.9	1A:538121164-538121183	
150bp_2A_F	TGCCGGCCATCAGTTGATGG	20	60.6	2A:722058206-722058225	150
150bp_2A_R	GTCTCTATGGCGTACCGGAC	20	57.0	2A:722058336-722058355	
200bp_3A_F	CACCACCTCGATGGATACGC	20	57.9	3A:381302864-381302883	200
200bp_3A_R	CACACGCTTGGTCTAAGGCC	20	58.6	3A:381303044-381303063	
250bp_4A_F	AGTACAGCGAACCTGGCAGC	20	60.1	4A:193096666-193096685	250
250bp_4A_R	CGCTGCTAGTTCAAGCAGCC	20	59.1	4A:193096896-193096915	
300bp_5A_F	AGGCTCACTCCAACACGACC	20	59.6	5A:32789504-32789523	300
300bp_5A_R	GAACAGAGTGTGGCGAGAGG	20	57.7	5A:32789784-32789803	
350bp_6A_F	TGCCGAGATTGGACGTCACG	20	59.9	6A:605537994-605538013	350
350bp_6A_R	GTGGTCTGTTGGTGGCCTTC	20	58.6	6A:605538324-605538343	
400bp_7A_F	TGGCGGCCATTAGACTAGCG	20	59.7	7A:433054330-433054349	400
400bp_7A_R	GGCGTCTCAGGTGTTGGACT	20	59.6	7A:433054710-433054729	
100bp_1B_F	CGTACCACTATCACGGCAGC	20	58.0	1B:102525056-102525075	100
100bp_1B_R	CTGGCATGAAGCCGTGTGCT	20	60.9	1B:102525136-102525155	
150bp_2B_F	TGGCCCTTTGG CTGGTGCTT	20	62.8	2B:649462075-649462094	150
150bp_2B_R	ACTATGTGCCGCTGATCGCC	20	60.2	2B:649462205-649462224	
200bp_3B_F	CGCTTCAACCCCGCCTATAGTA	22	58.6	3B:152216431-152216452	207
200bp_3B_R	AGACGGGTCTTTCACTCCCG	20	59.0	3B:152216618-152216637	

(continued)

Table 2
(continued)

Primer name	Sequence (5′-3′)	Length (nt)	Tm (°C)	Genomic location (on reference genome)	Product size (bp)
250bp_4B_F	CACCGACTAGGACCTCTGAC	20	56.4	4B:437626885-437626904	250
250bp_4B_R	GGTTCGGCGTTGTGCCGATT	20	61.3	4B:437627115-437627134	
300bp_5B_F	AGGAGTAGATGCTGCCTGCC	20	59.4	5B:63266364-63266383	300
300bp_5B_R	GACAGCTCACACGAGACACC	20	57.9	5B:63266644-63266663	
350bp_6B_F	CAGCTGGCCGGAATTCTTCC	20	58.8	6B:150479072-150479091	350
350bp_6B_R	GCGAGCAAGACTTGGCCAAC	20	59.5	6B:150479402-150479421	
400bp_7B_F	G CTAGGGGCGTAACCGTACT	20	58.9	7B:622973613-622973632	399
400bp_7B_R	GGCTGCATTGGCGCAGTCGC	20	65.1	7B:622973992-622974011	
100bp_1D_F	CCTGCGCGTTACAAGTCTCG	20	58.6	1D:112473637-112473656	100
100bp_1D_R	CCGAGACCGCCAGAAGTAAC	20	57.7	1D:112473717-112473736	
150bp_2D_F	AAGCACATG GATCGGGGCCT	20	61.9	20:27451470-27451489	150
150bp_2D_R	CTCCACTACGGGTCTCGCTA	20	58.0	2D:27370825-27370844	
200bp_3D_F	GCGGCGGCCTAAGATCTAAC	20	58.2	3D:103651721-103651740	200
200bp_3D_R	GCGCCATTCTGATCTCGGCT	20	60.2	3D:103651901-103651920	
250bp_4D_F	TCTAGACTCTGACCACGCGG	20	58.4	4D:1368997-1369016	250
250bp_4D_R	CTGCTCCAACCGCTATGACC	20	58.2	4D:1369227-1369246	
300bp_5D_F	GCCTTGGCAGAACTTCCTGC	20	59.2	5D:125349698-125349717	300
300bp_5D_R	GACGTACTCCAGTCACCGCA	20	59.1	5D:125349978-125349997	
350bp_6D_F	TGGATGCGCATGAGCGTACG	20	60.5	6D:20108101-20108120	350
350bp_6D_R	GAGAAGCGCACGATCGGTTG	20	58.9	6D:20108431-20108450	

400bp_7D_F	CGTACGCTAGGTCTGGATCC	20	56.8	7D:138350913-138350932	402
400bp_7D_R	CTCCACCTGAATGGCCCGAT	20	59.7	7D:138351295-138351314	
100bp_1H_F	AAGAAGGACGGGCAGTGTCC	20	59.7	1H:30181438-30181457	103
100bp_1H_R	GTCCAGCTGCATCTGCAGGT	20	60.3	1H:30181355-30181374	
150bp_2H_F	CGCGTCGATCGTGTCCATAC	20	58.1	2H:652788188-652788207	147
150bp_2H_R	G CG TG CAGG CCTAATCCATC	20	58.8	2H:652788315-652788334	
200bp_3H_F	CATGTGATCTCCCACGGGCT	20	59.6	3H:246129369 246129388	200
200bp_3H_R	GGAGCGACGTTGAGAACTGC	20	58.6	3H:246129549-246129568	
250bp_4H_F	ACAAGACCACCAACAGCAGTTCCGG	25	63.3	4H:113766350-113766374	250
250bp_4H_R	GGGACTAGCAGCACAGGACAGG	22	63.6	4H:113766578-113766599	
300bp_5H_F	CTCTCTCTGTGGGCGCTAAC	20	57.5	5H:408498106-408498125	304
300bp_5H_R	TAGGTGGGTCGATCCGAGGA	20	59.5	5H:408498390 408498409	
350bp_6H_F	GGTGCGACGTGATCTACACC	20	57.9	6H:56727617-56727636	352
350bp_6H_R	GCAATGCACAGGGTGTTACTCG	22	58.7	6H:56727947-56727968	
400bp_7H_F	ATGTCCTCGAGTGCACCTGG	20	59.4	7H:186518611-186518630	399
400bp_7H_R	TGCCTAGATACGCCTACGCG	20	59.0	7H:186518990-186519009	

9. Superfrost microscope slides.
10. Cutting needle.
11. Glass coverslips (18 × 18 mm and 22 × 30 mm).
12. Spirit lamp.
13. Phase contrast microscope.
14. Dry ice.
15. Razor blade.
16. Pressure cooker (household).
17. Orbital shaker.
18. Centrifuge.
19. Liquid nitrogen.
20. Hexadecyltrimethylammonium bromide (CTAB).
21. 2-Mercaptoethanol.
22. Chloroform-isoamyl alcohol mixture (24:1).
23. Isopropanol.
24. Primers for the amplification of the FISH probes (G+ C-rich microsatellite DNA sequence and the sequence of the Centromeric Retrotransposon of Wheat, CRW) diluted to a concentration of 10 μM, stored at –20°C.
25. 2× DreamTaq PCR Master Mix (Thermo Scientific).
26. Atto488 and AZDye594 Nick Translation Labeling Kits (Jena Bioscience) for the labeling of FISH/GISH probes.
27. Blocking DNA: total wheat DNA.
28. Pepsin stock solution (1 mg/mL): Dissolve 1 mg of pepsin powder (Sigma-Aldrich, 3200–4500/mg protein) in 1 mL Milli-Q water. Distribute the solution into 100 μL aliquots and store at –20°C for months.
29. Pepsin-HCl working solution (50 μg/mL): Dilute 50 μL of stock solution in 949 μL Milli-Q water and add 1 μL of 1 M HCl.
30. Plastic coverslips (22 × 22 mm).
31. Heated plate or PCR machine equipped with a stainless-steel plate.
32. Coplin jars, for hybridization and washing slides.
33. 10× PBS buffer (Phosphate-Buffered Saline): Dissolve 80 g NaCl, 2 g KCl, 14.4 g Na_2HPO_4, and 2.4 g KH_2PO_4 in 800 mL Milli-Q water and make up the solution to 1 L. Autoclave.
34. PBS working solution (1× PBS): Dilute 100 mL of 10× PBS to 1000 mL with sterile water. To prevent bacterial

contamination, avoid storing 1× PBS at room temperature (RT) for over 1 week.

35. Syringe and 0.22 μm cellulose acetate membrane filters.
36. Paraformaldehyde (PFA) (4% *w/v*): Add 10 mL of isotonic 16% (*w/v*) PFA (Thermo Scientific) to 26 mL Milli-Q water and 4 mL of 10× PBS. Add 50 μL of Igepal CA-630. Prepare 1 mL aliquots and store at −20°C (*see* **Note 5**).
37. Dextran sulfate 50% (*w/v*): Dissolve 5 g powder in 8 mL Milli-Q water and make it up to 10 mL. Sterilize the solution using a 0.22 μm syringe filter.
38. 20× SSC (Saline Sodium Citrate) buffer stock solution: Dissolve 175.3 g NaCl and 88.3 g sodium citrate in 800 mL Milli-Q water, adjust the pH to 7.0, and make up the solution to 1 L. Autoclave.
39. 2× SSC buffer working solution: Dilute 100 mL of 20× SSC to 1000 mL with sterile water. Store at RT.
40. Deionized formamide.
41. Hybridization mixture: Mix 600 μL of deionized formamide, 200 μL of 50% dextran sulfate solution, and 200 μL of 2× SSC. Add the labeled probes to 17 or 18 μL of hybridization mixture per slide.
42. Incubator set at 37°C.
43. Humidity chamber.
44. Fine forceps.
45. Vectashield antifade mounting medium with DAPI (4′,6-diamidino-2-phenylindole) (Vector Laboratories).
46. Epifluorescence light microscope or confocal laser scanning microscope equipped with ultraviolet (e.g., BP 360), blue (e.g., BP 450-490), and green (e.g., BP 575-625) excitation filter blocks.

3 Methods

3.1 Generation of F1 Hybrids

3.1.1 Plant Growing

1. Incubate the wheat and barley seeds at RT in the dark in a Petri dish containing one or two layers of filter paper saturated with about 5 mL of GA_3 solution to stimulate germination.
2. Transfer the seedlings to moistened peat pellets and raise them to about 5 cm tall.

3. Vernalize the plantlets for 6 weeks at 2–4°C and 8-h photoperiod with 50–100 μmol m^{-2} s^{-1}.
4. Transfer the plantlets to 1.2 L pots and grow them for 4 weeks in a growth cabinet at 15/12°C (day/night), 60 ± 10% RH, and 12-h photoperiod with 520 μmol m^{-2} s^{-1}.
5. Grow the plants by raising the temperature in the growth cabinet to 18°C and extending the photoperiod to 16 h until flowering.

3.1.2 Emasculation and Pollination

1. Emasculate wheat spikes manually 3–4 days before the anthesis by removing the apical and basal spikelets, then all the inner florets in the rest of the spikelets and keep the two outermost ones. Remove the three unripe green anthers from each floret with needle-tip tweezers and insulate the spikes with cellophane bags (Fig. 1A).
2. On the day of anthesis, either early morning or late afternoon, clip off one-third of the barley spike florets with scissors, then cut and place the spikes with their stems into a jar of warm water to induce the anthesis.
3. Pollinate the emasculated wheat spikes by directly shaking fresh barley pollen from the emerging anthers onto their stigmas (Fig. 1B) and close the bags again.
4. Inject 1 mL of fresh SDI solution into the first internode cavity under the pollinated spikes (Fig. 1C, *see* **Note 6**).

Fig. 1 Generation of wheat × barley F1 hybrid plants: (**A**) Close-up of an emasculated maternal wheat spike. (**B**) Pollination with barley pollen. (**C**) Injection of SDI solution in the uppermost internode. (**D**) Swollen ovaries (pseudoseeds) 10–14 days after fertilizationin a pollinated spike. (**E**) 14-day-old pseudoseeds collected from a pollinated spike. (**F**) Developing embryo in a dissected pseudoseed. (**G**) Isolated embryo transferred to N6D medium (scale bar = 1 mm). (**H**) Vigorously growing plantlets

5. Place the plants back in the growth cabinet and grow them at 23°C and 16-h photoperiod for 14 d.
6. One day after pollination, add one droplet (ca. 5 μL) of 2,4-D solution to the surface of all pollinated stigmas with a pipette.

3.1.3 Embryo Rescue and Plant Regeneration

1. Collect 14-day-old pseudoseeds from each spike with tweezers (Fig. 1D, E) and pack them in one layer of a gauze sheet. Close the packs with a needle and self-adhesive labels for records.
2. Presterilize the gauze packs by soaking them in 70% EtOH for 5 min, then rinse them thrice with sterile water.
3. Sterilize the gauze packs by soaking them in hypochlorite solution 1%–2% (*w/v*) for 30 min and then rinse them thrice with sterile water.
4. Transfer each gauze pack to a sterile Petri dish and prepare two sterilized preparation needles.
5. Position the pseudoseed with the ventral furrow down so that its micropylar end points towards 12 o'clock direction (Fig. 1F).
6. Make a longitudinal cut in the pericarp tissue and open the pseudoseed with the two needles to check for the presence of an embryo (Fig. 1F).
7. Carefully adhere the embryo to the tip of the needle by surface tension without damaging it, then transfer to N6D medium (Fig. 1G) and incubate at 18°C and 12-h photoperiod with 520 $\mu mol\, m^{-2} s^{-1}$ until rooted plantlet form.
8. Transfer the vigorously growing plantlets (Fig. 1H) to moistened peat pellets and incubate for 1 week at 18°C, 100% RH, and 12-h photoperiod with 520 $\mu mol\, m^{-2} s^{-1}$ before transferring to the soil in pots.

3.2 MPCR Analysis

3.2.1 Direct DNA Extraction

1. Add 100 μL of DNA extraction solution from the Extract-N-Amp™ Plant PCR Kit and one stainless-steel bead into a 1.5 mL microfuge tube on ice (*see* **Notes 2 and** 7).
2. Cut a leaf sample approximately 5 × 5 mm with a sterile blade and place in the tube.
3. Fully homogenize the leaf sample in a mixer mill (speed 8, 1 min).
4. Incubate the sample in a thermoblock at 95°C for 12 min.
5. Cool the sample on ice for 1 min.
6. Add 100 μL of kit dilution solution to the homogenized sample.

7. Vortex the sample and centrifuge at 18,000×*g* for 2 min.
8. Transfer 100 μL of the supernatant into a new 1.5 mL microfuge tube and store at −20°C until required.

3.2.2 Multiplex PCR

1. Use the already verified primer sets as listed (Table 2) for setting up MPCR analysis (*see* **Note 8**).
2. Prepare the MPCR reaction in a sterile 0.2 mL PCR tube for each subgenome (A, B, D, and H) in a final volume of 20 μL per reaction. For a single tube, the components are as follows: 10 μL of 2× Phusion U Green Multiplex PCR Master Mix, 0.84 μL of genome-specific primer mix, 8.16 μL sterile Milli-Q water, and 1 μL of direct DNA extract. Add all components to the master mix, except the DNA samples. Gently vortex the master mix and briefly centrifuge to collect the mixture at the bottom of the tube. Prepare four master mixes for each subgenome per candidate plant (*see* **Note 3**).
3. For all MPCR mixes, set the cycling conditions in the thermal cycler as follows: 98°C, 3 min; 32× (98°C, 10 s; 65°C, 15 s; 72°C, 10 s); 72°C, 10 min; 4°C hold.
4. Prepare a 2% (*w/v*) agarose gel in 1× TBE containing ethidium bromide.
5. Load 4 μL of each PCR reaction onto the gel and run the samples at 150 V for 25 min in parallel with 1 μL of DNA ladder.
6. Capture the results of MPCR with a gel documentation system (Fig. 2).

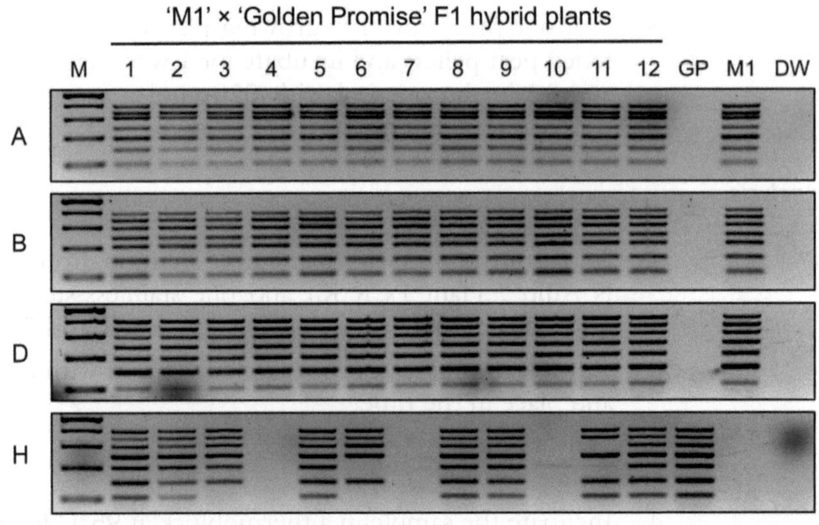

Fig. 2 Multiplex PCR analysis of 12 independent F1 ('M1' x 'Golden Promise') hybrid plants. The gel image shows bands for three maternal haploids (lanes 4, 7, 10), three partial hypoploids (lanes 3, 6, 11), and six full hybrids (lanes 1–2, 5, 8–9, 12). Markers for chromosomes 1–7 (bottom to top in the lanes) are amplified in subgenomes A, B, D, and H of wheat×barley F1 hybrid plants (lanes 1–12). Lanes include molecular size marker (M), 'Golden Promise' (GP), 'M1' wheat (M1), and no template control (DW)

3.3 Molecular Cytogenetic Analysis

3.3.1 Preparation of the Plant Material

1. For germination, arrange 10–15 seeds (wheat, barley) in a Petri dish on filter paper and add excess tap water to soak the paper. Incubate the seeds at RT in the dark for about 24 h until they initiate germination (*see* **Note 9**).
2. After germination, remove the excess water and transfer the Petri dish containing the seeds on the moist filter paper into a cold room (4°C) for 72 h.
3. Transfer the Petri dish to a dark box and leave it at RT (25–26°C) for 24–26 h.
4. Collect the fresh white roots of 1.5–2 cm that emerge.
5. To accumulate cells arrested in metaphase, place the collected roots in 10 mL glass vials filled with ice-cold Milli-Q water (containing approx. 1/3 volume of melting ice). Quickly transfer the vials to a thermocool box filled with ice and incubate in a cold room for at least 26 h.
6. After cold treatment, immediately place the roots in freshly prepared Clarke's fixative and incubate at 37°C for 5 days.
7. For chromosome observation, stain the fixed roots by immersing them in 1% acetocarmine at RT for 2 h.
8. Remove the acetocarmine solution, cover the stained roots with fresh Clarke's fixative, and transfer to –20°C for 14 days to ensure complete fixation (*see* **Note 10**).

3.3.2 Chromosome Preparation

1. After 2 weeks of storage in Clarke's fixative at –20°C, remove the roots from the fixative and place them into 45% acetic acid for 8–10 min (*see* **Note 11**).
2. Place one root on a microscopic slide, cutoff the root tip (2–3 mm), and discard the rest.
3. Add one drop of 60% acetic acid and gently squeeze out the cells from the root tip into the acetic acid drop.
4. Remove any visible root tissue and mix the squeezed cells with the acetic acid.
5. Carefully place an 18 × 18 mm glass coverslip on the top of the cell mixture and gently tap the loose coverslip by moving it a few mm.
6. Heat the slide for a few seconds over a spirit lamp to break down the cytoskeleton, flatten the cells, and facilitate the even distribution of the metaphase chromosomes on the slide. Heating also reduces the amount of cytoplasm around the chromosomes.

7. Observe the quality of the slide by using a phase contrast microscope.
8. If the slide is of good quality, turn it upside down and gently press it downwards on tissue paper, avoiding any side movement.
9. Place the slide on dry ice for at least 10 min, then remove the coverslip with a razor blade.
10. The preparations can be used for in situ hybridization or stored at –20°C for months.

3.3.3 Probe Preparation and Labeling

Total DNA Extraction (based on [11]):

1. Collect 5 g of fresh young leaves (use barley for extracting DNA for the GISH probe and wheat for the blocking DNA). Ground the leaves to a fine powder using liquid nitrogen. Transfer the powder into a 50 mL centrifuge tube.
2. Add 25 mL of prewarmed 2% (*w/v*) hexadecyltrimethylammonium bromide (CTAB) buffer containing 50 μL of 2-mercaptoethanol and gently shake the sample. The ground tissue must be kept frozen before extraction with the CTAB solution.
3. Incubate the mix for 30 min at 65°C and mix it twice during incubation.
4. Cool the sample to RT (1–2 min).
5. Add 12.5 mL of chloroform-isoamyl alcohol (24:1) to the tube and mix it gently.
6. Transfer the tube to an orbital shaker and mix at 30–50 rpm for 20 min.
7. Centrifuge the samples for 10 min at 11,000 × *g*.
8. Transfer the supernatant to a clean tube and repeat the chloroform extraction (Steps 5–8).
9. Precipitate the DNA by adding 15 mL of ice-cold isopropanol and shaking the tube gently. Leave the sample in a freezer overnight.
10. Pellet the DNA by centrifuging the tube for 5 min at 11,000 *g* Remove the supernatant and drain the tubes. Dry the DNA by leaving the tube on the bench overnight.
11. Dissolve the DNA in 1 mL of sterile Milli-Q water.

GISH Probe

1. Extract total DNA from 5 g of fresh, young barley leaves as above.

2. Fragment the DNA to a length of 300–500 bp in a pressure cooker for 6 min.
3. Label 1 μg of fragmented barley DNA by nick translation following a standard procedure (see Probe labeling).

FISH Probes

1. Amplify the barley centromere-specific G+C-rich repeat sequence and a part of the wheat centromeric retrotransposon CRW by PCR (based on [12] and [13], respectively).
2. Prepare the PCR reactions in sterile 0.2 mL PCR tubes at a final volume of 50 μL per reaction. Components for a single reaction: 25 μL of 2× DreamTaq PCR Master Mix, 3–3 μL of the forward and reverse primers, 18 μL sterile Milli-Q water, and 1 μL of barley (in the case of the G+C-rich probe) or wheat (CRW probe) total genomic DNA (50 ng/μL).
3. Cycling conditions for the amplification of the G+C-rich microsatellite repeat: 95°C, 2 min; followed by a cycle of 95°C, 30 s; 63°C, 20 s; and 72°C, 40 s. In the following five cycles, the annealing temperature is reduced by 1°C until it reaches 58°C. The subsequent 45 cycles are run at (95°C, 30 s; 58°C, 20 s; 72°C, 40 s) with a final extension step of 72°C, 2 min; 4°C hold. The CRW retrotransposon can be amplified by using the following cycling conditions: 95°C, 2 min; 35× (95°C, 20 s; 62°C, 15 s; 72°C, 20 s); 72°C, 2 min; 4°C hold.
4. For each probe, use 5 μL of the respective PCR reaction mixes for nick translation labeling by following a standard procedure. For simultaneous detection, label the two probes with different fluorochromes.

Probe Labeling

FISH and GISH probe labeling is carried out according to the manufacturer's recommendations for commercial Nick-Translation Kits (e.g., AZDye 594 Nick Translation Labeling Kit, Jena Bioscience). In brief:

1. The nick translation mixture contains: 1 μg of total genomic DNA (GISH probe) or amplified DNA sequences (FISH probe) 2 μL of 10× NT labeling buffer, 2 μL of AZDye594 NT labeling mix, and 2 μL of enzyme mix. When using two probes simultaneously for a double FISH, a different fluorochrome needs to be used for the second probe, e.g., Atto488.

2. Suspend the mix carefully until homogenous by using a pipette.
3. Place the tube into a thermocycler precooled to 15°C.
4. Incubate the reaction mixture for 90 min at 15°C.
5. Add 5 μL of Stop buffer and proceed to hybridization or store at −20°C.

Preparation of the Blocking DNA

1. Extract total DNA from 5 g of fresh, young wheat leaves (see DNA extraction above).
2. Shear the DNA in a pressure cooker (1 min) to obtain about 1000-bp fragments. Measure the DNA concentration and adjust it to 1 μg/μL. Store at −20°C.

3.3.4 *In Situ Hybridization*

Prehybridization Treatments

1. Remove the slides with chromosome preparations from the freezer and place them on the bench to quickly air dry (3–5 min).
2. To remove the cytoplasm and facilitate the probe access to the chromosomal DNA, add 50 μL of pepsin-HCl working solution to the slides while covering them with a plastic coverslip. Place the slides on a heated plate (37°C) for 1–2 min (*see* **Note 12**).
3. Remove the plastic coverslips and wash the slides in 1× PBS buffer (twice for 5 min).
4. Drain the slides and place them under a fume hood. Add 50 μL of 4% PFA to each slide and cover them with a plastic coverslip. Allow PFA postfixation at RT for 10 min (*see* **Note 5**).
5. Remove the plastic coverslip and wash the slides in 1× PBS buffer (twice for 5 min).

Hybridization

1. Incubate the hybridization mixture at 37°C for 30 min.
2. For simultaneous GISH-FISH (Fig. 3A), prepare the probe mix containing 1 μL of the barley centromeric G+C-rich repeat sequence probe (FISH probe), 1.2 μL of GISH probe (~50 ng labeled probe), and 1 μL of blocking DNA (1 μg/μL) (20× blocking versus the GISH probe) in 18 μL of hybridization mixture per slide. Mix the liquid well by repeatedly suspending it with a pipette.
3. For double FISH (Fig. 3B), prepare the probe mix by pipetting 1 μL (approx. 50 ng) of the barley centromeric G+C-rich repeat sequence probe and 1 μL (approx. 50 ng)

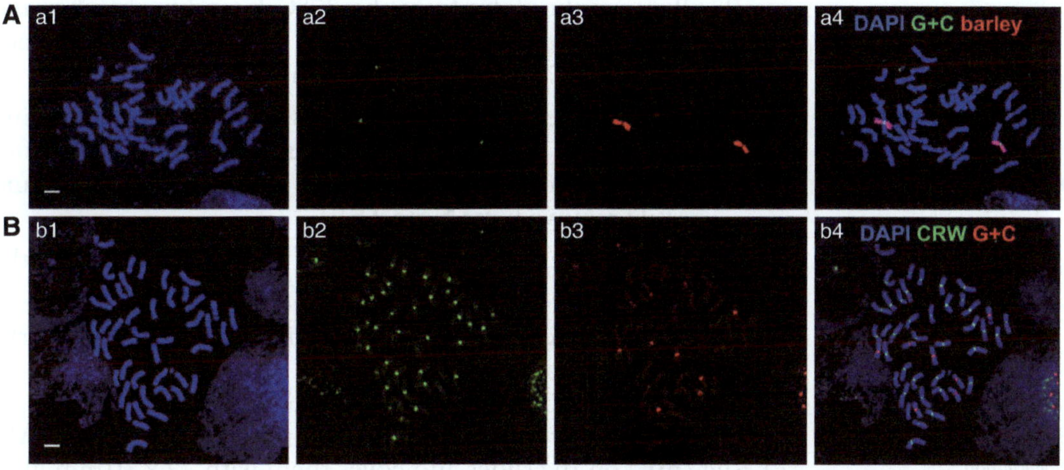

Fig. 3 Molecular cytological analysis of wheat × barley hybrid material. (**A**) Combined fluorescence- and genomic in situ hybridization (FISH–GISH) on somatic metaphase chromosomes of a wheat-barley disomic addition line 7 H ($2n = 44$). Chromosomes are counterstained with DAPI (blue, a1). The barley G+C-rich centromeric repeat sequences (FISH, green, a2) and barley DNA (GISH, red, a3) are visualized as well as their triple overlay (a4). (**B**) Fluorescence in situ hybridization on somatic metaphase chromosomes of a BC1 generation of a wheat × barley F1 hybrid line ($2n = 45$). Chromosomes are counterstained with DAPI (blue, b1). Wheat centromeres ($n = 38$) are shown by the CRW probe (green, b2), while barley centromeres ($n = 7$) are visualized by the G+C-rich repeat probe (red, b3) as well as their triple overlay (b4). Scale bar = 5 μm

of the wheat centromeric CRW retrotransposon sequence probe into 17 μL hybridization mixture per slide. Make sure the probes are well mixed in (*see* **Note 13**).

4. Denature the DNA in the probe mixture by placing the tubes into a PCR machine set at 85°C for 8.5 min.
5. Remove the tubes from the PCR machine and place them on ice immediately. Cover the tubes with ice and keep the denatured probe mixture on ice for at least 5 min.
6. Pipette the cold probe mixture on the pretreated slides and cover them with a 22 × 30 mm glass coverslip.
7. Place a stainless-steel plate on the heated surface of a PCR machine set to 75°C and wait for 2–3 min until the plate reaches the temperature. Preset a timer for 3 min and place the preparations on the hot plate. Denature the preparations for 3 min.
8. Remove the slides from the hot plate and place them into a humidity chamber set at 37°C (*see* **Note 14**). Incubate the slides for 18 h (overnight).

Posthybridization Washes

1. Prewarm 1× PBS buffer in a washing jar for 30 min in a water bath or incubator set to 37°C.

2. Remove the slides from the 37°C humidity chamber and immerse them in the prewarmed buffer. Place the jar back in the 37°C water bath or incubator for 5 min.
3. Remove the coverslips from the slides by carefully pushing them away from the preparation with fine forceps. After the 5-min soaking, the coverslips should become loose and easily removed (*see* **Note 15**).
4. Replace the buffer with fresh 1× PBS buffer and wash the slides twice for 5 min.
5. Remove the slides from the buffer and place them vertically on clean tissue paper. Allow to drain for 2–3 min.

3.3.5 Microscopic Observation

Pipette 15 μL of Vectashield antifade mounting medium with DAPI (1.5 μg/mL) on the slides and cover them with 22 × 30 mm glass coverslips. Turn the covered slides upside down on tissue paper and gently press them against the tissue to remove any excess liquid.

Analyze the slides using an epifluorescence or confocal laser scanning microscope (Fig. 3).

Explanation of Fig. 3.

Fig. 3A: A mitotic metaphase cell in a plant derived from a recurrent backcross generation of a wheat × barley F1 hybrid. This plant carries the complete chromosome complement of wheat (42 chromosomes stained by the DAPI counterstaining, blue), and a pair of barley chromosome 7 H hybridizing to the GISH probe (red signal) that uniformly covers the barley chromosomes. The centromeric DNA of the barley chromosomes is detected by the FISH probe containing the G+C-rich barley centromeric microsatellite repeats. The barley DNA and the centromeric regions can thus be monitored across various mitotic and meiotic developmental stages.

Fig. 3B: A mitotic metaphase cell in a plant derived from the first backcross generation of a full wheat × barley F1 hybrid. DAPI counterstaining (blue) uniformly labels both wheat and barley DNA. Double FISH specifically detects wheat (CRW probe, green signal) or barley (G+C-rich probe, red) centromeres. The method can be applied to cells at different developmental stages and combined with immunodetection to reveal epigenetic regulation of parental centromere activity

4 Notes

1. Other cultivars with similar results are: wheat—“Chinese Spring” (and its *Ph*^{I} derivative), “Sichuan,” and J-11; barley – “Morex.”
2. We use the Extract-N-Amp™ Plant PCR Kit (Merck, Sigma-Aldrich), but similar kits are available from other suppliers.

3. The PCR mix and conditions must be optimized for multiple primers. For convenience, we use the 2× Phusion U Green Multiplex PCR Master Mix (Thermo Scientific). Phusion Green Hot Start II High-Fidelity DNA Polymerase (Thermo Scientific) and its 5× Phusion Green HF Buffer can also be used with comparable MPCR results to reduce costs. However, in this case, a 10 mM dNTP mix must also be included in the master mix.
4. Other PCR cyclers may deliver different results, so it is necessary to test the reaction parameters beforehand.
5. Paraformaldehyde is a potential carcinogen; wear appropriate protective clothing and gloves to avoid inhalation and skin contact by handling it in a chemical fume hood. Dispose of wipes and liquids as hazardous waste in a dedicated container.
6. Pierce one-sided holes on the top and bottom of the internode to serve as indicator and injection points, respectively. Inject the SDI solution slowly into the cavity via the lower point until drops appear at the top hole.
7. Direct DNA extraction can be replaced by traditionally purified DNA, but such DNA may affect the number of off-target amplicons.
8. Alternatively, assemble a bioinformatic workflow as outlined [4] to identify unique chromosome-specific loci with defined lengths on reference genomes and full genomes of various wheat and barley cultivars.
9. Barley seeds may need more than 24 h for germination.
10. Material fixed for less than 2 weeks might result in more fragile tissues, in which the cells and the chromosomes are easily lost during manual preparation.
11. Soft tissues need shorter acetic acid treatment.
12. The treatment time depends on the amount of cytoplasm around the chromosome spreads as observed in a phase contrast microscope.
13. As multiple slides are usually used for hybridization, a master probe mix needs to be prepared. Due to the high viscosity and inevitable loss of a small proportion of the liquid during pipetting, it is advised to calculate with slightly larger quantities than the number of slides used (e.g., when using eight slides, calculate the probe mix for 8.5 slides). For the denaturation step, the volumes for two to three slides can be measured in 0.2 mL PCR tubes. The liquid can reach but not exceed the level of the wells in the PCR machine to ensure precise temperature control during the heating procedure.

14. High humidity is critical for successful hybridization.
15. Longer soaking times may be needed, especially when slides dry out during hybridization.

Acknowledgments

This chapter is based on publications supported by the Hungarian Research, Development and Innovation Office (grant numbers FK134264, FK134992, K125300, and K129221), the Hungarian National Laboratory Program (grant number RRF-2.3.1-21-2022-00007), and the Research Excellence Programme of the Hungarian University of Agriculture and Life Sciences.

References

1. Ishii T, Karimi-Ashtiyani R, Houben A (2016) Haploidization via chromosome elimination: means and mechanisms. Annu Rev Plant Biol 67:421–438
2. Polgári D, Cseh A, Szakács É et al (2014) High-frequency generation and characterization of intergeneric hybrids and haploids from new wheat–barley crosses. Plant Cell Rep 33:1323–1331
3. Polgári D, Mihók E, Sági L (2019) Composition and random elimination of paternal chromosomes in a large population of wheat × barley (*Triticum aestivum* L. × *Hordeum vulgare* L.) hybrids. Plant Cell Rep 38:767–775
4. Ali M, Polgári D, Sepsi A et al (2024) Rapid and cost-effective molecular karyotyping in wheat, barley, and their cross-progeny by chromosome-specific multiplex PCR. Plant Methods 20: 37
5. Makai D, Mihók E, Polgári D et al (2023) Rapid in-solution preparation of somatic and meiotic plant cell nuclei for high-quality 3D immunoFISH and immuneFISH-GISH. Plant Methods 19: 80
6. Sigurbjörnsson B, Micke A (1969) Progress in mutation breeding. Induced Mutation in Plants. Proceedings of an International Symposium on the Nature, Induction and Utilization of Mutation in Plants. International Atomic Energy Agency, Vienna, pp. 673–697
7. Kumlehn J, Lörz H, Kranz E (1998) Differentiation of isolated wheat zygotes into embryos and normal plants. Planta 205:327–333
8. Chu CC, Wang CC, Sun CS et al (1975) Establishment of an efficient medium for anther culture of rice through comparative experiments on the nitrogen sources. Sci Sin 18:659–668
9. Peacock AC, Dingman CW (1967) Resolution of multiple ribonucleic acid species by polyacrylamide gel electrophoresis. Biochemistry 6:1818–1827
10. Clarke JL (1851) Researches into the structure of the spinal chord. Phil Trans R Soc Lond 141:607–621
11. Saghai-Maroof MA, Soliman KM, Jorgensen RA, Allard RW (1984) Ribosomal DNA spacer-length polymorphisms in barley: mendelian inheritance, chromosomal location, and population dynamics. Proc Natl Acad Sci USA 81:8014–8018
12. Hudakova S, Michalek S, Presting GG et al (2001) Sequence organization of barley centromeres. Nucleic Acids Res 29:5029–5035
13. Mihók E, Polgári D, Lenykó-Thegze A et al (2024) Plasticity of parental CENH3 incorporation into the centromeres in wheat × barley F1 hybrids. Front Plant Sci 15:1324817

Check for updates

Chapter 8

Bread Wheat Anther Culture

Sue Broughton, Marieclaire Castello, Li Liu, Julie Killen and Christopher McMullan

Abstract

Doubled haploid (DH) technology can rapidly deliver fixed lines for breeding and research purposes, especially in crops like wheat (*Triticum aestivum* L.). In Western Australia, we have developed an anther culture protocol that generates more than 12,000 bread wheat DH lines annually. Key aspects of the protocol include the growth conditions of donor plants, microspore developmental stage, stress pretreatment, induction and regeneration medium and culture conditions. Using controlled environment rooms for donor plant growth is important in Australia, enabling year-round production of high-quality plants. We select spikes for anther culture when microspores are at the mid–late to late uninucleate stage. Several stress pretreatments are applied to excised anthers to reprogram microspores from a pathway of gametogenesis to embryogenesis and plant development. We use a combination of mannitol and *n*-butanol stress pretreatments to induce embryogenesis and these are critical steps in our protocol. Following stress pretreatments, anthers are coincubated with ovaries in a liquid induction medium (LIM) for 4–6 weeks. The inclusion of ovaries during the induction phase is another important step. Resulting embryos are then transferred to a solid regeneration medium for shoot and root development. Plants are subsequently transferred to a soil medium and grown to maturity. We rely on spontaneous chromosome doubling to restore fertility in our wheat DH lines.

Keywords Doubled haploid, Microspore, Anther culture, Ovary, Wheat

1 Introduction

Wheat is Australia's most important crop in terms of area and production. Annually, Australia produces approximately 25 million metric tons (MMT) of wheat from around 12 million ha. In recent years, production has reached 30–40 MMT, contributing about 20% of global wheat exports and 4% of global wheat production. Most of Australia's wheat (65%–75%) is exported, with Western Australia being the largest exporting state [1, 2]. Breeding new and improved wheat varieties remains a critical goal, and the production of doubled haploid (DH) lines is a valuable tool for rapidly generating fixed lines. In self-pollinating crops like wheat, DH lines can reduce the time to

Ana María Castillo et al. (eds.), *Doubled Haploid Methods in Cereals*, Methods in Molecular Biology, vol. 3029,
https://doi.org/10.1007/978-1-0716-5245-9_8,

variety release by several years. Moreover, phenotypic evaluation and selection are more reliable and accurate in DH populations than in other breeding populations. Doubled haploid populations are also helpful for gene mapping, genome-wide association studies, and genome sequencing and can be used in transformation and gene editing studies to generate homozygous transformed or edited plants in a single step [3–6].

A bread wheat anther culture protocol has been developed at the Department of Primary Industries and Regional Development (DPIRD) in Western Australia and is used to generate 12,000–16,000 wheat DH lines annually. The demand is primarily for spring wheat. Most DH lines are used for breeding (85%–90%) and the remainder for research [7–9]. Like many androgenesis protocols, key factors include donor plant growth conditions, microspore developmental stage, stress pretreatments, and induction/regeneration culture medium and conditions.

In our program, donor plants are always grown in controlled environment rooms, in relatively cool temperatures with good quality lighting, enabling year-round DH production. Spikes are selected when microspores are at the mid–late to late uninucleate stage. Stress pretreatments are required to reprogram microspores from a pathway of gametogenesis to embryogenesis. We apply two sequential pretreatments to excised anthers in our protocol; anthers are placed on a solid 1.0 M mannitol pretreatment medium for 5 days, followed by a 5-h treatment with *n*-butanol. The mannitol pretreatment, adapted from Cistué et al. [10], correlates with increased green plant production in barley anther culture. We found that pretreating barley anthers on solid 1.0 M mannitol medium yielded more green plants than cold pretreatment, particularly when combined with increased maltose levels in the induction medium (e.g., 100 g/L) [11]. We also use solid 1.0 M mannitol pretreatment medium in wheat, with the addition of calcium and macronutrients, which can increase embryo and green plant regeneration [12, 13]. Macronutrients are included at the same concentration as the induction medium, and calcium is added as per Soriano et al. [14].

Following mannitol pretreatment, anthers are treated with *n*-butanol for 5 h, which has been shown to stimulate embryogenesis in wheat, likely through the disruption of cortical microtubules [14, 15]. Recent research showed that mannitol pretreatment depolymerizes microtubules in wheat microspores, and when combined with *n*-butanol (mannitol followed by *n*-butanol), significant changes in microtubule organization occur, enhancing microspore embryogenesis [16]. After mannitol and *n*-butanol pretreatments, we generally perform a 48-h treatment with dimethyl sulfoxide (DMSO). Adding DMSO to solid mannitol pretreatment medium in barley and wheat anther culture has been shown to increase green plant numbers [17]. In our program, we have obtained the best results when DMSO was combined with our liquid induction medium (LIM) and applied at the start of the induction phase (unpublished data). Another critical step in our protocol is the inclusion of ovaries and ovary-conditioned medium during the induction phase [15, 18]. Following the coincubation of pretreated anthers and ovaries in ovary conditioned LIM, resulting embryos are transferred to solid regeneration medium for shoot and root development. When regenerated plants have sufficient root growth, they are transferred to soil, usually in seedling trays within a protected environment, and subsequently moved to a glasshouse or tunnel house to grow to maturity.

We rely on spontaneous chromosome doubling for routine wheat DH production to restore fertility. In cereals, such as wheat and barley, this mechanism is primarily the result of nuclear fusion in the early stages of microspore embryogenesis [19–21]. It avoids the need to screen regenerant plants for ploidy and treat haploid plants with colchicine via root wash (immersion) treatments, which can result in plant death and reduced and delayed seed set. In contrast, spontaneous doubling results in good seed volumes, which is useful for breeding companies requiring rapid seed increase before field testing. Morphological observations of mature plants are used to identify haploid and sterile lines. Our long-term average for spontaneous doubling in wheat is 30%–40%, with observed frequencies ranging from 14% to 80%.

The following sections describe the protocol used at the Western Australian DPIRD DH Program to produce wheat DHs for commercial breeding programs and researchers in Australia.

2 Materials

The materials and tissue culture media required for wheat anther culture are summarized below (*see* Sects. 2.1–2.5).

2.1 Plant Material

The DPIRD protocol has been used to generate 12,000–16,000 wheat DH lines annually in the last 6 years. The bulk of the lines are used for breeding purposes (85%–90%), while the remainder are used for research at DPIRD and other Australian institutions. In recent years, the program has processed approximately 70 wheat crosses per year, using 4–8 F_1 donor plants per cross. The program primarily focuses on spring wheat crosses.

2.2 Equipment and Materials for Plant Growth

1. Controlled environment rooms: Walk-in rooms with controlled temperature and high-quality lighting are required for year-round DH production in Australia. Sodium vapor lamps (Osram 400 W) provide approximately 450–500 μmol m^{-2} s^{-1} at the canopy level.

2. Greenhouses: We use polycarbonate greenhouses with large benches. Each bench is approximately 1.8 × 9 m. The benches are covered in capillary matting for watering. The greenhouses have either refrigerative or evaporative cooling systems that work well in cooler months in Western Australia (e.g., April–November).
3. Fertilizers:
   - Soluble fertilizer with 20:8.7:16.6 (N:P:K elemental analysis) and trace elements.
   - Iron chelate–EDTA (13% Fe soluble EDTA).
   - Slow-release fertilizer with 15:3.9:9.1 (N:P:K elemental analysis) plus 1.2 MgO and trace elements.
4. Pots and seed trays: Nursery pots (5.2 L) and Kwikpot trays (30-cell) (Garden City Plastics, https://www.gardencityplastics.com).
5. Soil mix: Commercial potting mix containing fine pine bark compost, coco peat, and river sand in a 0.5:0.25:0.25 ratio. Nutrients included per cubic meter are dolomite (1.4 kg), lime (1.0 kg), mineral mix (0.9 kg), calcium nitrate (0.4 kg), urea (0.2 kg), superphosphate (0.2 kg), and iron chelate (0.05 kg), pH 5.8–6.0.

2.3 Laboratory Equipment and Materials

1. Filter units: Steritop-GP Filter Unit 0.22 μm filter.
2. Autoclave.
3. Balance.
4. Bunsen burner.
5. Freezer (−20°C).
6. Fine forceps.
7. Glassware including Schott bottles.
8. Hot-plate/stirrer.
9. Fume cabinet.
10. Laminar flow cabinet.
11. Oven.
12. pH meter.
13. Optical microscope.
14. Refrigerator (4°C).
15. Scissors.
16. Scalpels.
17. Stereo or dissecting microscope.
18. Vacuum pump.
19. Water bath.

20. Incubators used for the pretreatment and induction phases at 25°C. These incubators do not have lights or humidity control.
21. "Culture room" for the regeneration phase with mesh shelving units. Fluorescent lamps are located on the underside of the shelves and provide lighting for cultures on the shelf below. The distance from the lamps to the cultures is approximately 25 cm. The lamps are a mixture of 36 W Sylvania Grolux T8 and 36 W Phillips or Osram cool white lamps and provide approximately 50–100 μmol m^{-2} s^{-1}. Two air conditioning units control the temperature (25 ± 2°C) and the culture room does not have humidity control.
22. Sterile 55 × 14 mm Petri dishes.
23. Sterile 92 × 16 mm Petri dishes.
24. Sterile 90 × 20 mm Petri dishes.
25. Parafilm™.

2.4 Laboratory Solutions

1. Chemicals: In addition to the media components (see Table 1), you will require the following chemicals:
   - Acetic acid (glacial).
   - Commercial bleach (sodium hypochlorite 42 g/L and sodium hydroxide 9 g/L).
   - Carmine.
   - *n*-Butanol.
   - Dimethyl sulfoxide (DMSO).
2. Fertilizer solution: Dissolve soluble fertilizer and iron chelate–EDTA (13% Fe soluble EDTA) together in tap water at approximately 5.2 g/L and 1.4 g/L, respectively.
3. Acetocarmine stain: Prepare 100 mL acetocarmine stain (1% or 2% w/v). Wearing PPE and working in a fume cabinet, prepare a 45% acetic acid solution by carefully adding 45 mL glacial acetic acid to 55 mL deionized (DI) water. Heat solution in a conical flask to boiling. Add the dye (1 or 2 g) slowly to the boiling solution, stirring with a glass rod. Boil gently until the dye dissolves. Cool to room temperature, filter and store in bottle with glass stopper [22].
4. Bleach solution: Prepare 1 L of a "1%" bleach solution (10.5 g/L sodium hypochlorite) by combining 250 mL commercial bleach with 750 mL DI water. Use within 1 month of preparation.

2.5 Tissue Culture Media

Four tissue culture media are required for this protocol (see Table 1) including:

- Pretreatment medium: 1.0 M mannitol (solid).
- 0.7 M mannitol medium (liquid).

Table 1
Media for wheat anther culture

Components	Pretreatment medium (mg/L)	0.7 M liquid mannitol medium (mg/L)	Liquid induction medium (LIM) (mg/L)	Regeneration medium (mg/L)
Macronutrients				
KNO_3	1400.0	1400.0	1400.0	1400.0
NH_4NO_3	300.0	300.0	300.0	300.0
$CaCl_2.2H_2O$	6340.0	6340.0	440.0	440.0
$MgSO_4.7H_2O$	370.0	370.0	370.0	370.0
KH_2PO_4	170.0	170.0	170.0	170.0
Micronutrients				
$MnSO_4.4H_2O$	–	–	22.3	22.3
$ZnSO_4.7H_20$	–	–	8.6	8.6
H_3BO_3	–	–	6.2	6.2
KI	–	–	0.83	0.83
$Na_2Mo_4.2H_2O$	–	–	0.25	0.25
$CoCl_2.6H_2O$	–	–	0.025	0.025
$CuSO_4.5H_2O$	–	–	0.025	2.5
Iron source				
FeNaEDTA (Sigma E6760)	–	–	40.0	40.0
Other components				
Glutamine	–	–	750.0	750.0
Casein hydrolysate	–	–	300.0	300.0
Myo-inositol	–	–	100.0	100.0
Proline	–	–	100.0	100.0
Citric acid	–	–	10.0	10.0
Pyruvic acid	–	–	10.0	10.0
Glycine	–	–	2.0	2.0
Nicotinic acid	–	–	2.0	2.0
Pantothenic acid	–	–	2.0	2.0
Pyridoxine HCl	–	–	2.0	2.0
Thiamine HCl	–	–	0.4	0.4
Biotin	–	–	0.02	0.02
Gum arabic (Sigma G9752)	–	–	10.0	–

(continued)

Table 1
(continued)

Components	Pretreatment medium (mg/L)	0.7 M liquid mannitol medium (mg/L)	Liquid induction medium (LIM) (mg/L)	Regeneration medium (mg/L)
+(−) Arabinogalactan (Sigma 10830)	–	–	10.0	–
2,4-D	–	–	2.0	–
Kinetin	–	–	0.5	–
Cefotaxime	–	–	100.0	100.0
Maltose	–	–	100.0 g/L	30.0 g/L
Mannitol	182.0 g/L	127.5 g/L	–	–
Agar (Sigma A7921)	16.5 g/L	–	–	–
Phytagel (Sigma P8169)	–	–	–	3.0 g/L
pH	not adjusted	not adjusted	5.8	5.8

- Liquid induction medium (LIM).
- Regeneration medium (solid).

Instructions required to prepare the media:

1. Stock solutions: Prepare stock solutions (see Table 2). Stocks are indicated in mg/L, except for 2,4-D and kinetin (mg/mL). Different stock volumes are prepared, depending on the volume of stock used in the media (see Tables 2 and 3) (*see* **Note 1**).
   - Biotin is dissolved in ammonium hydroxide, as per Sigma technical specifications. We dissolve 25 mg biotin in 1–2 mL of 2 M ammonium hydroxide, make up to 100 mL with DI water, then add 8 mL of biotin stock to 1 L of vitamin stock.
   - The 2,4-D stock is prepared by dissolving 8 mg/mL 2,4-D in absolute ethanol.
   - Kinetin is purchased as a liquid (1 mg/mL), frozen in single-use aliquots, and added during medium preparation.
   - Vitamins, glycine, gum arabic, and +(−) arabinogalactan are prepared as separate stocks; however, you may wish to combine them e.g., add glycine to the vitamin stock, and combine the gum arabic and the +(−) arabinogalactan stocks.
2. Pretreatment medium: 1.0 M mannitol (solid): To prepare 1 L medium, add ingredients, including macronutrient stock, $CaCl_2.2H_2O$, and mannitol, to a 1 L beaker (see Table 3). Add

Table 2
Stocks for wheat anther culture

Stocks	mg/L	Preparation and storage notes
Macronutrient Stock		
KNO_3	14,000	Prepare 2 L stock. Add KNO_3 and 500–1000 mL DI water to a volumetric flask. Mix until dissolved. Add each remaining ingredient with ~100 mL DI water, ensuring each one is dissolved before adding the next ingredient. Store refrigerated at 4°C, replace every 3 months.
NH_4NO_3	3000	
$CaCl_2.2H_2O$	4400	
$MgSO_4.7H_2O$	3700	
KH_2PO_4	1700	
Iron Stock		
FeNaEDTA (Sigma E6760)	800	Prepare 1 L stock. Store refrigerated at 4°C, replace every 3 months.
Micronutrient Stock		
$MnSO_4.4H_2O$	4450	Prepare 200 mL stock. Store refrigerated at 4°C, replace every 3 months.
$ZnSO_4.7H_20$	1700	
H_3BO_3	1250	
KI	170	
$Na_2Mo_4.2H_2O$	50	
$CoCl_2.6H_2O$	5	
$CuSO_4.5H_2O$	5	
Vitamin Stock		
Citric acid	1000	Prepare 500 mL or 1 L stock. Store frozen in 10 mL aliquots and replace every 6 months.
Pyruvic acid	1000	
Nicotinic acid	200	
Pantothenic acid	200	
Pyridoxine.HCl	200	
Thiamine.HCl	40	
Biotin	2	
Glycine Stock		
Glycine	200	As above
Gum Arabic Stock		
Gum arabic (Sigma G9752)	1000	As above
+(−) Arabinogalactan Stock		
+(−) Arabinogalactan (Sigma 10830)	1000	As above

(continued)

all dry ingredients have dissolved, adjust pH, then transfer to a volumetric flask and make up to 500 mL with DI water. Filter sterilize through a Steritop 0.22 μm filter. Combine the autoclaved and filter-sterilized portions (*see* **Note 3**), swirl thoroughly to combine, and pour into 90 × 20 mm Petri dishes (20 mL per dish). Store the dishes in sealed plastic sleeves at room temperature in the dark and use within 4 weeks.

3 Methods

3.1 Wheat Donor Plant Growth

3.1.1 Maintain controlled environment rooms for donor plant (F_1) growth at 18/13°C (day/night) with a 12 h photoperiod.

3.1.2 Sow F_1 seeds directly into 5.2 L pots (one plant per pot) containing commercial potting mix (*see* Sect. 2.2). Fertilize weekly with commercial soluble fertilizer and iron chelate–EDTA (*see* Sect. 2.2 and **Note 5**).

3.2 Wheat Spike Harvest, Storage, and Sterilization

3.2.1 Harvest wheat spikes for anther culture when the leaf sheath surrounding the spike is just beginning to open (approximately 1–2 mm), although this may vary with genotype (Fig. 1A). Cut tillers several nodes below the spike and place them in a beaker with approximately 100 mL tap water. To determine microspore developmental stage, remove three anthers from a central floret, squash anthers in a drop of

Fig. 1 Harvest and sterilization of wheat spikes. (**A**) Spikes with the leaf sheath opened approximately 1–2 mm. (**B**) Spikes ready for sterilization

acetocarmine stain (we have used 1% and 2% w/v acetocarmine stain), cover with a coverslip, and observe under 400× magnification. Microspores should be at the mid–late to late uninucleate stage of development (*see* **Note 6**).

3.2.2 After harvest, spikes are either sterilized the same day or stored in a beaker with water for 1–4 days at 4°C. This period can be extended to 5–7 days if necessary.

3.2.3 Sterilize the spikes as described by Kasha et al. [23] by trimming awns and placing 4–6 spikes into a 50 mL centrifuge tube (Fig. 1B). Fill tubes with bleach solution (10.5 g/L sodium hypochlorite) and sterilize spikes for 15 min. Rinse the spikes in sterile DI water for 5 min, twice. Spikes can be dissected immediately following sterilization or left at 4°C overnight and processed the next day (*see* **Note 7**).

3.3 Wheat Anther Pretreatment (Mannitol)

3.3.1 In a laminar flow cabinet, prepare a dissecting microscope and equipment, including fine forceps for anther and ovary excision. Prepare 92 × 16 mm Petri dishes containing pretreatment medium by marking the base of the dishes into quarters (one dish per four spikes). Prepare 55 × 14 mm Petri dishes with 4 mL LIM, one per spike. Label all dishes with relevant information (e.g., genotype and date).

3.3.2 Excise three anthers per floret from approximately 20 florets (ten from each side of the spike). We use a 92 × 16 mm Petri dish lid as a sterile surface for anther excision and replace it after each set of four spikes. Place the anthers on the surface of the wheat pretreatment medium, placing anthers from each spike on different quarters of the Petri dish. The three anthers from each floret can be left in a group; they need not be separated (Fig. 2A). Record anther numbers and other information as required.

3.3.3 Following anther excision, excise ten ovaries from each spike and place them in a 55 × 14 mm Petri dish containing LIM (Fig. 2B).

3.3.4 Cover the Petri dishes containing anthers and ovaries with foil until all spikes are processed.

3.3.5 When all spikes are processed, seal the Petri dishes with Parafilm™ and incubate in the dark at 25°C for 5 days (anthers) and 7 days (ovaries).

3.4 Wheat Anther Pretreatment (n-butanol)

3.4.1 After 5 days of mannitol pretreatment, prepare a fresh set of 55 × 14 mm Petri dishes with 4 mL of 0.7 M liquid mannitol medium. Transfer the anthers from each spike onto a dish containing 0.7 M liquid mannitol (*see* **Note 8**).

Fig. 2 Bread wheat anther culture. (**A**) Anthers from four spikes on solid pretreatment medium. (**B**) Ovaries from one spike in 4 mL liquid induction medium (LIM): **C**) Anthers and ovaries combined in ovary-conditioned LIM for 4–6 weeks induction phase. **D**) Visible embryos after 4–6 weeks culture

3.4.2 Once all anthers are transferred, add 8 μL *n*-butanol (0.2% v/v) to each dish.

3.4.3 Leave the Petri dishes unsealed, carefully wrap bundles of 16–24 dishes in foil and incubate in a fume cabinet at room temperature (approximately 22–25°C) for 5 h.

3.5 Wheat DMSO Treatment (Optional) and Embryo Induction

Anthers are treated with DMSO after the *n*-butanol pretreatment. We now routinely include this step, but it can be omitted (*see* **Note 9**).

3.5.1 After the 5-h *n*-butanol pretreatment, remove the 0.7 M liquid mannitol medium + *n*-butanol from each dish using a pipette, leaving the anthers behind.

3.5.2 Add 4 mL LIM supplemented with 0.6% DMSO (v/v) to each dish. Leave the dishes unsealed, carefully wrap bundles of 16–24 dishes in foil, and incubate in the dark at 25°C for approximately 48 h.

3.5.3 After 48 h, remove the LIM + DMSO from each dish using a pipette, leaving the anthers behind. Take the dishes of LIM + ovaries (incubated for 7 days) and transfer the contents into a dish containing anthers. The ovaries and anthers do not need to be matched to the same spike. If any ovaries are left behind, gently transfer them with forceps. The ovaries grow during the incubation in LIM; by the time they are added to the anthers, they have nearly doubled in size.

3.5.4 Gently swirl the dishes so that all anthers are floating and not stuck to the sides or bottom of the dish. Seal dishes with Parafilm™. Coincubate the anthers and ovaries in the dark at 25°C for 4–6 weeks for embryo induction (Fig. 2C).

3.6 Wheat Regeneration and Root Induction

3.6.1 Embryos should be visible after 4–6 weeks (Fig. 2D).

3.6.2 Transfer embryos to regeneration medium by tipping the embryos and LIM on to the regeneration medium. Typically, embryos from one Petri dish containing LIM medium are transferred onto one dish of regeneration medium. If embryo density is low, embryos from several dishes can be combined on one regeneration dish. While ovaries are typically removed before transferring the embryos to the regeneration medium, they can also be left and transferred along with the embryos. After transferring the embryos to regeneration medium, remove excess LIM by gently tipping the dish and pouring off the LIM (ensuring embryos remain on the medium) or by using a pipette.

Fig. 3 Wheat plant regeneration from anther culture. (**A**) Embryos tipped onto regeneration medium develop into plants. (**B**) Sub-cultured wheat plants on regeneration medium (10 per dish)

Fig. 4 Grow-out of wheat regenerant plants from anther culture. (**A**) Clear plastic covers are placed over the trays for the first week to maintain humidity. (**B**) Wire racks are added to the trays to prevent plants from falling over. (**C**) Plants are grown to maturity in 30-cell trays on capillary matting

3.6.3 Seal the Petri dishes with Parafilm™ and incubate them in the dark at 25°C for 7 days.

3.6.4 Transfer the Petri dishes to a 25°C culture room or growth cabinet with a 12 h photoperiod. After approximately 1–2 weeks, select individual green plants from the mass of embryos and plants (Fig. 3A). Place ten plants on each Petri dish containing fresh regeneration medium (Fig. 3B) and incubate the cultures under the same conditions (*see* **Note 10**). Continue incubation for 3 weeks or until sufficient root growth (at least 1.5–3 cm long on each plant), when the plants are ready to be transferred to soil.

3.7 Transfer Wheat Regenerant Plants to Soil

3.7.1 Prepare the Kwikpot trays with potting mix and a slow-release fertilizer (*see* Sect. 2.2 and **Note 11**). Make a hole in the corner of each cell and add 1.5–2.0 g slow-release fertilizer. Cover the fertilizer with soil. Lightly water trays prior to transplanting.

3.7.2 Remove the plants from Petri dishes and gently remove excess medium from the roots. Lightly trim any very long or yellowing leaves with scissors prior to transplanting. Place one plant into the center of each cell, ensuring that the roots are pressed firmly into the soil. Clear plastic covers should be placed on the trays for the first week after transplanting to maintain humidity and facilitate hardening-off (Fig. 4A).

3.7.3 Transfer the regenerant plants directly to a glasshouse if seasonally appropriate (e.g., April–November in Australia) or establish them in controlled environment rooms (e.g., 18/13°C) for 3–4 weeks before transferring to a glasshouse. Place Kwikpot trays on capillary matting in the glasshouse. Plants are grown to maturity in the trays, and since they send some roots into the matting, the trays cannot be moved once in position. Therefore, position the trays with space between them to allow the plants to develop to maturity. Add wire racks to the trays once the plastic covers are removed to support the plants as they mature (Fig. 4B, C).

3.7.4 Harvest fertile wheat DH lines at maturity (*see* **Note 12**).

4 Notes

1. All ingredients, except for biotin and 2,4-D, are water soluble. Deionized (DI) water is used for all stock and medium preparation.
2. Agar (Sigma A7921) is preferred for the pretreatment medium. This medium requires a higher concentration of agar to ensure the medium is a firm consistency (1.65%). However, the agar concentration can be adjusted if required. If the agar is too firm, the anthers tend to fly off and stick to the underside of the Petri dish lid. If the agar is too soft, it's not easy to place the anthers on the agar. Phytagel (Sigma P8169) is used for the regeneration medium. Phytagel

requires the presence of divalent cations to form a gel matrix; the calcium and magnesium contained in most plant tissue culture media are usually sufficient for gelation.

3. Schott bottles, autoclave times and cooling media (post-autoclaving): When autoclaving media in Schott bottles, we only fill the bottles to half or two thirds of their capacity to prevent media from boiling over during autoclaving. Autoclave times are adjusted based on volume. Volumes up to 500 mL are autoclaved for 20 min, while 750 mL or 1 L volumes are autoclaved for 30 min. Cooling methods (post-autoclaving) depend on the gelling agent and whether the media will be combined with a filter-sterilized portion. For example, for the pretreatment medium (*see* Sect. 2.5), we cool the medium to approximately 50°C before pouring into Petri dishes. The cooling can be hastened by placing the bottles in a laminar flow cabinet and removing the lids. This medium contains agar (Sigma A7921), which congeals at 30–35°C. For regeneration medium (*see* Sect. 2.5), where autoclaved and filter-sterilized portions are combined, and Phytagel is the gelling agent, we cool the autoclaved portions (but not too much) and warm the filter-sterilized portions. Phytagel congeals at 27–32°C, so if the filter-sterilized portions are not warmed, the medium will become lumpy and set rapidly once the portions are combined. For this medium, cool the autoclaved portion to 55–60°C by placing bottles in a water bath (or oven set to 60°C) for 15–20 min and warm the filter-sterilized portions to 55–60°C in the same water bath or oven for 15–20 min.

4. When preparing media that require frozen stocks (e.g., LIM or regeneration medium), we remove stocks from the freezer the day before and defrost them in the refrigerator. We also autoclave Schott bottles (for filter-sterilized media) the day before, so the bottles are cool and ready. For LIM, we often filter sterilize into a 1 L bottle for convenience, but then store in 2 × 500 mL bottles to reduce the risk of contamination. For this chapter, we have indicated how to prepare 1 L of each media. However, when the program is busy, we often prepare larger volumes, and amounts are scaled up accordingly. For example, we often prepare 2 L LIM and 4 L regeneration medium.

5. Once the wheat F_1 seedlings are established (after 2–3 weeks), apply fertilizer at approximately 125 mL per pot, increasing to 250 mL per pot over the following weeks until the spike harvest is complete.

6. At the beginning of each cross, check the developmental stage of wheat microspores in a few spikes and correlate with morphological indicators, such as the width of the leaf sheath opening surrounding the spike, to select spikes for the

remainder of the spike harvest. For anther culture, mid–late and late uninucleate microspores are preferred, but good responses have also been obtained from slightly older spikes where mitosis is visible.

7. If spikes are left overnight after sterilization, retain the spikes in the 50 mL centrifuge tubes with approximately 0.5 mL sterile DI water. When processing large numbers of spikes, we usually sterilize them the day before.

8. We apply *n*-butanol treatment in 0.7 M liquid mannitol medium. Wheat LIM can also be used at this step, but 0.7 M liquid mannitol medium is more cost-effective for the relatively short *n*-butanol treatment (5 h).

9. Our in-house experiments have demonstrated that a DMSO treatment, at the start of the induction phase, improves green plant regeneration, although the effect varies with genotype (unpublished data). We routinely include it in our protocol, but it could also be omitted. When DMSO treatment is omitted, after the 5-h *n*-butanol pretreatment, remove the 0.7 M liquid mannitol medium + *n*-butanol from each Petri dish using a pipette, leaving the anthers behind. Take the dishes of LIM + ovaries (incubated for 5 days) and transfer the contents into a dish containing anthers (*see* Sect. 3.5).

10. It is easier to select individual plants when they are small; hence, it is recommended to sub-culture regenerant plants (10 per dish) after 1–2 weeks of light incubation. If left too long, the plants can become tangled, making it more challenging to separate them. Selecting individual plants, and placing ten plants per dish, facilitates easier transplanting and reduces the risk of planting multiple plants together.

11. We generally use 30-cell Kwikpot trays for wheat, but larger cell trays (e.g., 20-cell Kwikpot trays) or pots, depending on glasshouse space and required seed volumes, can be used. Fill the trays with potting mix, wet them with water, and add slow-release fertilizer. This can be done the day before, or the day of transplanting.

12. For routine wheat DH production, we rely on spontaneous chromosome doubling and use morphological observations of mature plants to identify haploid and sterile lines.

Acknowledgments

This work has been supported by the Western Australian Department of Primary Industries and Regional Development (DPIRD).

References

1. ABARES (2024) Australian crop report: March 2024, Australian Bureau of Agricultural and Resource Economics and Sciences, Canberra, March. CC BY 4.0. https://doi.org/10.25814/n3jr-fh31
2. Kingwell R, Maharjan D (2022) Bulk and containerized grain exports from Australia's main grain producing states. AEGIC Industry Supplementary Report, August, 2022. https://www.aegic.org.au/wp-content/uploads/2022/08/AEGIC-Bulk-and-containerised-grain-exports-from-Australias-main-grain-producing-states-2022.pdf
3. Devaux P, Cistué L (2016) Wheat doubled haploids: production to sequencing. What makes them so appealing? In: Bonjean AP, Angus WJ, Ginkel van M (eds) The world wheat book, a history of wheat breeding, vol 3. Lavoisier Tec & Doc Publishers, Paris, pp 885–938
4. Weyen J (2021) Applications of doubled haploids in plant breeding and applied research. In: Segui-Simarro JM (ed) Doubled Haploid Technology. Methods in Molecular Biology, vol 2287. Humana, New York, pp 23–39. https://doi.org/10.1007/978-1-0716-1315-3_2
5. Brew-Appiah RAT, Ankrah N, Liu W et al (2013) Generation of doubled haploid transgenic wheat lines by microspore transformation. PLOS One 8:e80155. https://doi.org/10.1371/journal.pone.0080155
6. Bhowmik P, Ellison E, Polley B et al (2018) Targeted mutagenesis in wheat microspores using CRISPR/Cas9. Sci Rep 8:6502. https://doi.org/10.1038/s41598-018-24690-8
7. Zhang Y, Hu X, Islam S et al (2018) New insights into the evolution of wheat avenin-like proteins in wild emmer wheat (*Triticum dicoccoides*). PNAS 115:13312–13317. https://doi.org/10.1073/pnas.1812855115
8. Choudhury S, Hu H, Fan Y et al (2019) Identification of new QTL contributing to barley yellow dwarf virus-PAV (BYDV-PAV) resistance in wheat. Plant Dis 103:2798–2803. https://doi.org/10.1094/PDIS-02-19-0271-RE
9. Bhoite R, Smith R, Bansal U et al (2023) Exome-based new allele-specific PCR markers and transferability for sodicity tolerance in bread wheat (Triticum aestivum L.). Plant Direct 7:e520. https://doi.org/10.1002/pld3.520
10. Cistué L, Ramos A, Castillo AM, Romagosa I (1994) Production of large number of doubled haploid plants from barley anthers pretreated with high concentrations of mannitol. Plant Cell Rep 13:709–712. https://doi.org/10.1007/BF00231629 PMID: 24193525
11. Broughton S, Priest P (1997) Barley doubled haploids in Western Australia–improved green plant regeneration from anthers pretreated with high concentrations of mannitol. Eighth Australian Barley Technical Symposium, Gold Coast, Queensland, 7–12 Sep 1997
12. Hu TC, Ziauddin A, Simion E, Kasha KJ (1995) Isolated microspore culture of wheat (*Triticum aestivum* L.) in a defined media I. Effects of pretreatment, isolation methods, and hormones. In Vitro Cell Dev Biol 31:79–83
13. Hoekstra S, van Bergen S, van Brouwershaven IR et al (1997) Androgenesis in *Hordeum vulgare* L.: effects of mannitol, calcium and abscisic acid on anther pretreatment. Plant Sci 126:211–218
14. Soriano M, Cistué L, Castillo AM (2008) Enhanced induction of microspore embryogenesis after n-butanol treatment in wheat (*Triticum aestivum* L.) anther culture. Plant Cell Rep 27(5):805–811. https://doi.org/10.1007/s00299-007-0500-y
15. Broughton S (2011) The application of n-butanol improves embryo and green plant production in anther culture of Australian wheat (*Triticum aestivum* L.) genotypes. Crop Pasture Sci 62:813–822. https://doi.org/10.1071/CP11204
16. Dubas E, Castillo AM, Żur I et al (2021) Microtubule organization changes severely after mannitol and n-butanol treatments inducing microspore embryogenesis in bread wheat. BMC Plant Biol 21:586. https://doi.org/10.1186/s12870-021-03345-3
17. Echávarri B, Cistué L (2016) Enhancement in androgenesis efficiency in barley (*Hordeum vulgare* L.) and bread wheat (*Triticum aestivum* L.) by the addition of dimethyl sulfoxide to the mannitol pretreatment medium. Plant Cell Tissue Organ Cult 125:11–22. https://doi.org/10.1007/s11240-015-0923-z
18. Broughton S (2008) Ovary co-culture improves embryo and green plant production in anther culture of Australian spring wheat genotypes (*Triticum aestivum* L.). Plant Cell Tiss Org Cult 95:185–195. https://doi.org/10.1007/s11240-008-9432-7
19. Kasha KJ, Hu TC, Oro R et al (2001) Nuclear fusion leads to chromosome doubling during mannitol pretreatment of barley (*Hordeum vulgare* L.) microspores. J Exp Bot 52:1227–1238. https://doi.org/10.1093/jexbot/52.359.1227
20. Daghma DES, Hensel G, Rutten T et al (2014) Cellular dynamics during early barley pollen embryogenesis revealed by time-lapse imaging. Front Plant Sci 5:17–30. https://doi.org/10.3389/fpls.2014.00675
21. Griggs R, Zheng MY (2016) Nuclear fusion during early stage of microspore embryogenesis indicates

chromosome doubling in wheat (*Triticum aestivum*). Am J Plant Sci 7(3):489–499. https://doi.org/10.4236/ajps.2016.73043
22. Sharma AK, Sharma (1965) Chromosome. Techniques: theory and practise, 2nd edn. Butterworths, London
23. Kasha KJ, Simion E, Miner M et al (2003) Haploid wheat isolated microspore culture protocol. In: Malszynnski M, Kasha KJ, Forster BP (Eds.) Doubled haploid production in crop plants: a manual. Kluwer Academic Publishers, Dordrecht, pp 29–34 29

Check for updates

Chapter 9

Bread Wheat Microspore Culture

Hai Ying Yuan and Alison M.R. Ferrie

Abstract

Microspore culture is one of the doubled-haploid-producing techniques that can speed up the breeding process and shorten the time for releasing new cultivars. Growing world population and changing climate demand to expedite the development of new resilient and high-yielding cultivars for staple crops such as bread wheat (*Triticum aestivum* L.). Here, we describe a comprehensive protocol for obtaining bread wheat doubled haploids through microspore culture. Key steps including donor plant growth and spike pretreatment, microspore isolation and density-gradient-based purification, microspore-ovary coculture and developmental stage tracking, embryo transferring and plantlet regeneration, as well as ploidy level determination and chromosome doubling of microspore-developed plantlets are covered in detail.

Keywords Bread wheat, Doubled haploids, Embryogenesis, Microspore culture, *Triticum aestivum* L.

1 Introduction

Doubled haploid (DH) technology is an important speed breeding technology, since it can significantly reduce the time used for the development of new cultivars. Instead of five to six generations to stabilize the new traits through the traditional breeding method, DH only needs one generation. The benefit of instant homozygosity enables a wide range of applications for DH technology beyond breeding programs [1, 2]. Among different approaches for generating DHs, in vitro culture of the male gametophytes—microspores is the most feasible and frequent approach for most plant species including crops such as maize (*Zea mays* L.), oat (*Avena sativa* L.), barley (*Hordeum vulgare* L.), and *Brassica* species [3–6].

Bread wheat (*Triticum aestivum* L.) is one of the global staple cereal crops and an important part of the human diet, and is therefore crucial for maintaining worldwide food and nutrition security [7]. With the continued global population growth, as well as shifting growing environments, and emerging new pests and diseases caused by global climate change, rapid development of improved cultivars becomes critical to address these challenges. DH technology can

Ana María Castillo et al. (eds.), *Doubled Haploid Methods in Cereals*, Methods in Molecular Biology, vol. 3029,
https://doi.org/10.1007/978-1-0716-5245-9_9,

well alleviate these challenges through its advantages of shortened breeding periods.

We recently reported a wheat microspore culture methodology that optimized a combination of parameters and showed high embryogenesis efficiency in a diverse group of bread wheat genotypes including Canadian spring wheat and winter wheat [8]. In this chapter, we detail the steps regarding this improved methodology for bread wheat DH production through microspore culture. These include optimized factors such as spike pretreatment, maltose density gradient purification, as well as epigenetic chemicals along with the timelines required to generate homozygous DH plants (Fig. 1).

2 Materials

2.1 Plant Material

1. Spring wheat genotype AC Nanda [9].
2. Winter wheat genotype Radiant [10].

2.2 Donor Plant Growth

1. Plant growth cabinet for vernalization set at 4°C, with 12/12 h light/dark at 150–250 μmol/m^{2}s with a $R_{600\text{--}700\,nm}/FR_{700\text{--}780\,nm}$ ratio of 3.5.
2. Plant growth cabinet set at 15°C day/12°C night, with 18/6 h light/dark at 400–420 μmol/m^{2}s with a $R_{600\text{--}700\,nm}/FR_{700\text{--}780\,nm}$ ratio of 7.7.
3. Plant growth cabinet set at 20°C day/18°C night, with 18/6 h light/dark at 350–420 μmol/m^{2}s with a $R_{600\text{--}700\,nm}/FR_{700\text{--}780\,nm}$ ratio of 7.7.
4. Root trainers for easy transplanting of winter wheat after vernalization.
5. 20 cm Ø pots.
6. Potting mix (such as Sunshine® mix #4, which is composed of Canadian sphagnum peat moss, perlite, dolomite lime, RESiLIENCE®, as well as a wetting agent).
7. Slow-release fertilizer 14-13-13 (14% nitrogen, 13% phosphorus, 13% potassium).
8. All-purpose fertilizer 20-20-20 (20% nitrogen, 20% phosphorus, 20% potassium).

2.3 Spike Staging and Collection

1. 4°C refrigerator.
2. Inverted epifluorescence microscope and a digital camera (such as a Zeiss Axio Observer 7 with a 350 nm excitation filter for DAPI staining observation and a Zeiss Axiocam 305 color camera).
3. Scissors.

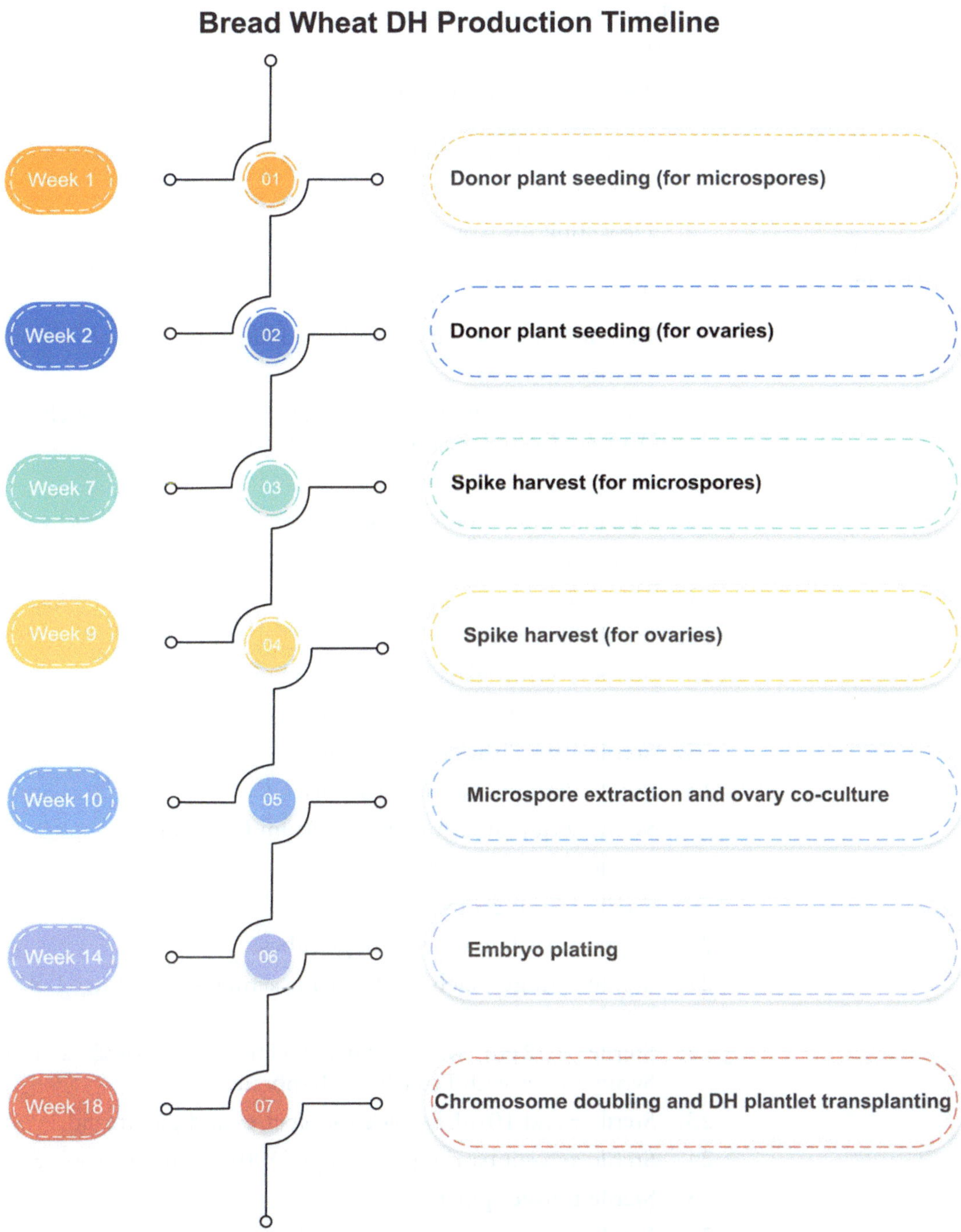

Fig. 1 Timeline for generating bread wheat doubled haploids through microspore culture. The spring wheat genotype AC Nanda is used as an example

4. Erlenmeyer flasks (500 mL).
5. Aluminum foil.
6. Glass slides and coverslips.

2.4 Spike Sterilization, Microspore Isolation, Culture, and Assessment

1. Laminar air flow hood.
2. Refrigerated benchtop centrifuge.
3. Dissecting microscope.
4. Inverted epifluorescence microscope and the digital camera described in Sect. 2.3 (item 2).
5. Inverted microscope (Leica Leitz Fluovert).
6. Hemocytometer.
7. Blender and mini blender containers (37–100 mL) (such as a Waring® Commercial Blender 7011S).
8. Benchtop micro incinerator.
9. Medium-pointed dissecting forceps.
10. Tweezers ultra-fine tip #5.
11. 10 mL syringe.
12. Sterile filter paper.
13. Sterile 500 mL glass bottle.
14. Sterile Petri dishes (100 × 25 mm).
15. Sterile Petri dishes (60 × 15 mm).
16. Sterile Petri dishes (35 × 10 mm).
17. Sterile 90 μm filter funnels (or 90 μm Cell Strainers).
18. Sterile 50 mL centrifuge tubes.
19. Sterile 15 mL centrifuge tubes.
20. Sterile 1.5 mL microcentrifuge tubes.
21. Sterile syringe (such as BD 5 mL syringes with Luer-Lok™ Tips).
22. Sterile 0.22 μm syringe filter (such as EZFlow® 25 mm Syringe filter with Foxx PES Membrane).
23. Sterile 5 and 10 mL serological pipettes and pipette aid.
24. Sterile aerosol barrier pipette tips (100 μL) and micropipette.
25. Sterile transfer pipette.
26. Parafilm.
27. Razor blade.
28. 30 μm nylon mesh filter.
29. Surface-sterilizing agents (e.g., commercial bleach solution).
30. Sterile double-distilled water.
31. Magenta boxes.

32. Autoclave (such as a Market Forge Sterilmatic STMEL Steam Sterilizer).
33. Incubators for microspore culture set at 28°C under dark conditions.
34. Tissue Culture room for plantlet growth set at 22°C with 16/6 h light/dark at 60–90 µmol/m²s with a $R_{600–700\,nm}$/ $FR_{700–780\,nm}$ ratio of 8.9.

2.5 Ploidy-Level Determination and Chromosome Doubling

1. Flow cytometer (such as a Sysmex Partec Cell Counter Analyzer).
2. Razor blade.
3. 30–µm nylon mesh filter.
4. Designated space for colchicine treatment, containing a fume hood.
5. Pots with soil or small cells for planting DH plants. Coverings (plastic cups or domes) for covering plants while hardening off.

2.6 Solutions

1. 0.3 M mannitol solution: Weigh 54.65 g mannitol and 500 mg MES • H_2O (2-(N-morpholino)ethanesulfonic acid monohydrate), dissolve in double-distilled water, adjust pH to 6.5, and final volume to 1 L. Store at 4°C. MES • H_2O is used as a pH buffering agent in the solution.
2. 1 mg/mL DAPI (4′,6-diamidino-2-phenylindole) stock solution: Dissolve 5 mg DAPI in 5 mL DMSO [dimethyl sulfoxide, ≥99.5% (GC), Sigma cat # D4540], aliquot, protect from the light, and store at −20°C. Mix 25 µL DAPI stock solution with 1 mL 0.3 M mannitol to get 1 mL of 25 µg/mL DAPI working solution, and store at 4°C (*see* **Note 1**).
3. Modified Carnoy's solution (ethanol and glacial acetic acid, 3:1 v/v): Mix 75 mL absolute ethanol and 25 mL glacial acetic acid to get 100 mL, and store at 4°C (*see* **Note 2**).
4. 50 µM TSA stock solution: Add 10 µL 5 mM Trichostatin A (Sigma T1952) to 990 µL DMSO (Sigma D4540).
5. 5 µM Trichostatin A (TSA) working solution: To obtain 1 mL 5 µM TSA working solution, add 100 µL 50 µM TSA stock solution to 900 µL DMSO, and aliquot into microcentrifuge tubes. Store at −20°C.
6. The CyStain UV Precise P kit from Sysmex buffers to study ploidy analysis.
7. 0.3% Colchicine solution: Add 0.3 g colchicine in 100 mL double-distilled water and store the solution in a brown bottle at room temperature (*see* **Note 3**).
8. 0.755% sodium hypochlorite: Add 50 mL of commercial bleach (7.55% w/w sodium hypochlorite) into 450 mL cold double-distilled water. Prepare fresh before each usage.

2.7 Plant Culture Media

The composition of culture media used in bread wheat microspore culture and plantlet regeneration is shown in Table 1.

1. Macrosalts stock solution for FHG-2 microspore isolation medium: prepare a stock solution (10×).
2. Stock solutions for NPB-99+ medium: Prepare a macro salts stock solution (10×), a micronutrients stock solution (1000×), and a vitamins stock solution (1000×).
3. Plant growth regulators: Prepare stock solutions of 2,4-D, kinetin, PAA and IAA, at a concentration of 0.5 mg/mL, and sterilize by filtration using a syringe and a 0.22 μm pore size filter.
4. Preparation of FHG-2 medium: For 1 L of FHG-2 medium, add the respective amounts of mannitol, MES, and 100 mL of the macro salts stock solution (10×). Bring the total volume to 1 L using double-distilled water (Table 1). Adjust the pH to 6.5 using 1 M KOH, filter sterilize through a 0.22 μm pore size filter, and store at 4°C.
5. Preparation of NPB-99+ medium: For 1 L of NPB-99^{+} medium combine the respective amounts of maltose, glutamine, cefotaxime, larcoll, glutathione, 100 mL of the macro salts stock solution (10×), 1 mL of the micronutrients stock solution (1000×), 1 mL of the vitamins stock solution (1000×), 0.4 mL of 2,4-D (0.5 mg/mL), 0.4 mL of kinetin (0.5 mg/mL), and 2 mL of PAA (0.5 mg/mL) (Table 1). Adjust the final volume to 1 L using double-distilled water, set the pH to 6.5 using 1 M KOH, filter sterilize, and store at 4°C.

Table 1
Media composition used in bread wheat microspore culture and plantlet regeneration

	Media composition (mg/L)		
Macros	**FHG-2[a]**	**NPB-99[+b]**	**B5-5[c]**
KNO_3	1900	1415	2500
KH_2PO_4	170	200	
NaH2PO4 • H2O			150
NH_4NO_3	165		
$(NH_4)_2SO_4$		232	134
CaCl2 • 2H2O	440	83	150
MgSO4 • 7H2O	370	93	250
FeSO4 • 7H2O	27.85	27.85	27.85
NA_2EDTA	37.25	37.25	37.25

(continued)

Table 1
(continued)

	Media composition (mg/L)		
Macros	**FHG-2[a]**	**NPB-99+[b]**	**B5-5[c]**
Micros			
H_3BO_3		5.0	3.0
$MnSO4 \cdot H2O$		5.0	10.0
$ZnSO4 \cdot 7H2O$		5.0	2.0
KI		0.4	0.75
$Na2MoO4 \cdot 2H2O$		0.0125	0.25
$CuSO4 \cdot 5H2O$		0.0125	0.025
$CoCl2 \cdot 6H2O$		0.0125	0.025
Vitamins, etc			
Nicotinic acid		0.5	1.0
Pyridoxine HCl		0.5	1.0
Thiamine-HCl		5.0	10.0
myo-inositol		50	100.0
Glutathione		0.61	
Glutamine		500	
$MES \cdot H2O$	1950		1000
Cefotaxime		100	
Larcoll		40	
Plant Growth Regulators			
2,4-D		0.2	
Kinetin		0.2	1.0
PAA		1.0	
IAA			1.0
	Media compositions (g/L)		
Mannitol	72.87		
Sucrose			20
Maltose		90	
Difco® agar			7
pH	6.5	6.5	5.8

[a]FHG-2 solution is modified and based on FHG [11]
[b]NPB-99+ medium is modified and based on NPB-99 medium [12] with glutathione [13], cefotaxime [14], and Larcoll [15]
[c]B5-5 medium is modified and based on B5 medium [16]

6. Preparation of B5-5 medium: For 1 L of B5-5 medium, add 3.21 g of B5 medium (Phytotech G398), 20 g sucrose and 1 g MES (Table 1). Bring the volume to 1 L using double-distilled water and adjust the pH to 5.8 using 1 M KOH. Add 7 g Difco agar and autoclave at 121°C and 0.1 MPa for 20 min. Let the medium cool to approximately 60°C, then add 2 mL of sterile Kinetin (0.5 mg/mL), and 2 mL of sterile IAA (0.5 mg/mL). Mix well and pour into 100×25 mm Petri dishes. When the medium solidifies, seal the dishes with Parafilm, and store them at room temperature.

3 Methods

3.1 Establishment of Donor Plants

1. Prepare 20 cm pots using potting mix (e.g., Sunshine mix #4) with 4.5 g of slow-release fertilizer 14-13-13 (N-P-K) (*see* **Note 4**).
2. Sow four seeds per pot and thin plants to three per pot after germination; repeat sowing every 2 weeks if needed (*see* **Note 5**). Donor plants for ovaries are planted 1 week later than the donor plants for microspores.
3. Set plant growth cabinets at 20°C day/18°C night, 18/6 h light/dark, at 350–420 μmol/m^{2}s (see Sect. 2.2), then transfer the pots to the cabinets.
4. Use low dosage (0.22 g/L) all-purpose fertilizer 20-20-20 (N-P-K) when watering, and water as needed.
5. Set a vernalization cabinet at 4°C with 12/12 h light/dark, at 150–250 μmol/m^{2}s. For winter wheat, 2 weeks after sowing, transfer plants to the cabinet for a 9-week vernalization period (*see* **Note 6** and Sect. 2.2).
6. After vernalization, transfer plants to growth cabinets with growing conditions set at 15°C day/12°C night, 18/6 h light/dark, at 400–420 μmol/m^{2}s (see Sect. 2.2).

3.2 Spike Collection and Pretreatment

1. Collect spikes for microspore culture when the majority of the microspores have reached the mid-late uninucleate stage after cold treatment at 4°C (*see* **Note** 7, Fig. 2A). Staging can be done through acetocarmine or DAPI. Use an inverted microscope for acetocarmine staining, or an epifluorescence microscope with a 350 nm excitation filter for DAPI to observe the microspores.
2. Trim off leaves from the collected spikes while keeping 3–5 cm of the flag leaf.
3. Place trimmed spikes for microspore culture into Erlenmeyer flasks containing 200 mL water at 4°C, and label clearly.
4. Store spikes at 4°C for 21–24 days for spring wheat before the microspore isolation, and 28–30 days for winter wheat.

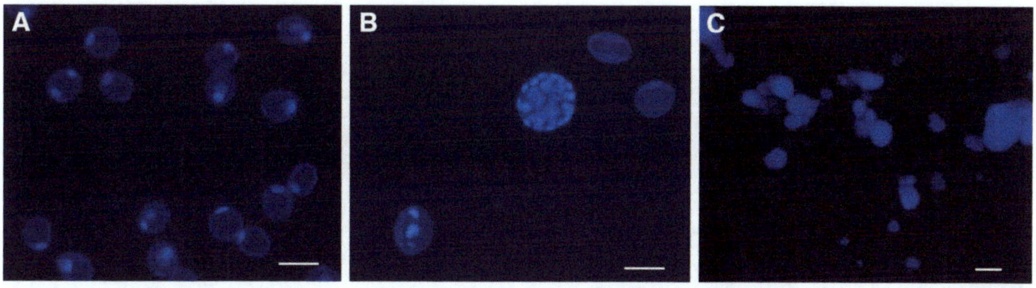

Fig. 2 DAPI staining of bread wheat microspore cultures: (**A**) Most microspores were at the mid-late uninucleate stage just after microspore isolation. (**B**) Most induced microspores reached the multinuclei stage 1 week after culture initiation. (**C**) Embryo-like structures at various developmental stages can be observed around 2 weeks of culture. The images were taken using a Zeiss Axio Observer 7 microscope with a 350 nm excitation filter and a Zeiss Axiocam 305. Image C was generated using the tiles module (3 × 3). The spring wheat genotype AC Nanda is used as an example. Scale bar: 50 μm (**A** and **B**), and 200 μm (**C**)

5. Collect spikes for ovary coculture when ovaries are at the elongated/cleft stage (i.e., stigma is elongated or just split but not fully branched). Staging can be done through the dissection of ovaries under a dissecting microscope (*see* **Note 8**).
6. Trimmed spikes for ovary coculture can be cut fresh or stored in Erlenmeyer flasks containing 200 mL water at 4°C for 1–2 weeks.

3.3 Microspore Isolation and Purification

1. For spike sterilization, remove the flag leaf and trim off the awns of the spikes (Fig. 3A, C: spikes used for ovary dissection; Fig. 3B, D: spikes used for microspore isolation).
2. Surface-sterilize the spikes by immersing for 3 min in 500 mL of 0.75% sodium hypochlorite with three drops of Tween-80 by gentle shaking, followed by 1.51% sodium hypochlorite for 5 min for spikes destined for ovaries (*see* **Note 9**).
3. Rinse four times in 500 mL cold, sterile double-distilled water. Keep the spikes for ovaries aside while using the rest for microspore isolation.
4. Position a mini blender container on ice and fill it with 50 mL cold, sterile FHG-2 medium.
5. Under sterile conditions, remove glumes from either side of the spikelets.
6. Take spikelets from the rachis and place them into the blender container. After adding two or three spikes, use sterile forceps to immerse the spikelets in the FHG-2 medium. Repeat until all the spikes are processed.
7. Blend the spikelets at low speed (about 18,000 RPM when using a Waring® commercial blender model 7011S) for 10 s and repeat once.

Fig. 3 Spikes for extracting microspores and ovaries used in bread wheat microspore culture. (**A**) Spikes for ovaries. (**B**) Spikes for microspores. (**C** and **D**) Spikes from **A** and **B** after removing the flag leaves, respectively. The spring wheat genotype AC Nanda is used as an example. Scale bar: 2.0 cm

8. Pour the blended suspension into a 500 mL sterile bottle through a 90 μm filter funnel, set on ice.
9. Rinse the blender container with cold FHG-2 medium, then pour through the filter funnel. Continue running cold FHG-2 medium until the bottle contains 300 mL.
10. Fill six 50 mL-centrifuge tubes with the filtrate, then centrifuge at 200 × *g* at 4°C for 5 min. Decant all tubes.
11. Add 20 mL of cold FHG-2 medium to one tube. Rinse each tube with the same 20 mL aliquot and keep the microspore solution in the last tube. Repeat the rinse with another 20 mL of cold FHG-2 medium.
12. Bring the volume to 50 mL by adding cold FHG-2 medium and centrifuge at 200 × *g* at 4°C for 5 min.
13. Decant and resuspend the microspore pellet in 5 mL of cold NPB-99$^+$ medium.
14. Transfer the microspore suspension into a 15 mL tube. Rinse the 50 mL tube with 5 mL of cold NPB-99$^+$ medium and transfer to the 15 mL tube. Repeat the rinse step until reaching a final volume of 15 mL.
15. Centrifuge at 200 × *g* at 4°C for 5 min and then decant.
16. Gently resuspend the microspores in 10 mL of cold 23% maltose and carefully place 1 mL cold NPB-99$^+$ medium over the maltose solution.
17. Centrifuge at 200 × *g* at 4°C for 13 min. A distinct microspore layer should form at the interface.
18. Carefully transfer the microspores to a clean sterile 15 mL tube using a sterile transfer pipette.
19. Fill the tube with cold NPB-99$^+$ medium to 6 mL and then add cold 0.3 M mannitol to 15 mL. Invert to mix.
20. Centrifuge at 200 g at 4°C for 5 min, and then decant.
21. Add 3 or 6 mL of cold NPB-99$^+$ medium depending on the pellet size. Resuspend the microspore pellet before the hemocytometer count.
22. Load 12 μL microspore suspension onto a hemocytometer and count the microspores in four 1 × 1 mm^2 squares using an inverted microscope. Repeat once and average all eight counts (*see* **Note 10**).
23. Calculate microspore yield and determine the number of Petri dishes required based on 1.5 mL microspore suspension per 35 × 10 mm Petri dish with a density of 60,000 microspores per mL (*see* **Note 11**).
24. Based on the calculation, add cold NPB-99$^+$ medium into the tube to reach the microspore density of 60,000 per mL.

25. Dispense 1.5 mL microspore suspension to each 35 × 10 mm Petri dish; add 2.4 μL TSA to each dish and gently swirl to mix (*see* **Note 12**).

3.4 Microspore-Ovary Coculture

1. Aseptically isolate ovaries from previously sterilized spikes. Only use ovaries at the elongated/cleft stage.
2. Place three ovaries into each 35 × 10 mm Petri dish of microspore suspension and wrap with Parafilm.
3. Briefly observe the microspores under an inverted microscope and take images as part of the record-keeping process (Fig. 4).
4. Place six Petri dishes (35 × 10 mm) in a large 150 × 25 mm Petri dish with an open 60 × 15 mm water-filled Petri dish in the center.
5. Incubate the cultures at 28°C in the dark for 28–35 days.

3.5 Assessment of Microspore Development (Optional)

1. Using a microscope, assess microspore development periodically (such as weekly) under brightfield. Microspore-derived embryos can be observed around 2 weeks after initiating the culture (Fig. 5A); while the embryos are ready to be transferred to the regeneration medium around 4 weeks (Fig. 5B).
2. For assessment using DAPI, pipette 50 μL subsample from each Petri dish to a microcentrifuge tube and add 100 μL Carnoy's solution. Store at 4°C for 2–3 h before processing to the next

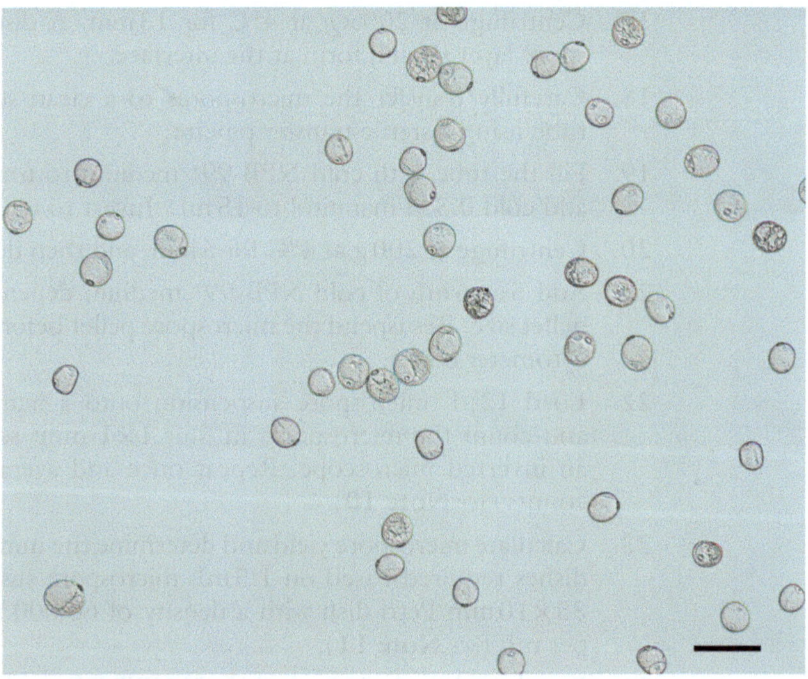

Fig. 4 Freshly isolated microspores from a spring wheat genotype AC Nanda. The image was generated using the tiles module (3 × 3) through a Zeiss Axio Observer 7 Microscope and a Zeiss Axiocam 305. Scale bar: 100 μm

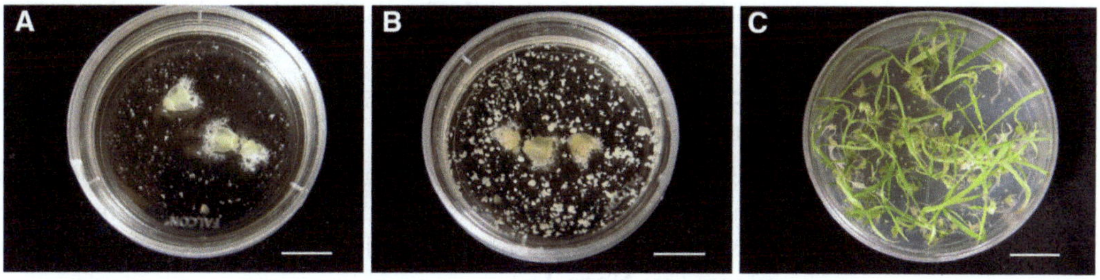

Fig. 5 Bread wheat microspore culture. (**A**) Microspore-derived embryos at 2 weeks of culture. (**B**) Microspores-derived embryos at 4 weeks of culture. (**C**) Plantlets regenerated after 4 weeks in the regeneration medium. The spring wheat genotype AC Nanda is used as an example. Scale bar: 1.2 cm (**A** and **B**) and 2.0 cm (**C**)

step. If the samples cannot be processed, remove Carnoy's solution, add 70% ethanol and store at 4°C in the fridge.

3. Remove Carnoy's solution or 70% ethanol and add 20 μL DAPI working solution. Mix well and store samples for at least 5 min in the dark (*see* **Note 13**).
4. Transfer the entire sample to a glass slide, cover it with a cover glass, and then observe under an epifluorescence microscope. Figure 2B and C show DAPI staining of microspore culture at 1 week (B) and 2 weeks (C) after culture initiation.

3.6 Embryo Transferring and Plantlet Regeneration

1. After 4–5 weeks of microspore culture initiation, remove and dispose of the ovaries aseptically in a laminar flow hood.
2. Transfer the embryos onto 100 × 25 mm Petri dishes with solid B5-5 medium and maintain roughly 100 embryos per plate.
3. Use a sterile transfer pipette to remove the liquid media if there is any.
4. Gently spread the embryos across the medium surface using a sterile transfer pipette, then seal the Petri dish with Parafilm.
5. Place cultures at 22°C with 16/8 h light/dark at 60–90 μmol/m^2s. Embryos usually germinate within 1–2 weeks.
6. After 2–3 weeks, plantlets can be transferred to Magenta boxes containing B5-5 medium or they can be left in the Petri dish if the plant density is sufficient to allow further growth and development (Fig. 5C, *see* **Note 14**).

3.7 Ploidy-Level Determination

1. Collect 0.5 cm^2 young green leaf tissue from each plantlet, and place in a plastic Petri dish or weighing boat.
2. Add 400 μL CyStain UV Precise P nuclei extraction buffer and finely chop the leaf tissues.
3. Incubate for a maximum of 5 min in the extraction buffer before filtering through a 30 μm nylon filter.

4. Add 1.6 mL CyStain UV Precise P staining buffer into the filtrate and incubate for 30–60 s.
5. Prepare a control sample using leaf tissues from donor bread wheat plants.
6. Measure the ploidy level through a flow cytometer. DNA content of the tested plantlet is determined by comparing the peak position of the plantlet with the peak position of the donor plants (*see* **Note 15**).

3.8 Chromosome Doubling

1. Remove haploid plantlets having three to four leaves, as determined through the previous step, from Magenta boxes and gently wash off the agar from the roots.
2. Submerge the roots and crown of the plantlets in a 0.3% colchicine solution for 4 h at room temperature.
3. Wash the roots in water and transfer the plantlets to pots with Sunshine mix #4.
4. Cover the plantlets initially and go through the hardening process by gradually lifting the cover and reducing the humidity while using the same growing conditions set for donor plant growth. The plants can be maintained until maturity.

4 Notes

1. Store the DAPI staging solution in a brown bottle or wrapped with aluminum foil to protect it from light. Generally, this solution is stable for 6 months at 4°C. In addition, DAPI is a known mutagen so handle with care. Wear personal protective equipment including safety glasses, gloves, and a lab coat.
2. Microspores treated with modified Carnoy's solution can increase the permeability of DAPI into microspores.
3. Colchicine is corrosive and toxic when inhaled, swallowed, or comes into contact with the skin. Follow laboratory safety precautions and use inside a fume hood. Wear personal protective equipment including safety glasses, gloves, and a lab coat.
4. 10–12 pots per 16 sqft growth cabinet is recommended.
5. Sowing at 2-week intervals provides continuous access to spikes for microspores and ovaries at the proper developmental stages.
6. Alternatively, root trainers can be used to start winter wheat and then transferred to 20 cm pots with three plants per pot after the vernalization period.
7. Microspore developmental stage is critical and can be determined through the correlation between the stage and the

distance of the spike head below the flag leaf. During the cold treatment period, continued development occurs. Therefore, the distance at the spike collection will be the staging distance plus 1 cm in general (Fig. 3B). We usually use 10–12 spikes per experiment.

8. Ovary developmental stage can be determined through the correlation between the stage and the distance of the spike head above the flag leaf. For Spring wheat genotype AC Nanda, the spike is usually harvested when there are three to four spikelets above the flag leaf (Fig. 3A).
9. For microspore isolation, work quickly under cold conditions to get microspores into the culture medium. All equipment and media are stored at 4°C and/or placed on ice during isolation. In addition, make a fresh bleach solution before each experiment.
10. Gently swirl the tube to ensure microspores are evenly distributed. Take the sub-sample for counting before microspores start to settle.
11. If the microspore pellet was resuspended in 3 mL of NPB-99^+ medium, the number of plates for the experiment can be obtained by multiplying the average count by 0.311; if in 6 mL of NPB-99^+ medium, then the number of plates doubles. Using AC Nanda as an example, microspores extracted from ten spikes can usually be plated in 30–40 Petri dishes (35 × 15 mm) containing 1.5 mL of microspore suspension at a density of 60,000 microspores/mL.
12. TSA is effective for microspore embryogenesis in multiple spring and winter bread wheat genotypes from our study [8]. The optimal concentration is 0.008 μM. Using the formula $C_1V_1 = C_2V_2$ (C1 refers to TSA stock concentration and C2 refers to TSA concentration in the culture; while V1 is the volume of TSA stock and V2 is the volume of the culture), 2.4 μL of 5 μM TSA can be added to a 1.5 mL culture plate to reach the desired concentration of 0.008 μM.
13. Use aluminum foil to wrap each single tube after adding DAPI working solution to protect it from light. The samples can be stable for hours before observing using a fluorescence microscope under dark conditions.
14. Albino plant development is common in bread wheat microspore culture and the frequency varies depending on the genotypes.
15. Spontaneous doubling can be revealed through the process and the frequency varies depending on the genotypes. We identified no more than 20% spontaneous doubling in microspore-developed plantlets from the genotypes that we evaluated.

Acknowledgments

We greatly acknowledge HM Wang, JL Enns, KL Nelson, JM Brost, and TD Orr for their contribution to the development of the bread wheat microspore culture protocol. We appreciate I Smith, K Irmen, and K Kjargaard from the Plant Growth Facility at the Aquatic and Crop Resource Development Research Centre, National Research Council of Canada for their help related to this work.

References

1. Ferrie AMR, Möllers C (2011) Haploid and doubled haploids in Brassica spp. for genetic and genomic research. Plant Cell Tiss Organ Cult 104:375–386
2. Hale B, Ferrie AMR, Chellamma S et al (2022) Androgenesis-based doubled haploidy: past, present, and future perspectives. Front Plant Sci 12:751230. https://doi.org/10.3389/fpls.2021.751230
3. Vergne P, Gaillard A (2021) Isolation of staged and viable maize microspores for DH production. In: Segui-Simarro JM (ed) Doubled haploid technology: methods in molecular biology. Humana, New York, NY, pp 281–293
4. Kasha KJ, Simion E, Oro R et al (2001) An improved in vitro technique for isolated microspore culture of barley. Euphytica 120:379–385
5. Ferrie AMR (2003) Microspore culture of Brassica species. In: Maluszynski M, Kasha K, Forster B, Szarejko I (eds) Doubled haploid production in crop plants. Kluwer Academic Publishers, Dordrecht, pp 205–214
6. Sidhu PK, Davies PA (2009) Regeneration of fertile green plants from oat isolated microspore culture. Plant Cell Rep 28:571–577. https://doi.org/10.1007/s00299-009-0684-4
7. Erenstein O, Jaleta M, Mottaleb KA et al (2022) Global trends in wheat production, consumption and trade. Wheat Improvement. Springer International Publishing, Cham, pp 47–66
8. Wang HM, Enns JL, Nelson KL et al (2019) Improving the efficiency of wheat microspore culture methodology: evaluation of pretreatments, gradients, and epigenetic chemicals. Plant Cell, Tissue Organ Cult 139:589–599. https://doi.org/10.1007/s11240-019-01704-5
9. Sadasivaiah RS, Perkovic SM, Pearson DC, Postman B (2000) Registration of 'AC Nanda' wheat. Crop Sci 40:579–580. https://doi.org/10.2135/cropsci2000.0029rcv
10. Thomas JB, Conner RL, Graf RJ (2012) Radiant hard red winter wheat. Can J Plant Sci 92:169–175. https://doi.org/10.4141/CJPS2011-082
11. Hunter CP (1988) Plant regeneration from microspores of barley *Hordeum vulgare* L. (Doctoral dissertation, Imperial College London), University of London
12. Zheng M, Liu W, Weng Y et al (2001) Culture of freshly isolated wheat (*Triticum aestivum* L.) microspores treated with inducer chemicals. Plant Cell Rep 20:685–690. https://doi.org/10.1007/s00299-001-0393-0
13. Asif M, Eudes F, Goyal A et al (2013) Organelle antioxidants improve microspore embryogenesis in wheat and triticale. Vitr Cell Dev Biol-Plant 49:489–497. https://doi.org/10.1007/s11627-013-9514-z
14. Asif M, Eudes F, Randhawa H et al (2013) Cefotaxime prevents microbial contamination and improves microspore embryogenesis in wheat and triticale. Plant Cell Rep 32:1637–1646. https://doi.org/10.1007/s00299-013-1476-4
15. Letarte J, Simion E, Miner M, Kasha KJ (2006) Arabinogalactans and arabinogalactan-proteins induce embryogenesis in wheat (*Triticum aestivum* L.) microspore culture. Plant Cell Rep 24:691–698. https://doi.org/10.1007/s00299-005-0013-5
16. Gamborg OL, Miller RA, Ojima K (1968) Nutrient requirements of suspension cultures of soybean root cells. Exp Cell Res 50:151–158. https://doi.org/10.1016/0014-4827(68)90403-5

Chapter 10

Durum Wheat Anther Culture: a Complete Protocol

Ioannis N. Xynias and Theano B. Lazaridou

Abstract

Durum wheat is one of the most recalcitrant species among cereals for anther culture, and is characterized by a low embryogenesis rate and a high percentage of albinism (95%). This chapter describes key features of durum wheat anther culture protocol. This includes growing donor plants under the most appropriate conditions, selecting spikes with microspores at the mid- to late-uninucleate stage of development, establishing aseptic conditions for dissection, and culturing anthers in appropriate media and conditions. A modified W14 medium is described that performs well in durum wheat anther culture. The green plants produced are transplanted into pots under optimal temperature and light conditions. The low percentage of spontaneous chromosome doubling obtained makes the application of antimitotic agents necessary. The low number of fertile DH plants can be faced in the future by developing new protocols or more suitable media for durum wheat, by applying wide hybridization methods, or even more by using speed breeding to accelerate the production of doubled haploid plants.

Keywords Durum wheat, Anther culture, Recalcitrance, Embryoid production, Albino plants, Green plants

1 Introduction

Durum wheat (*Triticum turgidum* subsp. *durum* Desf.) is a very important crop for Mediterranean countries, contributing to the production of high-value products. Its significance lies in past making and cooking quality, dietary fiber content, improved beta-glycan content, and the proportion of soluble arabinoxylans in semolina, which determines the nutritional and health properties of the crop [1]. However, there is an urgent need to develop new cultivars better adapted to the Mediterranean region, characterized by high temperatures and low rainfall, due to the prevailing climate change, which is leading to uncontrollable, extremely dry periods, followed by short but intense precipitation.

Various approaches have been proposed to reduce the time needed to release new varieties. A very easy and effective one is the use of anther culture, which leads to the production of 100%

Ana María Castillo et al. (eds.), *Doubled Haploid Methods in Cereals*, Methods in Molecular Biology, vol. 3029,
https://doi.org/10.1007/978-1-0716-5245-9_10,

homozygous lines in a single generation. However, durum wheat is a highly recalcitrant crop to anther culture [2], due to its low frequency of embryoid production and the high rate of albino plant: i.e., 95% of the plants produced are albino [3]. Many efforts have been attempted to reduce the recalcitrance in durum wheat, a trait that is influenced by both genetic and physiological factors. Genetic factors play an important role in the efficiency of anther culture and its potential for widespread application [4, 5]. For example, it has been reported that B genome genes in bread wheat significantly affect anther culture response [6, 7]. Based on this assumption, Tersi et al. [2] attempted to transfer B genome genes from bread wheat into durum genome, by crossing the two species. However, this approach was not successful and only one hexaploid green plant was produced. Lazaridou et al. [8] concluded that the D genome is responsible for anther culture response. If this is true, it may be a challenge for durum wheat cultivars to respond well to this methodology.

Physiological factors affecting anther culture include the microspore developmental stage, culture media composition, duration of a low-temperature pretreatment [9, 10], and culture conditions [11, 12]. In an attempt to improve the androgenic response in durum wheat, Cistué et al. [12] reported that subjecting anthers to a stress pretreatment with 0.7 M mannitol for 5 days promoted green plant regeneration. Furthermore, Labbani et al. [13] demonstrated that a combined pretreatment of anthers with 0.3 M mannitol and low temperature for 7 days had a crucial effect on the number of embryos and regenerated green plants in durum wheat. Lazaridou et al. [14] assayed different anther culture media and produced green plants in FHG [15] and a modified W14 medium [16, 17].

The aim of this chapter is to describe a complete approach to anther culture in durum wheat.

2 Materials

2.1 Plant Material

This protocol has been applied to the cultivars Anna, Sifnos, Mexikali 81, and Elpida that were developed at the Cereal Institute, Thessaloniki (Greece).

2.2 Plant Growth Materials

1. Pots (2.5 L).
2. Soil mixture (2:1:1 soil:sand:peat).
3. Fertilizer (N:P:K, 20:10:0).
4. Trays.
5. Plastic bags.
6. Frames.
7. Mouldboard plough, disc harrow, and cultivator.

2.3 Equipment

1. Laminar flow hood.
2. Autoclave.
3. Magnetic stirrer.
4. pH meter.
5. Stereoscope.
6. Optical microscope.
7. Analytical balances.
8. Refrigerator.
9. Distiller/deionizer.
10. Growth chamber set at 28 °C without light.
11. Growth chamber set at 26 °C with 50 μmol m^{-2} s^{-1}, provided by cool-white fluorescent lamps and 16 h light.
12. Fume hood.
13. Air pump.
14. Slide press equipment.

2.4 Laboratory Materials

1. Filter paper.
2. Forceps.
3. Scissors.
4. Sterile Petri dishes (100 × 20 mm).
5. Pasteur pipettes.
6. Plastic pipettes.
7. Erlenmeyer flasks.
8. Measuring cylinders.
9. Vials (30 mL).
10. Beakers.
11. Microscope slides and coverslips.
12. Pencil.
13. Scalpel.
14. Gloves.
15. Masks.

2.5 Solutions

1. Mannitol 0.7 M: Dissolve 127 g of mannitol in 1000 mL of water, put it on the magnetic stirrer until the solution becomes clear.
2. Orcein acetate (1.5%): Weigh 1.5 g of orcein acetate into an Erlenmeyer flask containing 45 mL of glacial acetic acid. Seal the flask tightly to avoid evaporation and transfer it to a water bath at 80 °C. Keep the flask in the water for 3 h, shaking periodically to improve dissolution. Once the orcein is fully

dissolved, gradually add 55 mL of distilled water to make a 1.5% orcein acetate solution (*see* **Note 1**).

3. Sodium hypochlorite (1.65%) solution: Take 30 mL of commercial bleach (5.5% sodium hypochlorite) and make up to 100 mL.
4. Colchicine solution (0.06%): Dissolve 0.3 g of colchicine in 500 mL distilled water, add 15 mL of dimethylsulfoxide (DMSO), and adjust the pH to 5.5 (*see* **Note 2**).

2.6 Culture Medium

In the present protocol, three different media are used: The induction medium is a modified W14 medium (mW14) [16, 17] for callus induction, the regeneration medium is 190-2 [18] for plantlet regeneration, and the rooting medium is a MS medium [19] without growth regulators to facilitate the production of roots (*see* **Note 3**). The composition of these three media is described in Table 1.

2.6.1 Induction Medium (mW14)

- *Macronutrients stock solution (10×)*: Add 400 mL of distilled water to an Erlenmeyer flask and place it on a magnetic stirrer. Then, add the required amounts of KNO_3, $NH_4H_2PO_4$, $MgSO_4.7H_2O$, $CaCl_2.2H_2O$, and K_2SO_4 with continuous stirring until dissolved. Take the volume to 1000 mL and store at 4 °C.
- *Micronutrients stock solution (100×)*: Add 400 mL of distilled water to an Erlenmeyer flask and place it on a magnetic stirrer. Then, add the required amounts of $MnSO_4.H_2O$, H_3BO_3, $ZnSO_4.7H_2O$, $Na_2MoO_4.2H_2O$, KI, $CuSO_4.5H_2O$, and $CoCl_2.2H_2O$ with continuous stirring until dissolved. Take the volume to 1000 mL and store at 4 °C.
- *Vitamins and myoinositol stock solution (10×)*: Add 400 mL of distilled water to an Erlenmeyer flask and place it on the magnetic stirrer. Then, add the required amounts of thiamine chloride, pyridoxine chloride, nicotinic acid, glycine, L-cystine, and myo-inositol with continuous stirring until dissolved. Take the volume to 1000 mL and store at 4 °C (*see* **Note 4**).
- *Iron source stock solution (10×)*: Add 400 mL of distilled water to an Erlenmeyer flask and place it on the heated stirrer. Add the required amounts of $FeSO_4{\cdot}7H_2O$ and Na_2EDTA with continuous stirring until dissolved. Take the volume to 1000 mL and store at 4 °C.
- *2.4-Dichlorophenoxyacetic acid (2.4-D) stock solution (5 mg/L)*:

  Dissolve 50 mg of 2.4-D in a minimal amount of 96% ethanol under continuous stirring. Once dissolved, add deionized water slowly and take the volume to 100 mL. Store the solution at 4 °C.

Table 1
Culture media composition for durum wheat anther culture

	W14 (mg/L)	190-2 (mg/L)	MS (mg/L)
KNO_3	2000	1000	1900
$NH_4H_2PO_4$	380	-	1,650
KH_2PO_4	-	300	170
$MgSO_4{\cdot}7H_2O$	200	200	370
$CaCl_2{\cdot}2H_2O$	140	-	440
KCl	-	40	-
$Ca(NO_3)_2{\cdot}4H_2O$	-	100	-
$(NH_4)SO_4$	-	200	-
K_2SO_4	700		-
$MnSO_4{\cdot}2H_2O$	8	8	16.9
H_3BO_3	3	3	6.2
$ZnSO4{\cdot}7H_2O$	3	3	10.6
$Na_2Mo_4{\cdot}2H_2O$	0.005	-	0.25
KI	0.5	0.5	0.83
$CuSO_4{\cdot}5H_2O$	0.025	-	0.025
$CoCl_2{\cdot}2H_2O$	0.005	-	0.025
$FeSO_4{\cdot}7H_2O$	27.8	27.8	27.8
Na_2EDTA	37.2	37.2	37.2
Myo-inositol	500	100	100
Thiamine chloride	2.0	1.0	1.0
Pyridoxine chloride	0.5	0.5	0.5
Nicotinic acid	0.5	0.5	0.5
Glycine	2.0	2.0	2.0
L-cysteine	50	-	-
Glutamine	-	-	730
2,4-D	5	-	-
BAP	0.5	-	-
Gibberellic acid (GA3)	-	0.1	-
IBA	-	5	-
Maltose	100.000	-	-
Sucrose	-	30.000	30.000
Gelrite	6.000	6.000	6.000
pH	5.8		

- *Benzylaminopurine (BAP) stock solution (0.5 mg/L)*:
  Dissolve 5 mg of BAP in a minimal amount of 1 N NaOH or KOH under continuous stirring. Once dissolved add deionized water slowly and take the volume to 100 mL. Store the solution at 4 °C.
- *To prepare 1 L of W14 induction medium*: Transfer 100 mL of macronutrients, vitamins and myoinositol and iron source stock solutions, and 10 mL of micronutrients stock solution to a 1 L Erlenmeyer flask with constant stirring. Add sucrose with continuous stirring to dissolve it. Add 10 mL of 2,4-D, and BAP stock solutions, and take the volume to 1 L with distilled water. Adjust the pH to 5.8. Add 6 g Gelrite (Sigma G1910) and autoclave at 121 °C for 21 min. Pour 30 mL of medium into 100 mm sterile Petri dishes.

2.6.2 Regeneration Medium (190-2)

- *Macronutrients stock solution (10×)*:
  Add 400 mL of distilled water to an Erlenmeyer flask and place it on a magnetic stirrer. Then, add the required amounts of KNO_3, KH_2PO_4, $MgSO_4{\cdot}7H_2O$, KCl, $Ca(NO_3)_2{\cdot}4H_2O$, and $(NH_4)SO_4$, with continuous stirring until dissolved. Take the volume to 1000 mL and store at 4 °C.
- *Micronutrients stock solution (100×)*:
  Add 400 mL of distilled water to an Erlenmeyer flask and place it on a magnetic stirrer. Then, add the required amounts of $MnSO_4{\cdot}H_2O$, H_3BO_3, $ZnSO_4{\cdot}7H_2O$, and KI with continuous stirring until dissolved. Take the volume to 1000 mL and store at 4 °C.
- *Iron source stock solution (10×)*:
  Add 400 mL of distilled water to an Erlenmeyer flask and place it on a heated stirrer. Then, add the required amounts of $FeSO_4{\cdot}7H_2O$ and Na_2EDTA with continuous stirring until dissolved. Take the volume to 1000 mL and store at 4 °C.
- *Vitamins and myo-inositol stock solution (10×)*:
  Add 400 mL of distilled water to an Erlenmeyer flask and place it on a magnetic stirrer. Then, add the required amounts o f thiamine chloride, pyridoxine; nicotinic acid; glycine, and myo-inositol with continuous stirring until dissolved. Take the volume to 1000 mL and store at 4 °C.
- Gibberellic acid (GA3) *stock solution (0.1 mg/L)*:
  Dissolve 1 mg of Gibberellic acid in a minimal amount of distilled water under gentle heating and continuous stirring. Once dissolved, add deionized water slowly and take the volume to 100 mL. Store the solution at 4 °C.
- *Indole-3-butyric acid (IBA) stock solution (5 mg/L)*:
  Dissolve 50 mg of IBA in a minimal amount of ethanol 96% under continuous stirring. Once dissolved, add deionized wa-

ter slowly and take the volume to 100 mL. Store the solution at 4 °C.

- *To prepare 1 L of the 190-2 regeneration medium*: Transfer 100 mL of macronutrients, iron source and vitamins and myo-inositol stock solutions, and 10 mL of micronutrients stock solution to a 1 L Erlenmeyer flask with constant stirring. Then, add sucrose with continuous stirring until dissolved. Next, add 10 mL of GA_3 and IBA stock solutions and take the volume to 1 L. Adjust the pH of the medium 5.8. Add 6 g Gelrite (Sigma G1910) and autoclave at 121 °C for 21 min. Pour 30 mL medium into 100 mm sterile Petri dishes.

2.6.3 Rooting Medium (MS)

- *Macronutrients stock solution (10×)*:
Add 400 mL of distilled water to an Erlenmeyer flask and place it on a magnetic stirrer. Then, add the required amounts of KNO_3, $NH_4H_2PO_4$; $MgSO_4 \cdot 7H_2O$, $CaCl_2 \cdot 2H_2O$, and KH_2PO_4 with continuous stirring until dissolved. Make the volume to 1000 mL and store at 4 °C.
- *Micronutrients stock solution (100×)*:
Add 400 mL of distilled water to an Erlenmeyer flask and place it on a magnetic stirrer. Then, add the required amounts of $MnSO_4 \cdot H_2O$, H_3BO_3, $ZnSO_4.7H_2O$, $Na_2MoO_4 \cdot 2H_2O$, KI, $CuSO_4 \cdot 5H_2O$, and $CoCl_2 \cdot 2H_2O$ with continuous stirring until dissolved. Make the volume to 1000 mL and store at 4 °C.
- *Iron source stock solution (10×)*: Add 400 mL of distilled water to an Erlenmeyer flask and place it on the heated stirrer. Then, add the required amounts of $FeSO_4 \cdot 7H_2O$ and Na_2EDTA with continuous stirring until dissolved. Make the volume to 1000 mL and store at 4 °C.
- *Vitamins and myo-inositol stock solution (10×)*: Add 400 mL of distilled water to a Erlenmeyer flask and place it on the heated stirrer. Then, add the required amounts of thiamine chloride, pyridoxine, nicotinic acid, glycine, glutamine, and myo-inositol with continuous stirring until dissolved. Make the volume to 1000 mL and store at 4 °C (*see* **Note 4**).
- *Preparation of 1 L of the rooting medium*:

  Transfer 100 mL of macronutrients, iron source and vitamins and myo-inositol stock solutions, and 10 mL of micronutrients stock solution to a 1 L Erlenmeyer flask with constant stirring. Add sucrose with constant stirring until it is completely dissolved. Take the volume to 1 L with distilled water. Adjust the pH of the medium to 5.8. Add 6 g Gelrite (Sigma G1910) and autoclave at 121 °C for 21 min. Pour 30 mL medium into 100 mL sterile Erlenmeyer flasks.

3 Methods

3.1 Growing Conditions of Donor Plants

1. Prepare the field for sowing using a mouldboard plough, a disc harrow and a cultivator. Apply the fertilizer N:P:K (20:10:0) at 80 and 40 kg/ha, respectively, before sowing.
2. Sow the donor plant seeds in mid-November in a well-prepared field.
3. Irrigate the crop when rainfall is deficient and keep the crop free of weeds by hand hoeing when necessary.

3.2 Harvest of Spikes and Stress Treatment

1. Harvest the spikes from the described cultivars when the awns have just emerged, stage 48 according to the Zadoks scale [20].
2. Placed the spikes in Erlenmeyer flasks containing a 0.7 M mannitol solution at 4 °C for 7 days for osmotic stress and low temperature pretreatment [14].
3. Pretreat 100 spikes from each cultivar.

3.3 Determination of Microspore Developmental Stage

1. Isolate the anthers from flowers at the central part of the spikes and place them on the center of a slide. Cut the anthers vertically with a scalpel and add a drop of 1.5% orcein acetate. Cover the preparation with a coverslip and gently squeeze it with the opposite end of a pencil to spread the microspores.
2. Determine the developmental stage of the microspores by observation under an optical microscope.
3. Select spikes with microspores at the mid- to late-uninucleate stage (Fig. 1).

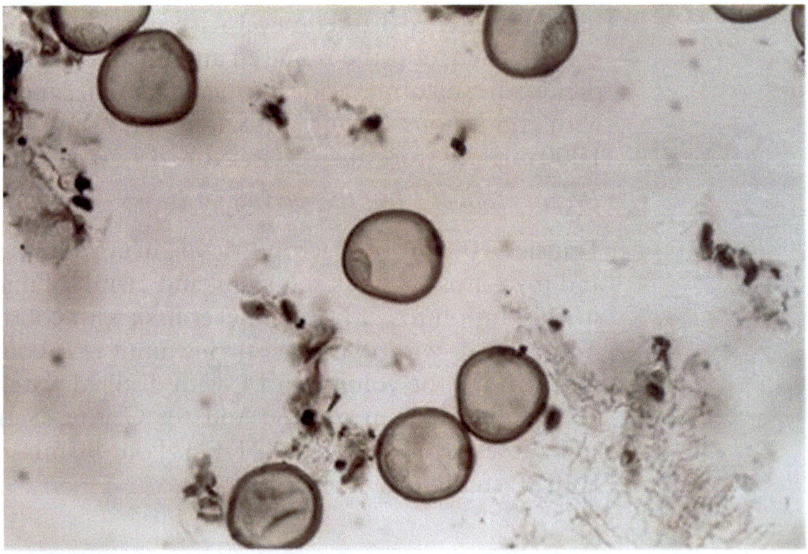

Fig. 1 Durum wheat microspores at the mid- to late-uninucleate stage

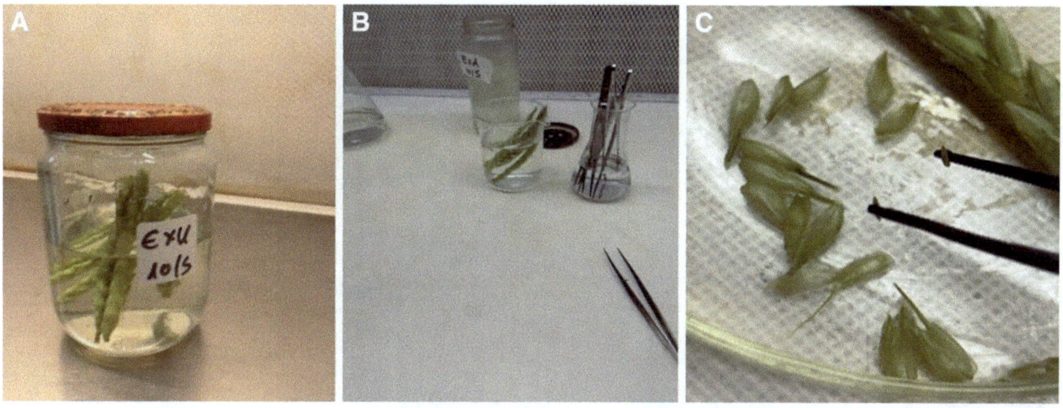

Fig. 2 Spike sterilization and anther isolation. **(A)** Spikes in sodium hypochorite solution. **(B)** Sterilized spikes after washing. **(C)** Anther dissection from spikes containing microspores at the mid- to late-uninucleate stage

3.4 Anther Sterilization

1. Sterilize the spikes in a sodium hypochloride solution (1.65%) for 20 min [21, 22] (Fig. 2A).
2. Rinse the spikes three times for 5 min in sterilized distilled water inside a laminar flow hood (Fig. 2B).

3.5 Anther Culture

1. Aseptically dissect the anthers containing microspores at the mid- to late-uninucleate stage with forceps (Fig. 2C).
2. Inoculate the anthers in Petri dishes containing about 30 mL of modified W14 medium [16, 17] (Table 1). Culture approximately 60 anthers from each spike in a Petri dish (Fig. 3A).
3. Incubate the cultures at 28 °C in the dark for more than 5 weeks (Fig. 3B, C).

3.6 Plant Regeneration

1. Transfer the calli or embryoids (Fig. 3B) to the 190-2 regeneration medium (Table 1).
2. Incubate the cultures at 26 °C with 16 h photoperiod for 4 weeks until green or albino plants develop (Fig. 3D).
3. Transfer the green plants to MS medium without growth regulators for rooting (Table 1), and incubate them at 26 °C with 16 h photoperiod for 4 weeks (Fig. 3E).
4. Roots emerge after 2–3 days, depending on the genotype, and young plants have well-developed roots after 5 weeks.

3.7 Plant Acclimatization

1. Transplant the green plantlets into small pots containing a mixture 2:1:1 soil sand: peat mixture and cover them with plastic bags in order to maintain high humidity. Place the small pots in a growth chamber at 26 °C with a 16 h photoperiod (Fig. 4A).
2. Gradually acclimatize the plants by opening and reclosing the plastic bag for longer periods each time after the second day:

Fig. 3 Durum wheat anther culture. (**A**) Anthers aseptically placed in the induction medium. (**B**) Calli (or embryo-like structures) developed after five weeks of culture. (**C**) Anthers cultured in the dark in a growth chamber. (**D**) Albino plants. (**E**) Green plants

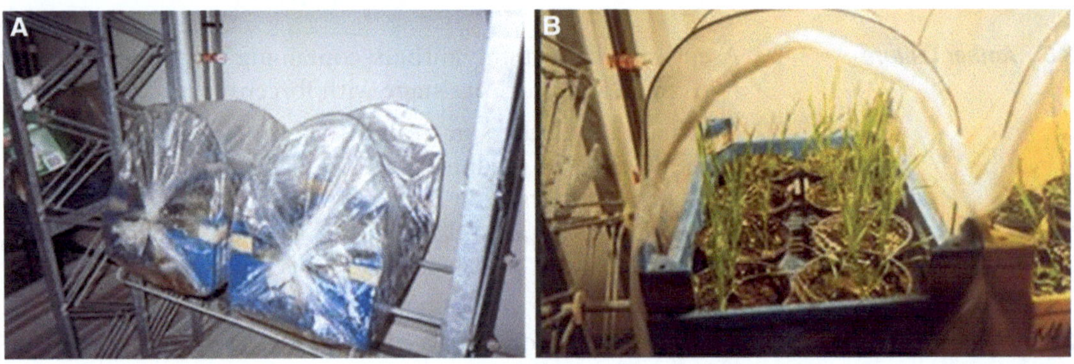

Fig. 4 Acclimatization of durum wheat plantlets: (**A**) The plantlets are placed in pots and covered with plastic bags to protect them from excessive transpiration. (**B**) The bags are opened after the third day for gradual acclimatization of the plantlets

i.e., open for 1 h on the third day, 2 h on the fourth day, and so on (Fig. 4B).

3.8 Chromosome Doubling and Seed Production

1. Cultivate the seedlings in pots at a temperature of 26 °C and 16h of light (*see* **Note 5**).
2. Take the plants out of the pots at the 3–4 tillering stage.
3. Wash the roots to remove the soil (Fig. 5A), cutoff the end of the roots, leaving about 5 cm from the crown (Fig. 5B).

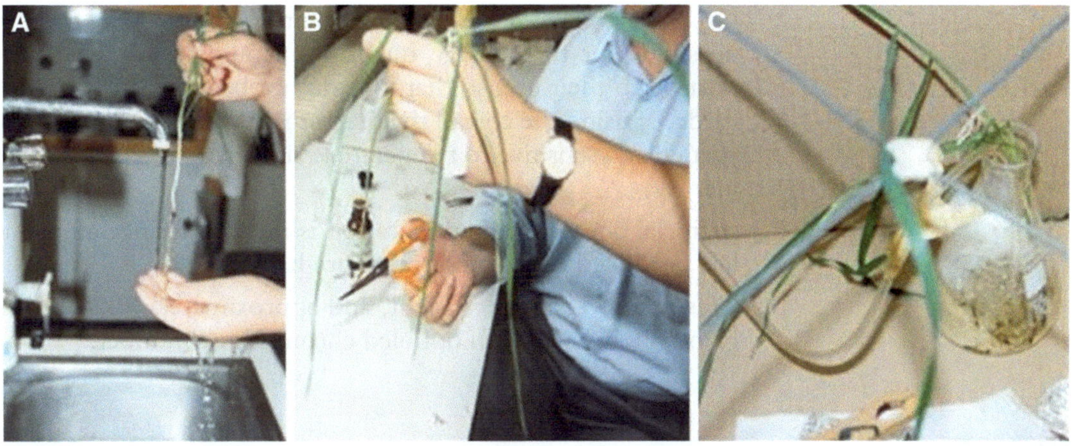

Fig. 5 Colchicine treatment of durum wheat haploid plants. (**A**) Washing of the roots. (**B**) Cutting of root tips. (**C**) Colchicine treatment with continuous aeration

4. Place the plants (i.e., their roots) in the colchicine solution for 5 h inside a fume hood. Use an air pump to aerate the solution (Fig. 5C) (*see* **Note 2**).
5. Repot the plants, cover them with plastic, and place them in the growth chamber with a day/night temperature of 21/19 °C and a 16 h photoperiod. Keep the space under the plastic saturated with moisture to prevent the plants from dying due to evapotranspiration. Acclimatize the plants as previously described (Sect. 3.7.2, Fig. 4). Successful diploidization is confirmed by the production of seeds from the treated plants (*see* **Note 6**).

4 Notes

1. Warm glacial acetic acid before adding orcein to dissolve it.
2. DMSO helps colchicine to enter the treated cells. Colchicine is very toxic, and care must be taken to avoid accidents. For this, gloves, masks, and eye protection glasses must be used during colchicine manipulation to avoid the risk of exposure.
3. The culture media are composed of various nutrients and are prepared in the form of stock solutions: macronutrients (10×), micronutrients (100×), iron source (10×), and vitamins (10×). The preparation of stock solutions simplifies the work because it allows for quick and consistent media preparation, reduces the chances of errors in weighing individual components, and ensures uniformity across multiple experiments.

4. Caution: For best results, stock solutions must be as fresh as possible. For this reason, quantities should be made in proportion to their use, so that they do not become stale and discarded, raising the cost of the medium excessively. Vitamins are particularly sensitive to temperature and have the shortest shelf life.
5. Artificial chromosome duplication is inevitable since all plantlets produced from microspores are haploids.
6. This method is very efficient, and approximately 80%–90% of plants show areas with doubled chromosome number.

References

1. Xynias IN, Mylonas I, Korpetis EG et al (2020) Durum wheat breeding in the mediterranean region: current status and future prospects. Agronomy 10:432. https://doi.org/10.3390/agronomy10030432
2. Tersi M, Xynias I, Gouli-Vavdinoudi E, Roupakias D (2005) Effect of genome, induction medium and temperature pretreatment on green plant production in durum (*Triticum turgidum* var *durum*) × bread wheat (*Triticum aestivum* L em Thell) crosses. Acta Physiol Plant 27(4):641–649
3. Agache S, Bacheller B, De Buyser J et al (1989) Genetic analysis of anther culture in wheat using aneuploid, chromosome substitution and translocation lines. Theor Appl Genet 77:7–11
4. Barnabas B, Szakacs E, Karsai I, Bedo Z (2001) *In vitro* androgenesis of wheat: from fundamentals to practical application. Euphytica 119:211–216
5. Lantos C, Weyen J, Orsini JM et al (2013) Efficient application of in vitro anther culture for different European winter wheat (*Triticum aestivum* L) breeding programs. Plant Breed 132:149–154
6. Löschenberger F, HänseL H, Heberle-Bors E (1993) Plant formation from pollen of tetraploid wheat species and *Triticum tauschii*. Cereal Res Commun 21(2/3):133–139
7. Szakacs E, Kovacs G, Pauk J, Barnabas B (1988) Substitution analysis of callus induction and plant regeneration from anther culture in wheat (*Triticum aestivum* L.). Plant Cell Rep 7:127–129
8. Lazaridou TB, Pankou CI, Xynias IN, Roupakias DG (2016) Effect of D genome on wheat anther culture response after cold and mannitol pre-treatment. Acta Biol Crack Ser Bot 58:95–102
9. Lazaridou TB, Sistanis I, Lithourgidis AS et al (2011) Response to in-vitro anther culture of F3 families originating from high and low yielding F2 barley (*Hordeum vulgare* L) plants. Aust J Crop Sci 5:262–267
10. Kaushal J, Sharmai R, Balachandran SM et al (2015) Effect of cold pretreatment on improving anther culture response of rice (*Oryza sativa* L). J Exp Biol Agric Sci 2:233–242
11. Patel M, Darvey NL, Marshal DR, Berry JO (2004) Optimization of culture condition for improved plant regeneration efficiency from wheat microspore culture. Euphytica 140:197–204
12. Cistué L, Soriano M, Castillo AM et al (2006) Production of doubled haploids in durum wheat (*Triticum turgidum* var *durum*) through isolated microspore culture. Plant Cell Rep 25:257–264. https://doi.org/10.1007/s00299-005-0047-8
13. Labbani Z, De Buyser J, Picard E (2007) Effect of mannitol pre-treatment to improve green plant regeneration on isolated microspore culture in *Triticum turgidum* ssp *durum* cv 'Jennah Khetifa'. Plant Breed 126:565–568
14. Lazaridou TB, Mavromatis AG, Xynias IN (2021) Effect of culture medium and mannitol pre-treatment on durum wheat anther culture response. Cytol Genet 55(6):590–597
15. Hunter CP (1987) European Patent Application No 0245898 A2, Shell Int. Res: 1–8
16. Ouyang JW, Liang H, Zhanf C et al (2004) The effect of potato extract used as additive in anther culture medium on culture responses in *Triticum aestivum*. Cereal Res Commun 32:501–508. https://doi.org/10.1007/BF03543341
17. Ponitka A, Slusarkiewicz-Jarzina A (2009) Regeneration of oat androgenic plants in relation to induction media and culture conditions of embryo-like structures. Acta Soc Bot Pol 8(3):209–213. https://doi.org/10.5586/asbp.2009.026
18. Zhuang J, Jia X (1983) Increasing differentiation frequencies in wheat pollen callus. In:

Hu H, Vega MR (Eds.) Cell and tissue culture techniques for cereal crop improvement. Sinica Science Press, Beijing, 413–432
19. Murashige T, Skoog F (1962) A revised medium for rapid growth and bioassays with tobacco tissue cultures. Physiol Plant 15:473–497
20. Zadoks JC, Chang TT, Konzak CF (1974) A decimal code for the growth stages of cereals. Weed Res 14:415–421
21. Zamani I, Gouli–Vavdinoudi E, Kovacs G et al (2003) Effect of parental genotypes and colchicine treatment on the androgenic response of wheat F1 hybrids. Plant Breed 122:1–4
22. Barnabas B (2003) Protocol for producing doubled haploid plants from anther culture of wheat (*Triticum aestivum* L). In: Maluszynski M, Kasha KJ, Forster BP, Szarejko I (Eds.) Doubled haploid production in crop plants. A manual. Kluwer, Dordrecht, 65–70

Chapter 11

Barley Doubled Haploid Plant Production by Anther Culture

Ana María Castillo, Begoña Echávarri, Patricia Fustero, Aimar Navarro, María Asunción Costar, Nuria Isach and María Pilar Vallés

Abstract

Barley is the fourth most cultivated cereal worldwide, traditionally used for animal feed and malt production. Its high adaptability to diverse environmental conditions is particularly evident in drought- and salt-prone areas. The development of resilient varieties is a priority to support food security and agricultural sustainability in the context of climate change. An important tool in barley breeding is the production of doubled haploid (DH) plants. This process enables the rapid development of homozygous plants in a single generation, significantly reducing the time required to release new varieties. DH technology has been widely adopted in the field of barley breeding, leading to the development of numerous new varieties. The two main methods for DH plant production in barley are interspecific crosses with *Hordeum bulbosum* and microspore embryogenesis (ME). The former involves the selective elimination of *H. bulbosum* chromosomes in the hybrid. In ME, microspores are reprogrammed to develop into embryos and finally into haploid or DH plants. Despite its high efficiency, ME faces challenges with certain genotypes, such as spring and six-row types. The barley anther culture protocol described in this chapter is based on the work of Cistué et al. (2003) with modifications to improve plant regeneration and root development. This protocol involves an osmotic and starvation treatment with mannitol for four days, followed by culture on an FHG solidified medium for embryo development, and a modified J25-8 medium for plant regeneration. A spontaneous doubling rate of 85%–90% has been achieved. This approach has enhanced the efficiency and cost-effectiveness of DH plant production in barley breeding programs.

Keywords Barley, Microspore embryogenesis, Anther culture, Mannitol, Doubled haploid

1 Introduction

Barley (*Hordeum vulgare* L.) is the fourth most cultivated cereal in the world, with a total area of 46.7 Mha (million hectares) and an annual production of approximately 150 Mt (million tons) in 2022–2023 [1]. The primary use of barley is as a livestock feed, in addition to its role in malt production [2]. However, it is now attracting considerable attention as a foodstuff due to its health benefits [3]. This crop is notable for its versatility and remarkable adaptability to various environmental conditions, particularly in areas affected by drought and salinity, where the growth of other cereals

Ana María Castillo et al. (eds.), *Doubled Haploid Methods in Cereals*, Methods in Molecular Biology, vol. 3029,
https://doi.org/10.1007/978-1-0716-5245-9_11,

is hindered [4]. Agriculture, and food production in particular, is becoming increasingly vulnerable to climate change. According to the latest climate projections, a significant decrease in barley yields is expected [5]. There is an urgent need for the development of new barley varieties to ensure global food security and agricultural sustainability in the context of climate change.

One of the tools available to plant breeders to accelerate the development of new varieties is the production of doubled haploid (DH) plants. DH technology allows the obtention of completely homozygous plants in a single generation, as opposed to the six to eight generations of selfing required by traditional breeding methods. Furthermore, phenotypic selection is more efficient on homozygous lines than in early generations of segregating material (traditional methods). Consequently, the release of a new variety onto the market can be accomplished in 4–5 years. The impact of this technology in barley is evidenced by the registration of a large number of barley varieties selected from DH lines [6]. For example, it was estimated that 50% of new commercial varieties in Europe were produced via a DH system [7]. Additional applications of DH in barley include genetic mapping, QTL analyses, genome-wide association studies, etc. ([8], see Chap. 1).

The most efficient methods for DH plant production in barley are interspecific crosses with *H. bulbosum* and microspore embryogenesis (ME) [9]. In the first method, after the fertilization of barley with *H bulbosum*, the chromosomes from *H. bulbosum* are selectively eliminated during the early stages of development in the hybrid embryo, resulting in endosperm abortion. Therefore, the embryo needs to be rescued and cultured in vitro for haploid plant production [10]. The main advantage of this method over ME is its lower genotype dependency [11].

ME has the highest potential for haploid production in barley. In this method, microspores are reprogrammed to follow an embryogenic pathway. This developmental change is triggered by the application of a stress treatment, followed by a culture phase that leads to the development of embryos and, finally, haploid or DH plants [12]. One of the main advantages of ME over intergeneric crosses with *H bulbosum* is the direct recovery of a high percentage of spontaneous DH plants [13]. Barley is considered a model species for ME among cereals. However, there are still genotypes that are recalcitrant to this methodology, especially spring and six-rowed types [14, 15]. This fact hinders the full application of this technology in barley breeding.

This chapter describes a protocol for anther culture in barley, based on that of Cistué et al. [16]. Their approach included the application of a starvation and osmotic treatment with 0.7 M mannitol, the incorporation of Ficoll into the FHG induction medium, and the use of FHG medium for plant regeneration and rooting. The modifications presented in this protocol include the replacement of Ficoll with agarose in the FHG induction medium and the use of a modified J25-8 medium [17] for plantlet regeneration and rooting. The J25-8 medium is supplemented with 3.8 mg/L $CuSO_4.5H_2O$, 0.5 mg/L indoleacetic acid (IAA), and 1 mg/L benzylaminopurine (BAP) for plant regeneration. In addition, 100 mg/L cefotaxime is incorporated into the culture media. These implementations reduce costs and improve plant regeneration and root development, favouring plant survival during the acclimatization process. The percentage of spontaneous doubling with this protocol is approximately 85%–90%, thus eliminating the need for chromosome doubling agents.

2 Materials

2.1 Plant Material

This protocol has been successfully applied to: a highly ME-responsive DH barley line (DH6148), the low ME-responsive cultivar Noga, which is an elite cultivar in Israel, and three CMPP (BC2DH) lines generated from *H. vulgare* ssp. *spontaneum* × Noga crosses [18]. DH6148 was selected from a DH population derived from a cross between the model cultivar Igri and DH46 (a DH line with a high rate of albinism selected from a DH population of Igri × Dobla) [14].

2.2 Equipment

1. Growth chamber set at 6°C with 8 h photoperiod and a light intensity of 100 µmol $m^{-2}s^{-1}$, provided by Philips Master Daylight TLD 36W/865 fluorescent lamps, for plant vernalization.
2. Growth chamber set at 14/12°C (day/night) with 12 h photoperiod and 250 µmol $m^{-2}s^{-1}$ light intensity at the canopy level, provided by high-pressure metal halide lamps (Philips Powertone HPI-T plus 400W/645), and relative humidity (RH) set at 60%, for vegetative growth of the mother plants.
3. Growth chamber set at 21/18°C (day/night) with 16 h photoperiod and 450 µmol $m^{-2}s^{-1}$ at the canopy level, provided by high-pressure metal halide lamps (Philips Powertone HPI-T plus 400W/645) and 60% RH, for plant flowering.
4. Growth chamber set at 14/12°C (day/night) with 12 h photoperiod, and 180 µmol $m^{-2}s^{-1}$ at the canopy level provided by OSRAM Lumilux Cool White L58W/840 fluorescent lamps and 70% RH, for plant hardening-off.
5. Greenhouse set at 25/20°C (day/night), supplemented with high pressure sodium lamps (Philips Son-T AGRO 400W) providing 300 µmol $m^{-2}s^{-1}$, with 16 h photoperiod, for DH seed production.
6. Incubator set at 25°C without light, for stress treatment and embryo induction.

7. Incubator set at 25°C with 16 h photoperiod and a light intensity of 120 µmol $m^{-2}s^{-1}$, supplied by Philips Master TLD 58W/865 and GE LED BrightStick 15W 3000K, for plant regeneration.
8. Inverted microscope (Nikon TE300 or similar) with epifluorescence and filters EX 365/DM 400/BA, along with a digital sight DS-Ri1 and NIS-Elements D software for DAPI analysis and image processing.
9. Stereo microscope (Nikon SMZ750 or similar) with a digital camera and NIS-Elements F software.
10. Flow cytometer (CyFlow™ Ploidy Analyser with UV-LED—365 nm, Sysmex).
11. Laminar flow hood.
12. Bunsen-type burner, or similar, for flame sterilization.
13. Precision balance.
14. pH meter.
15. Autoclave.
16. Agitator.
17. Refrigerator (4°C).
18. Freezer.
19. Vacuum pump.
20. Milli-Q water production system.

2.3 Plant Growth and Laboratory Materials

1. Peat, vermiculite (size 3), and fine sand.
2. Soil fertilizers: Osmocote Pro (N:P:K, 17:11:10) and Fertipron (N:P:K, 20:20:20; boron 0.2%, copper 0.05%, zinc 0.05%, iron 0.05%, manganese 0.05%, and molybdenum 0.0025%).
3. Plastic seedbeds (trays with 96 sockets) and pots (17 cm Ø × 15.5 cm height and 8 cm Ø × 6 cm height).
4. Plastic propagator (62 × 42 × 21 cm height).
5. Plastic spray bottles.
6. 96 % (*v/v*) ethanol.
7. Beakers and magnetic stirrer.
8. Glass bottles (100 mL, 500 mL, 1 L, and 2 L).
9. Graduated cylinders.
10. Automatic pipettes (0-20 µL, 0–200 µL, and 0-1 mL), and manual propipettes and pipettes (1, 5, and 10 mL).
11. Zip-closure plastic bags (150 mL).
12. Plastic tubes (25 mL).
13. Glass jars (250 mL).

14. Plastic containers (1 L).
15. Polystyrene sterile Petri dishes (6 and 9 cm Ø).
16. Sterile 0.22 μm pore size membrane filters (500 mL capacity).
17. Parafilm ®.
18. Microscope slides and coverslips.
19. Scissors and gauze.
20. Micro-dissecting stainless-steel tweezers (110- and 200-mm length).
21. Stainless-steel scissors.
22. Magenta vessels.
23. Filter paper.
24. Small plastic boxes with wet filter paper.
25. Polystyrene Petri dish lids (3 cm Ø, non-sterile).
26. Polypropylene tube (3.5 mL, 55 × 12 mm, Sysmex).
27. Celltrics® 30 μm pore-size (Sysmex).
28. Stainless-steel scalpel and razor blade.

2.4 Solutions for Cytological Studies, Ploidy Determination, and Plantlet Fertilization

1. 0.3 M mannitol: Dissolve 5.5 g mannitol in 100 mL of Milli-Q water and store at 4°C.
2. 4,6-diamidino-2-phenylindole (DAPI) stock solution (1 mg/mL, 250×): Dissolve 5 mg DAPI in 5 mL Milli-Q water, aliquot, and store in the dark at −20°C.
3. DAPI working solution (4 μg/mL): Take 40 μL of the DAPI stock solution to 10 mL with Milli-Q water, add four drops of Tween 80, aliquot into Eppendorf tubes, cover with aluminum foil, and store at −20°C. This solution might also be stored at 4°C for several weeks.
4. Commercial CyStain® UV ploidy solution (Partec-Sysmex) containing lysis buffer and DAPI.
5. Modified Hoagland's salt fertilization stock solution (5×) (Table 1):
   - Prepare stock solutions separately for each macronutrient, and for Sequestrene:
     - Solution A: $Ca(NO_3)_2{\cdot}4H_2O$ (200×, 236 g/L).
     - Solution B: KNO_3 (200×, 101 g/L).
     - Solution C: $MgSO_4{\cdot}7H_2O$ (500×, 246 g/L).
     - Solution D: KH_2PO_4 (1000×, 136 g/L).
     - Solution E: $CaCl_2{\cdot}2\,H_2O$ (1000×, 184 g/L).
     - Solution F: Sequestrene (250×, 25 g/L).
   - Prepare the micronutrient stock solution (Solution G) (1000×) by dissolving 572 mg of H_3BO_3, 362 mg of $MnCl_2{\cdot}4H_2O$, 110 mg of $ZnCl_2$, 50 mg of $CuCl_2{\cdot}2H_2O$, and 25 mg of $Na_2MoO_4{\cdot}2H_2O$ in 1 L of distilled water.

Table 1
Composition of Hoagland's solution

Compounds	Hoagland salts
Macroelements [mg/L]	
$Ca(NO_3)_2 \cdot 4H_2O$	1180.70
KNO_3	505.50
$MgSO4 \cdot 7H_2O$	492.5
KH_2PO_4	136.1
$CaCl_2 \cdot 2\ H_2O$	183.73
Iron source [mg/L]	
Sequestrene EDTA	100
Microelements [mg/L]	
H_3BO_3	0.572
$MnCl_2 \cdot 4H_2O$	0.362
$ZnCl_2$	0.111
$CuCl_2 \cdot 2H_20$	0.05
$Na_2MoO_4 \cdot 2H_2O$	0.025
pH	6.0–6.5

- To prepare 2 L of Hoagland's solution (5×): Mix 50 mL each of solutions A and B, 20 mL of Solution C, 10 mL each of solutions D, E, and G, and 40 mL of solution F. Bring the volume up to 2 L and adjust pH to 6.0–6.5. Autoclave the solution at 121°C for 23 min and store at room temperature (RT) in the dark.

6. Hoagland's final working solution: Dilute the 5× Hoagland's stock solution 1:25 with water to obtain the final 0.2× watering solution.

2.5 Tissue Culture Media

Culture media composition for stress treatment, embryo induction, plant regeneration, and plant rooting are shown in Table 2. All media and stock solutions are prepared with Milli-Q water.

1. FHG [19] micronutrient stock solution (500×): Dissolve 83 mg of KI, 25 mg of $NaMoO_4 \cdot 2H_2O$, 2.5 mg of $CuSO_4 \cdot 5H_2O$, and 2.5 mg of $CoCl_2 \cdot 6H_2O$ in 200 mL of water (Table 2).
2. Thiamine-HCl stock solution (1000×): Dissolve 40 mg of Thiamine-HCl in 100 mL of water.
3. FHG medium stock solution (10×): Dissolve 19 g of KNO_3, 1.65 g of NH_4NO_3, 1.7 g of KH_2PO_4, 4.4 g of $CaCl_2 \cdot 2H_2O$,

3.7 g of $MgSO_4 \cdot 7H_2O$, 169 mg of $MnSO_4 \cdot H_2O$, 86 mg of $ZnSO_4 \cdot 7H_2O$, 62 mg of BO_3H_3, and 400 mg of FeNa(EDTA)$_2$ in 800 mL of water. Add 20 mL of micronutrient stock solution (500×), and 10 mL of Thiamine-HCl stock solution (1000×), and bring the volume up to 1 L (Table 2). Aliquot 100 mL into zip-closure plastic bags, and store at −20°C.

4. J25-8 medium [17] M1 macronutrient stock solution (20×): Dissolve 44 g of NO_3K, 12 g of NH_4NO_3, 6.2 g of $MgSO_4 \cdot 7H_2O$, 1.34 g of $(NH_4)_2SO_4$, and 5.9 g $CaCl_2 \cdot 2H_2O$ in 800 mL of water. Once dissolved, bring the volume up to 1 L (Table 2), aliquot 100 mL into zip-closure plastic bags, and store at −20°C.
5. J25-8 M2 macronutrients stock solution (20×): Dissolve 3 g of KH_2PO_4 and 1.5 g of $NaH_2PO_4 \cdot H_20$ in 800 mL of Milli-Q water. Once dissolved, bring the volume up to 1 L (Table 2), aliquot 100 mL into zip-closure plastic bags, and store at −20°C.
6. J25-8 micronutrient stock solution 1 (1000×): Dissolve 25 mg of $Na_2MoO_4 \cdot 2H_2O$, 2.5 mg of $CuSO_4 \cdot 5H_2O$, and 2.5 mg of $CoCl_2 \cdot 6H_2O$ in 100 mL of water (Table 2).
7. $CuSO_4 \cdot 5H_2O$ (2.5 mg/mL): Dissolve 250 mg of $CuSO_4 \cdot 5H_2O$ in 100 mL of water.
8. J25-8 micronutrient stock solution 2 (100×): Dissolve 50 mg of $MnSO_4 \cdot H_2O$, 30 mg of H_3BO_3, and 50 mg of $ZnSO_4 \cdot 7H_2O$ in 80 mL of water. Add 10 mL of micronutrient stock solution 1, and bring the volume up to 100 mL (Table 2). Aliquot 20 mL into plastic tubes and store at −20°C.
9. J25-8 vitamin stock solution (100×): Dissolve 10 mg of nicotinamide, 100 mg of thiamine-HCl, and 10 mg of pyridoxine in 80 mL of water. Once dissolved, bring the volume up to 100 mL (Table 2). Aliquot 20 mL into plastic tubes, and store at −20°C.
10. J25-8 5 iron stock solution (100×): Dissolve 400 mg of $FeNa_2$ EDTA in 80 mL of water, and bring the volume up to 100 mL (Table 2). Aliquot 20 mL into plastic tubes, and store at −20°C.
11. J25-8 amino acid stock solution (100×): Dissolve 1.2 g of L-glutamine, 0.50 g of L-asparagine, 0.25 g of L-threonine, 0.25 g of L-arginine, and 0.5 g of L-proline in 80 mL water, and bring the volume up to 100 mL (Table 2). Aliquot 20 mL into plastic tubes, and store at −20°C.
12. Benzylaminopurine (BAP) stock solution (1 mg/mL): Dissolve 50 mg of BAP in a minimal volume of 1 N HCl, add Milli-Q water to bring the volume up to 50 mL, and store at 4°C for 2–3 months in a glass bottle (*see* **Note 1**).

Table 2
Media composition for barley anther culture (concentration expressed in mg/L)

Compounds	Stress 0.7 M Mannitol	Embryo induction (FHGI)	Plant regeneration (J25-8 Reg)	Plant rooting (J25-8 Root)
Macronutrients (mg/L)				
KNO_3		1900	2200	2200
NH_4NO_3		165	600	600
KH_2PO_4		170	150	150
$CaCl_2 \cdot 2H_2O$	5800	440	295	295
$MgSO_4 \cdot 7H_2O$		370	310	310
$(NH_4)_2SO_4$			67	67
$NaH_2PO_4 \cdot H_2O$			75	75
Micronutrients (mg/L)				
$MnSO_4 \cdot H_2O$		16.9	5	5
$ZnSO_4 \cdot 7H_2O$		8.6	5	5
H_3BO_3		6.2	3	3
KI		0.83		
$CuSO_4 \cdot 5H_2O$		0.025	3.825	0.025
$CoCl_2 \cdot 6H_2O$		0.025	0.025	0.025
$Na_2MoO_4 \cdot 2H_2O$		0.25	0.25	0.25
Iron source (mg/L)				
$FeNa_2EDTA$		40	40	40
Vitamins (mg/L)				
Nicotinamide			1	1
Tiamine-HCl		0.4	10	10
Piridoxine-HCl			1	1
Aminoacids				
L-Glutamine		730	120	120
L-Asparragine			50	50
L-Threonine			25	25
L-Arginine			25	25
L-Proline			50	50
Other Organic compounds (mg/L)				
Myo-inositol		100	100	100
Casein			125	125
Malic acid			100	100

Table 2
continued

Compounds	Stress 0.7 M Mannitol	Embryo induction (FHGI)	Plant regeneration (J25-8 Reg)	Plant rooting (J25-8 Root)
Coconut milk			25[a]	25[a]
Carbohydrates (mg/L)				
Mannitol	127,000			
Maltose		62,000		
Sucrose			20,000	20,000
Glucose			7000	7000
Growth regulators and antibiotic (mg/L)				
BAP		1	1	
IAA			0.5	
ANA				2
Cefotaxime		100	100	100
Solidifying agents and others (mg/L)				
Agarose	8,000	6,000		
Gelrite			3000	3000
Activated charcoal			4000	4000
pH		5.8	5.8	5.8

[a]25 mL of coconut milk

13. IAA stock solution (1 mg/mL): Dissolve 50 mg of IAA in a minimal volume of 1 N NaOH, add Milli-Q water to bring up to 50 mL. Store at 4°C in a glass bottle covered with aluminium foil to protect from light-induced degradation (*see* **Note 1**).
14. NAA stock solution (1 mg/mL): Dissolve 50 mg of NAA in a minimal volume of 1 N NaOH, add Milli-Q water to bring up to 50 mL, store at 4°C in a glass bottle (*see* **Note 1**).
15. Stress treatment medium: 0.7 M mannitol supplemented with 40 mM $CaCl_2{\cdot}H_20$ and 8 g/l Sea Plaque agarose (Table 2). To prepare 1 L of medium:
    - Solution A: Dissolve 127 g of mannitol in 400 mL of Milli-Q water by heating at 50°C (*see* **Note 2**), add 5.88 g of $CaCl_2{\cdot}2H_2O$, bring the volume up to 500 mL, and sterilize by filtration through a sterile 0.22 μm pore membrane filter in a sterile glass bottle inside a laminar flow hood.
    - Solution B: Add 8 g of agarose to Milli-Q water, bring up to 500 mL, and autoclave at 121°C for 23 min.

- Mix both solutions in the laminar flow hood and pour 10 mL into 6 cm Ø Petri dishes.

16. Embryo induction medium (FHGI): FHG medium supplemented with 1 mg/L of BAP, 100 mg/L of cefotaxime sodium, and 6 g/L of SeaPlaque agarose (Table 2). To prepare 1 L of medium:
    - Solution A: Defrost 100 mL of FHG stock solution (100×), add 62 g of maltose, 100 mg of myoinositol, 730 mg of glutamine, 1 mL of BAP stock solution (1 mg/mL), and 100 mg of cefotaxime sodium, and bring the volume up to 500 mL. Adjust pH to 5.8 with 1 N NaOH and sterilize by filtration through a 0.22 μm pore membrane filter in a sterile glass bottle inside the laminar hood.
    - Solution B: Add 6 g of Sea Plaque agarose to 500 ml of Milli-Q water, and autoclave at 121°C for 23 min.
    - Mix both solutions in the hood and pour 10 mL into 6 cm Ø Petri dishes.
17. Plantlet regeneration medium (J25-8 Reg): J25-8 medium supplemented with 3.825 mg/L of $CuSO_4{\cdot}5H_2O$, 1 mg/L of BAP, 0.5 mg/L of IAA, 100 mg/L of cefotaxime sodium, 4 g/L activated charcoal, and 3 g/L of Gelrite (Table 2). To prepare 2 L medium:
    - Solution A: Defrost 100 mL of M1 and M2 macronutrient stock solutions (20×), and 20 mL each of micronutrient stock solution 2, iron source, vitamin, and amino acid stock solutions (all at 100×), and 50 mL of coconut milk (*see* **Note 3**). Add the stock solutions to 500 mL Milli-Q water and mix. Then, add 3.04 mL of $CuSO_4{\cdot}5H_2O$ (2.5 mg/mL) (*see* **Note 4**), 2 mL of BAP, and 1 mL of IAA stock solutions (1 mg/mL), followed by 200 mg of myo-inositol, 250 mg of casein, 40 g of sucrose, 14 g of glucose, and 200 mg of cefotaxime. Separately, dissolve 200 mg of malic acid in 100 mL of Milli-Q water containing 6 drops of NH_4OH, and add it to the main solution. Adjust the volume to 1 L and the pH to 5.8 with 1 N KOH. Sterilize by filtration through a 22 μm pore membrane filter in a sterile glass bottle.
    - Solution B: Add 8 g/L of activated charcoal and 6 g of Gelrite to 1 L of Milli-Q water in a 2 L bottle and autoclave at 121°C for 23 min.
    - Mix solutions A and B in a laminar hood and transfer 10 mL into 6 cm Ø Petri dishes.
18. Rooting medium (J25-8 Root): J25-8 medium supplemented with 2 mg/L of NAA, 100 mg/L of cefotaxime sodium, and 3 g/L of Gelrite (Table 2). The medium is prepared as J25-8 Reg with some modifications in solution A: no extra supply of 3.8 mg /L of $CuSO_4{\cdot}5H_2O$, and the addition of 2 mg/L of NAA instead of 1 mg/L of BAP and 0.5 mg/L of IAA. After autoclaving, pour 50 mL of the medium into Magenta boxes.

3 Methods

3.1 Growth Conditions of the Mother Plants

1. Sow seeds of donor plants in a seedbed with a mixture of peat:vermiculite:fine sand (40 kg:8 kg:50 kg). Fertilize the soil with 2.9 g of Osmocote Pro (N:P:K, 17:11:10)/kg of soil mixture.
2. Vernalize spring and winter genotypes for 5 and 7 weeks, respectively, in a growth chamber at 6°C with 8 h photoperiod and a light intensity of 100 μE/m²/s.
3. After vernalization, transplant two plantlets into 17 cm Ø × 15.5 cm height pots with the same soil mixture described above, and cultivate in a growth chamber at 14/12°C (day/night) with a 12 h photoperiod and 250 μmol $m^{-2}s^{-1}$ light intensity (Fig. 1A).
4. After 4 weeks, transfer the pots to a growth chamber at 21/18°C (day/night) with a 16 h photoperiod and 450 μE/m²/s light intensity.

Fig. 1 Growth of barley donor plants, spike disinfection, and anther excision. **A**: Donor plants growing in a controlled-environment growth chamber. **B**: Disinfection of spikes by ethanol spraying. **C**: Disinfected spikes plated in a Petri dish can be used directly for stress treatment or stored in a refrigerator at 4°C. **D**: Anther excision with fine forceps under a stereoscopic microscope, and culture in 0.7 M mannitol stress medium

5. Fertilize each pot three times with 200 mL of a 3 g/L of Fertipron solution during the vegetative growth phase (short photoperiod), and twice with 250 mL after increasing the photoperiod. Water plants every 2–3 day as required. Plants should be free of pests and fungal infections (*see* **Note 5**).

3.2 Determination of Microspore Developmental Stage

The microspore developmental stage is determined by DAPI staining (Fig. 2).

1. Harvest spikes at different developmental stages in order to establish a correlation between the stage of microspore development and the distance between the ligules of the flag leaf and the second leaf in each genotype (Fig. 2A, B, *see* **Note 6**).

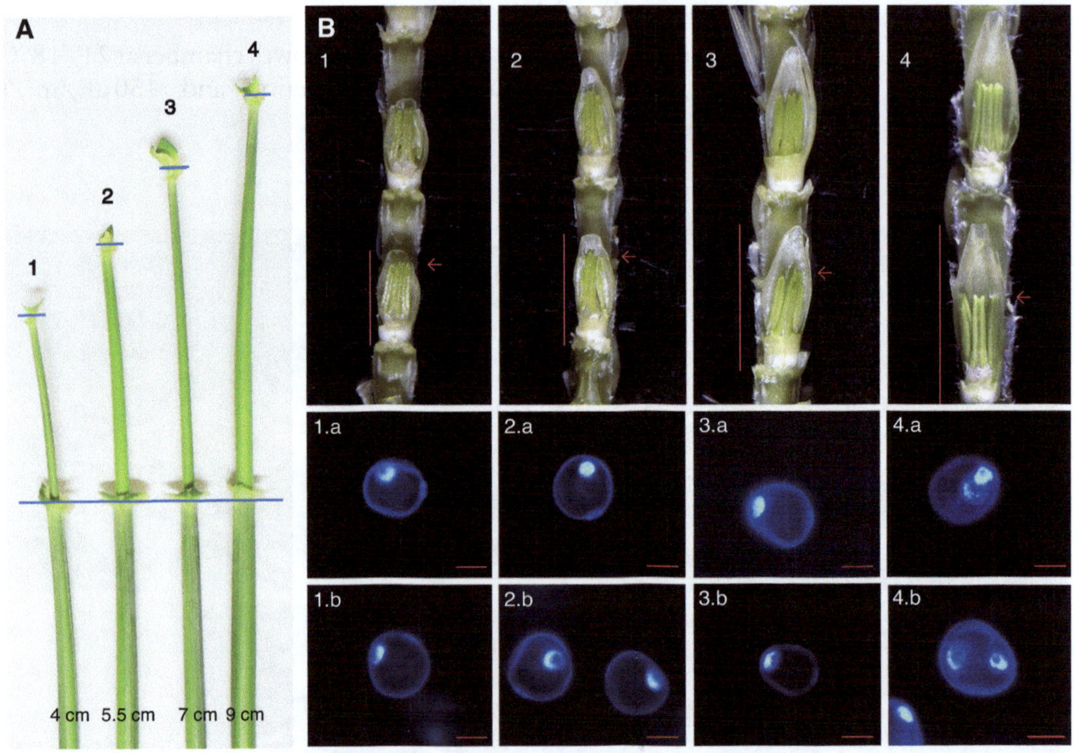

Fig. 2 Correlation between the microspore developmental stage and morphological characteristics of tillers and florets in the genotype DH6148. **A**: Tillers with 4.0 cm (1), 5.5 cm (2), 7.0 cm (3), and 9.0 cm (4) distance between the ligules of the flag leaf and the second leaf. **B**: Florets and microspores stained with DAPI from tillers 1-4 (1, 1a, 1b; 2, 2a, 2b; 3, 3a, 3b; 4, 4a, 4b). Note the color and brightness of the anthers and the position of the anther tips in relation to the distal end of the palea (1, 2, 3, 4). Early and mid-uninucleate microspores (1b and 1a, respectively), mid-late uninucleate microspores (2a, 2b), late uninucleate microspores (3a, 3b), and binucleate pollen (4a, 4b) are shown. Tillers with a 5.5-7.5 cm distance between the flag leaf and the second leaf contain florets with pale green anthers and anther tips in the upper third of the palea, corresponding to microspores at the optimal stage of development. Blue bars in A show the position of the ligules of the flag leaf and the second leaf of the tillers. Red bars and arrows in B:1, 2, 3, and 4 mark the total length of the palea and the position of the anther tip, respectively. Scale red bars in B 1.a, 1b, B 2.a, 2b, B 3a, 3b, B 4.a, 4.b: 25 μm

2. Excise four to six anthers from the central flowers of a spike and place them on a microscope slide with 15 μL of DAPI working solution. Chop the anther with a scalpel, cover with a coverslip, and incubate in the dark for 10 min.
3. Observe the microspores under an epifluorescence microscope with a 365 nm excitation filter and a 420 nm band-pass filter.
4. The optimal developmental stages for anther culture are vacuolated microspores at mid-late uninucleate or late uninucleate stages (Fig. 2B, *see* **Note 6**).

3.3 Harvest of Spikes and Disinfection

1. Harvest the spikes early in the morning when approximately 70% of the microspores are at the mid-late uninucleate or late uninucleate stage, according to the correlation established previously (see Sect. 3.2).
2. Remove the flag leaf and the remaining leaves. Place the spikes, still covered by their sheaths, in a plastic container with water and transfer them to the laboratory for sterilization.
3. Place the spikes on aluminum foil inside a laminar flow hood and spray with 96% ethanol for disinfection (Fig. 1B). After drying for 3 min, remove the spikes from the sheath, place them in 9 cm Ø Petri dishes (4–5 spikes/dish). Spikes can be used directly for anther stress treatment, or Petri dishes can be sealed with Parafilm® and stored at 4°C in the dark for no more than 4–5 days before stress treatment (Fig. 1C).

3.4 Anther Stress Treatment

1. Extract anthers from the spike under a stereoscopic microscope and place them on a 6 cm Ø Petri dish containing 0.7 M mannitol stress medium (Table 1) (Figs. 1D and 3A).
2. Select flowers at the correct developmental stage, characterized by pale green anthers with their distal portion positioned in the upper third of the palea (Fig. 2B, *see* **Note 7**).
3. Culture the anthers from four to five spikes of the same genotype or F_1 in the same Petri dish. Seal the dish with Parafilm® and incubate at 25°C in darkness for 4 days.

3.5 Anther Culture

1. After mannitol treatment, select white or yellowish swollen anthers (Fig. 3B, *see* **Note 8**) and transfer them to a 6 cm Ø Petri dish containing FHGI medium for embryo development (Table 2). Plate 50–60 anthers per dish, placing them close together in the centre of the dish. Incubate the cultures at 25°C in darkness.
2. Monitor culture development under a dissecting microscope every week. After 7 days in culture, the anthers dehisce, and some microspores appear outside the anthers. After 14 days in culture, globular and coleoptilar embryos can be observed

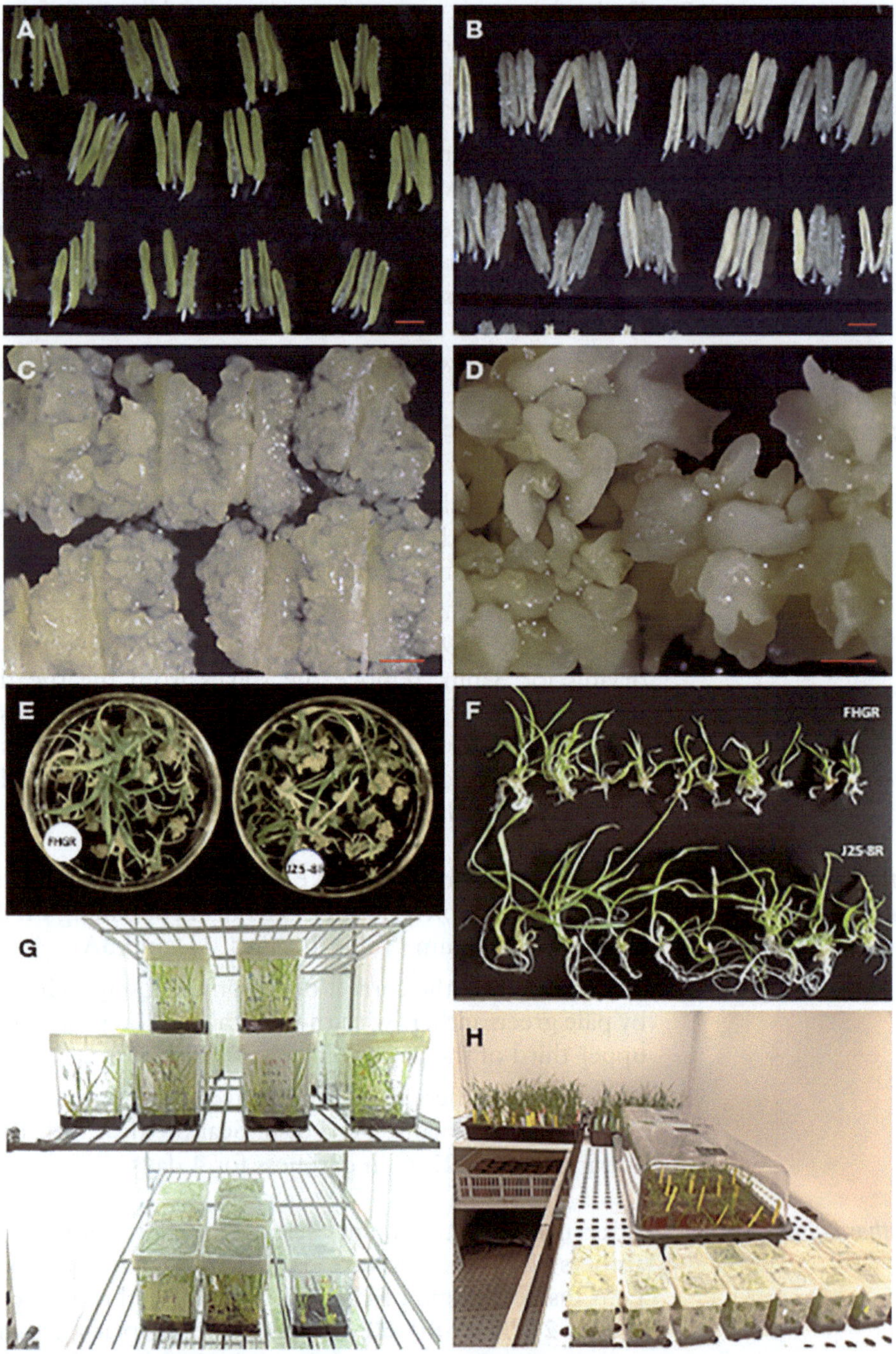

Fig. 3 Barley anther culture of the genotype DH6148. **A**: Anthers at 0 days of stress treatment. **B**: Anthers after 4 days in 0.7 M mannitol stress treatment; white swollen anthers are selected for culture. **C and D**: Embryogenic structures and embryos after 14 and 30 days in FHGI medium, respectively. **E**: Green and albino plant regeneration after 21 days in FHGR (on the left) and J25-8 Reg (on the right) regeneration media. **F**: Green plants developed in FHGR (upper) and J25-8 Reg (lower) regeneration media. Note the stronger plant development in J25-8 Reg medium. **G**: Plant vernalization in J25-8 Root medium. **H**: Acclimatization of the plants before transfer to the greenhouse

(Fig. 3C), and after 30 days, embryos with a well-developed embryo axis are formed (Fig. 3D).

3.6 Plant Regeneration and Rooting

1. After 30–35 days and 45–50 days of culture, transfer mature embryos with a well-developed embryo axis to J25-8 Reg medium (Table 2) for plant regeneration (*see* **Notes 9–11**).
2. Transfer approximately 20 embryos to one Petri dish, incubate at 25°C in darkness for 2–3 days, and then transfer to light (120 µmol $m^{-2}s^{-1}$ and 16 h photoperiod).
3. After 60 days in culture, tip all embryo-like structures over the regeneration medium (*see* **Note 9**).
4. Green and albino plants develop after 15–20 days in J25-8 Reg (Fig. 3E, F, *see* **Note 12**). Transfer the green plants to Magenta boxes containing J25-8 Root medium (Table 2) to strengthen the root system and incubate at 25°C with 120 µmol $m^{-2}s^{-1}$ light intensity and 16 h photoperiod for 5 days.

3.7 Vernalization and Transfer to Soil

1. Transfer the Magenta boxes to a vernalization growth chamber for 5–7 weeks, depending on the vernalization requirements of the genotypes, under the same conditions as described in Sect. 3.1.2 (Fig. 3G).
2. After vernalization, transfer the Magenta boxes to a growth chamber for hardening, set at 14/12°C (day/night) with 180 µmol $m^{-2}s^{-1}$ provided by LED lamps and 70% humidity for five days (Fig. 3H).
3. Transfer the rooted plants to small pots (8 cm Ø × 6 cm height) with a mixture of peat:vermiculite (3:2). Water the plants with the Hoagland's working solution (Table 1) and place them inside a propagator for 1 week. Remove the lid of the propagator and water the plants with the same solution as needed (Fig. 3H).

3.8 Ploidy Analysis

1. On the day the plants are transferred to soil, cut pieces of young leaves and place them in a plastic box with wet filter paper and store at 4°C until further use. Cut pieces of leaves from seed-derived plants to use as controls.
2. Place a 0.5 cm^2 section of the collected leaf on the lid of a 3 cm Ø Petri dish. Add 100 µL of Cystain® UV ploidy solution and finely chop the leaf with a razor blade to release as many nuclei as possible. Add 300 µL of the same solution to homogenize the sample.
3. Filter the homogenate through a 30 µm CellTrics® filter and collect the nuclei in a 3.5 mL polypropylene tube. Add 1 mL of Milli-Q water and mix thoroughly.

4. Insert the tube into the flow cytometer loading port and analyze the DNA content. Fix the fluorescence peak of the control samples at 200 on the X-axis and infer the ploidy level of the samples by comparison with the control. A minimum of 6000 counts per sample should be performed.

3.9 Seed Production

1. Transplant the plants to 17 cm Ø × 15.5 cm height pots with a peat:vermiculite:fine sand mix fertilized as described in Sect. 3.1.1.
2. Transfer the plants to a greenhouse at 25/20°C with a 16 h photoperiod.
3. Water the plants with the diluted modified Hoagland's stock solution once a week for the first 2 weeks, and then with water as needed.
4. Fertilize the plants five times before heading, as described in Sect. 3.1.3 (*see* **Note 13**).

4 Notes

1. To prepare the growth regulator stock solutions (1 mg/mL): Weigh 50 mg of the corresponding growth regulator on an analytical balance, put it in a small beaker with a magnetic bar, stir, and add 1 N HCl or 1 N NaOH (as indicated in sect. 2.5), drop by drop, until it is completely dissolved. Then, add Milli-Q water very slowly to bring the final volume to 50 mL.
2. Dissolve the mannitol in less than half of the final volume. For example, to prepare 1 L of medium, add 127.54 g mannitol to 400 mL of Milli-Q water, heat it at 50°C (avoid overheating), add 5.88 g of $CaCl_2 \cdot 2H_2O$, and then bring the volume to 500 mL with Milli-Q water.
3. This medium contains coconut milk, which is prepared in the laboratory from mature coconuts. Select heavy coconuts with high milk content. Poke a hole through the two softest of the three eyes of each coconut with a clean screwdriver. Pour the liquid from each coconut separately into a glass jar, making sure it smells good. Mix the milk from different coconuts and boil for 10 min to precipitate the proteins. Allow the solution to cool to RT and pass through a paper filter. Subsequently, aliquot 50 mL into 100 mL plastic bags and store at −20°C.
4. The final concentration of $CuSO_4 \cdot 5H_20$ in our modified J25-8 Reg is 3.825 mg/L, whereas the amount of this compound in the original J25-8 is 0.025 mg/L. Therefore, an additional amount of 3.8 mg/L of $CuSO_4 \cdot 5H_20$ is added to this medium.

5. It is extremely important that the mother plants grow free from pests and fungal infections, as the application of pesticides or fungicides reduces significantly the efficiency of ME.
6. The selection of the spikes at the correct stage of microspore development is one of the most critical points to achieve high efficiency in ME. The stage of development is determined by the position of the nucleus in the microspore. In the early uninucleate stage, the nucleus is close to the operculum (Fig. 2B: 1b); in the mid uninucleate stage, the nucleus is halfway between the pore and the opposite side (Fig. 2B: 1a); in the mid-late stage, the nucleus is close to the opposite side of the operculum (Fig. 2B: 2a, 2b); in the late stage, the nucleus is located on the opposite side of the operculum (Fig. 2B: 3a, 3b); and in bicellular pollen, a vegetative nucleus and a generative nucleus can be observed (Fig. 2B: 4a, 4b). To establish a correlation between the stage of the microspore development and the distance between the ligule of the flag leaf and the second leaf, four to five spikes of different distances are selected from each batch of plants and genotype for microspore DAPI staining (Fig. 2A, B). Since the optimal stage for anther culture is mid-late or late uninucleate, spikes with a 5.5–7 cm distance between the ligule of the flag leaf and the second leaf for genotype DH6148 should be harvested (Fig. 2).
7. Microspore developmental stage among the flowers varies within a spike. Therefore, the selection of the flowers at the correct developmental stages should be performed in the laboratory. The color and position of the anther tips relative to the palea correlate with the microspore developmental stage. Thus, the anthers should be pale green (Fig. 2B: 2, 3), neither too bright (Fig. 2B: 1), nor too dark (Fig. 2B: 4). In addition, the anthers tips should be located in the upper third of the palea (mid-late or late uninucleate) (Fig. 2B: 2, 3). The flowers with the anther tips below the upper third are discarded (binucleate stage) (Fig. 2B: 4).
8. During the stress treatment, anthers change their color and morphology, becoming white to yellow and swelling at different degrees. Whitish or yellowish swollen anthers are selected for culture (Fig. 3B).
9. There is a large variation in embryo quality, which is dependent on the genotype. Mature embryos are characterized by having a well-developed embryo axis and a white color due to starch accumulation in their cells. Embryos lacking a clear embryo axis are considered embryo-like structure.
10. Embryo development is asynchronous, therefore mature embryos or embryo-like structures are transferred to regeneration medium every 12–14 days.

11. Two different regeneration media were tested: FHG Reg [16] and the J25-8 Reg (Table 2). Similar regeneration and green plants rates were obtained in four barley genotypes; however, these variables were significantly higher in the genotype DH6148. Furthermore, plants regenerated on J25-8 Reg medium had larger size and a better root system in all five genotypes (Fig. 3F), ensuring higher survival rates during acclimatization (unpublished data).
12. The regeneration of albino plants is one of the bottlenecks in some barley genotypes. This factor is controlled by both genetic and physiological factors [14, 20]. Albino plants are discarded, as they are unable to survive after being transferred to soil.
13. If possible, transfer the plants to the greenhouse at the end of winter to avoid high temperatures during the seed-filling stage, which can cause shriveled grains. During winter, supplemental lighting with Philips Son-T AGRO lamps (400 W) is used to provide 300 μmol $m^{-2}s^{-1}$ light intensity.

Acknowledgments

We would like to thank Dr Eyal Fridman from the Volcani Agricultural Research Organization (ARO, Israel) for providing seeds from Noga cultivar and the three CMPP (BC2DH) lines. This research was supported by Projects PID2020-119540RB-I00 funded by MICIU/AEI/10.13039/501100011033, the grant PID2023-150342OB-100 funded by MICIU/AEI/10.13039/501100011033 and by ERDF/UE, and a grant of Reference Research Groups recognized by Diputación General de Aragón, Spain (Group A08_23R). Aimar Navarro was supported by the grant PRE2021-099847 funded by MICIU/AEI/10.13039/501100011033. Nuria Isach was supported by the grant JAEICU_24_01253 (linked to Conexión Trigo, WheatNet) funded by the Spanish National Research Council (CSIC).

References

1. Food and Agriculture Organization of the United Nations. FAOSTAT statistical database. https://www.fao.org/faostat/en/#home. Accessed 3 Jan 2025
2. Friero I, Martínez-Subirà M, Romero MP, Moralejo M (2024) Improving functional and nutritional profiles of barley flours with diverse starch types through pearling. Food Chem 460:140611
3. Raj R, Shams R, Pandey VK et al (2023) Barley phytochemicals and health promoting benefits: a comprehensive review. J Agric Food Res 14:100677
4. Rasane P, Singh J, Kaur S et al (2024) Barley. In: Singh J, Kaur S, Rasane P (eds) Cereals and nutraceutics. Springer Nature, Singapure, pp 23–46
5. Cammarano D, Ceccarelli S, Grando S et al (2019) The impact of climate change on barley

yield in the Mediterranean basin. Eur J Agron 106:1–11
6. Weyens J (2009) Barley and wheat doubled haploids in breeding. In: Touraev A, Forster BP, Jain Mohan (eds) Advances in haploid production in higher plants. Springer-Science +BussinessMedia BV, United Kingdom, pp 179–187
7. Forster BrianP, Kasha KJ, Touraev A (2007) The resurgence of haploid in plant breeding. Trends Plant Sci 12(8):368–375
8. Hale B, Ferrie AM, Chellamma S et al (2022) Androgenesis-based doubled haploidy: past, present, and future perspectives. Front Plant Sci 12:751230. https://doi.org/10.3389/fpls.2021.751230
9. Patial M, Chauhan R, Chaudhary HK et al (2023) Au-courant and novel technologies for efficient doubled haploid development in barley (*Hordeum vulgare* L.). Crit Rev Biotechnol 43(4):575–593
10. Kasha KJ, Kao KN (1970) High frequency haploid production in barley (*Hordeum vulgare* L). Nature 225:874–876
11. Devaux P, Kasha KJ (2009) Overview of barley doubled haploid production. In: Touraev A, Forter BP, Jain M (eds) Advances in haploid production in higher plants. Kluwer Academic Press, Netherlands, pp 47–63
12. Soriano M, Li H, Boutilier K (2013) Microspore embryogenesis: establishment of embryo identity and pattern in culture. Plant Reprod 26(3):181–196. https://doi.org/10.1007/s00497-013-0226-7
13. Castillo AM, Cistué L, Vallés MP, Soriano M (2010) Chromosome doubling in monocots. In: Touraev A, Forter BP, Jain M (eds) Advances in haploid production in higher P. Kluwer Academic Press, Netherlands, pp 329–338
14. Muñoz-Amatriain M, Castillo AM, Cistué L, Vallés MP (2008) Identification and validation of QTLs for green plant percentage in barley (*Hordeum vulgare* L.) anther culture. Mol Breed 22:119–129. https://doi.org/10.1007/s11032-008-9161-y
15. Makowska K, Oleszczuk S, Zimny A et al (2015) Androgenic capability among genotypes of winter and spring barley. Plant Breed 134:668–674
16. Cistué L, Vallés MP, Echávarri B et al (2003) Barley anther culture. In: Malupzinski M, Kasha K, Foster B (eds) Doubled haploid production in crop plants. a manual. FAO/IAEA, Austria, Viena, pp 29–34
17. Jensen CJ (1977) Monoploid production by chromosome elimination. In: Reinert J, Bajaj YPS (eds) Applied and fundamental aspects of plant cell, tissue and organ culture. Springer-Verlag, Berlin, pp 299–330
18. Bodenheimer S, Bdolach E, Be'ery A et al (2025) Cytonuclear analysis of barley spike traits using a cytoplasm-aware population. bioRxiv, 2025-04. https://doi.org/10.1101/2025.04.08.647843v1
19. Hunter CP (1988) Plant regeneration from microspores of barley, *Hordeum vulgare* L. PhD Thesis, Wye College University of London, London
20. Chen XW, Cistué L, Muñoz-Amatriaín M et al (2007) Genetic markers for doubled haploid response in barley. Euphytica 158:287–294

Chapter 12

Monitoring Barley Microspore Embryogenesis: In Vitro Culture and Immunolocalization Analyses

Yolanda Pérez-Pérez and Pilar S. Testillano

Abstract

The agricultural sector has a strong interest in developing new barley varieties to enhance productivity, quality, and resistance to pests and diseases. The production of doubled haploid (DH) plants is crucial in genetic improvement, as it enables the rapid and precise development of homozygous lines, accelerating the creation of enhanced varieties and consolidating new genetic combinations within a single generation. Microspore embryogenesis provides the most efficient tool for generating DH embryos and plants in barley (*Hordeum vulgare* L.). An appropriate in vitro protocol for inducing microspore embryogenesis must be accompanied by effective methods to monitor the cellular events and molecular determinants underlying the process. This combination is essential to elucidate the regulatory mechanisms involved and enable efficient manipulation. This chapter first presents a detailed, updated protocol for inducing microspore embryogenesis in barley using a 4°C cold stress treatment. It then describes methodologies for sample processing from embryogenic cultures to analyze cellular organization changes and localize key molecular determinants, such as auxin, through immunofluorescence assays. The described in vitro method incorporates several modifications to enhance culture efficiency, including reduced maltose and mannitol volumes during microspore isolation, optimized gradient preparation for clearer separation of viable microspores, improved cleaning to prevent aggregate formation that interferes with reprogramming, lower cell density in culture, or the use of larger Petri dishes to promote embryo differentiation. The cellular analysis protocol encompasses fixation, preembedding in gelatin, acrylic resin embedding, semithin sectioning, cytochemical staining, and immunofluorescence. This approach enables precise observation of cellular structures under light microscopy, offering detailed insights into embryogenic progression and allowing to carry out immunofluorescence studies using a wide range of antibodies. By integrating protocols for in vitro induction and cellular analysis, this chapter provides a comprehensive guide for monitoring barley microspore embryogenesis. It assists researchers in optimizing culture outcomes and advancing cellular studies, including structural analyses and the localization of key molecules like auxin.

Keywords *Hordeum vulgare*, Microspore embryogenesis, Cell reprogramming, Doubled haploid, Microspore culture, Plant breeding, Light microscopy, Sample processing, Semithin sectioning, Immunofluorescence

Ana María Castillo et al. (eds.), *Doubled Haploid Methods in Cereals*, Methods in Molecular Biology, vol. 3029,
https://doi.org/10.1007/978-1-0716-5245-9_12,

1 Introduction

The development of new barley varieties has gained particular importance within the agricultural sector, motivated by the need to produce high-quality crops with increased yield and enhanced resistance to pests and diseases. Barley is one of the most important cereal crops in the world. It is widely grown and ranks among the top ten crop plants in the world. As a globally significant cereal, barley holds forth in world production, with a total output of 155 million tons [1]. This crop is primarily used as animal feed, especially in colder climates where other cereals cannot thrive. Additionally, barley plays a crucial role in malt production, serving as a key ingredient in the brewing industry and the distillation of alcoholic beverages such as whisky and gin. The ability to obtain new barley varieties through advanced breeding techniques is, therefore, essential for meeting the growing demands of both the agricultural and brewing sectors.

In vitro embryogenesis is a powerful technique with significant applications in research, crop plant commercial micropropagation, and breeding programs. Among its most important contributions is the generation of doubled haploids (DH), which significantly reduces the time required to develop homozygous lines in comparison to traditional breeding methods. This technique enables the production of homozygous lines in a single generation, providing a rapid and efficient pathway to fix new genetic variability. However, finding efficient conditions for the in vitro production of DH plants remains a significant challenge in many agricultural and forestry species of interest.

In 1970, a novel technique for producing DH barley plants was developed through the interspecific hybridization of *Hordeum vulgare* and *Hordeum bulbosum*. Later, DH barley plants were successfully obtained through anther [2–4] and isolated microspore cultures [4, 5]. In both methods, the process entails the regeneration of DH plantlets from microspores. These microspores were initially destined to develop into pollen grains but undergo a reprogramming process that enables them to follow an embryogenic pathway upon the application of a stress treatment (Fig. 1). Among all methods for DH production in barley, the isolated microspore culture has proven to be the most efficient for large-scale production [6]. Additionally, this species exhibits a spontaneous diploidization rate of nearly 90% [7], which provides a significant advantage by eliminating the need for antimitotic or diploidizing agents.

This report describes, on the one hand, a detailed and updated protocol, originally outlined by Kumlehn and collaborators [8], to induce microspore embryogenesis in isolated microspore cultures using a cold stress treatment at 4°C, incorporating some modifications that improved culture efficiency. The protocol describes the selection of the most responsive microspore developmental stage to

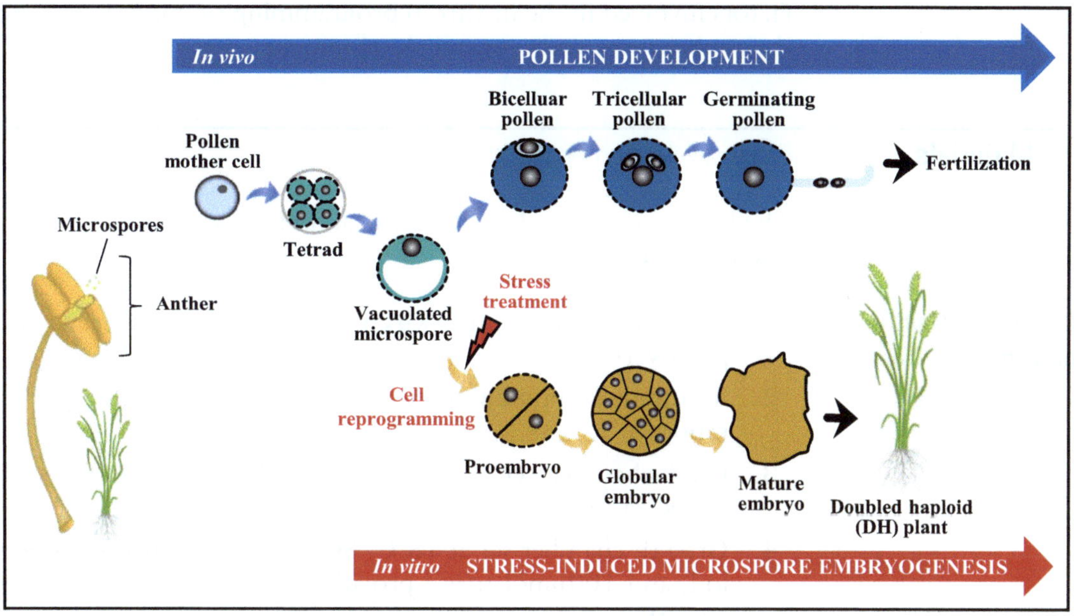

Fig. 1 Microspore developmental pathways in barley. Illustration depicting the main stages of pollen development (pathway occurring in vivo within anthers) and microspore embryogenesis (pathway occurring in vitro following stress-induced cell reprogramming)

initiate the culture, followed by the pretreatment of spikes at 4°C, and the in vitro culture of microspores in an appropriate medium under optimal conditions for embryogenesis induction. This protocol was initially developed for the winter barley variety Igri, which shows a good embryogenesis response, but recent results from our group indicate its effectiveness in other barley varieties with lesser embryogenic response, including Reinette and Hop.

On the other hand, this chapter also describes a detailed protocol for the processing of culture samples for light microscopy analysis. The procedures outlined in this protocol encompass fixation, preembedding in gelatin, resin embedding, semithin sectioning, and cytochemical staining of the different developmental stages of the in vitro isolated microspore culture.

Recent studies have described that the initiation and progression of in vitro induced embryogenesis involve a complex network of factors, such as hormonal regulation, cell wall remodeling, cell death, and epigenetic control [9–16]. This protocol, in addition to enabling a thorough cellular analysis of the various stages of in vitro microspore culture, can also be applied to immunofluorescence localization studies of specific epitopes of interest, such as auxin, pectins, autophagy-related proteins, cysteine proteases, or epigenetic marks, as demonstrated in previous studies [9–16]. Therefore, the protocol described in this report, represents a useful tool for monitoring and understanding, through optical microscopy techniques, the multiple factors involved in the in vitro reprogramming of microspores and their subsequent embryogenic development.

2 Materials

2.1 Plant Material

The barley genotypes Igri, Reinette, and Hop were tested with this microspore culture protocol.

2.2 Donor Plant Growth

1. Trays with cells 4 cm Ø × 4 cm height for sowing seeds and plant vernalization.
2. Substrate composed of peat, sand, and vermiculite (2:1:1 ratio).
3. Cold chamber set at 4°C under dark conditions for seed vernalization.
4. Growth chamber set at 18°C, 80% humidity, 16 h photoperiod, and 100 μmol $m^{-2}s^{-1}$ provided by LED lamps, for plant growth.
5. Pots 18 cm Ø × 15 cm height for plant growth.
6. Greenhouse with a controlled temperature of 19°C, 12 h photoperiod, and 200 μmol $m^{-2}s^{-1}$ provided by fluorescent tubes.
7. Liquid NPK (20-20-20) universal fertilizer.

2.3 Laboratory Material

1. Sterile Petri dishes (60 × 15 mm and 100 × 20 mm).
2. Sterile Whatman filter paper.
3. Parafilm.
4. Waring blender container (Eberbach).
5. Funnel with a 100 μm nylon filter (Scharlab) into a beaker.
6. Sterifil Aseptic System and Holder unit (Millipore) with a 0.22 μm polyvinylidene fluoride membrane filter.
7. Bunsen burner.
8. Tweezers.
9. Scissors.
10. Eppendorf tubes (1.5 mL).
11. Falcon tubes (15 and 50 mL).
12. Glass bottle (Pyrex).
13. Beaker.
14. Glass pipettes.
15. Pasteur pipettes.
16. Micropipettes for handling volumes of 10 and 100 μL.
17. Pipette tips.

18. Neubauer counting chamber.
19. Gelatin capsules (size 2, Agar Scientific).
20. Rak for gelatin capsules.
21. Glass slides, multiwell slides, and coverslips.
22. Toothpicks.
23. Single-edged razor blades (RB-009).
24. Glass knife (Taab Superglass).
25. Minisart® Syringe Filters (0.22 μm, Sartorius).
26. Syringe.
27. Nitrile gloves.
28. Safety glasses.
29. Lab coat.

2.4 Equipment

1. Laminar flow hood.
2. Swinging-bucket centrifuge.
3. Magnetic stirrer.
4. Waring blender (Eberbach).
5. Incubator at 25°C.
6. Refrigerator set at 4°C.
7. Freezer set at −20°C.
8. Light microscope and fluorescence microscope, equipped with a UV light for DAPI visualization (excitation wavelength range: 340–360 nm). Fluorescence emission was collected within the 450–460 nm spectral detection window. A digital camera (DFC42 C, Leica Microsystems) coupled to the microscope was used to acquire images.
9. Stereomicroscope.
10. Light inverted microscope.
11. Autoclave.
12. Precision balance.
13. pH meter.
14. Vacuum pump (Boeco R-300).
15. Swinging-bucket centrifuge (Eppendorf 5702 RH).
16. Rotary shaker.
17. Ultramicrotome (LKB Ultramicrotome III).
18. Knifemaker (LKB Bromma 7800).
19. Transilluminator.
20. Fume cupboard (Burdinola ST-1500).

21. Confocal laser scanning microscope, equipped with appropriate laser excitation lines for fluorophore visualization, DAPI was excited using a laser excitation wavelength range: 340–360 nm), and fluorescence emission was collected within the 450–460 nm spectral detection window. Alexa Fluor 488 was excited using a laser in the 495–505 nm range, and fluorescence emission was collected within the 510–530 nm spectral detection window.

2.5 Solutions and Culture Medium for In Vitro Culture

1. Autoclaved distilled water.
2. Sterilization solution (0.18% sodium hypochlorite): Take 5 mL of commercial bleach (3.7% sodium hypochlorite) to 100 mL of distilled water.
3. DAPI (4′,6-Diamidino-2-phenylindole) working solution (10 μg/mL) in distilled water with 1% (*v/v*) Triton X-100 (*see* **Note 1**).
4. 0.4 M mannitol solution: Prepare the solution by dissolving 72.9 g mannitol in 1 L of distilled water and autoclave at 121°C for 15 min.
5. 0.55 M maltose solution: Prepare the solution by dissolving 1.98 g maltose in 100 mL of distilled water and autoclave at 121°C for 15 min.
6. KBP medium previously described by Kumlehn et al. [8] (Table 1). To simplify the process, use previously prepared and stored stock solutions to formulate the KBP culture medium, as it contains a large number of components. Prepare stock solutions of macronutrients 10×, micronutrients 1000×, vitamins 1000×, L-glutamine 100×, and BAP 100×. Into a 1 L beaker, add the appropriate volumes of the macronutrient (100 mL), micronutrient (1 mL), vitamin (1 mL), L-glutamine (10 mL), and BAP (10 mL) stock solutions, followed by the addition of 100 mg myo-inositol (a vitamin not included in the stock solution), 438 mg of glutamine, and 90 g of maltose. Adjust the medium to a final volume of 1 L with sterile distilled water, and the pH to 5.8.

Filter-sterilize the culture medium using the Sterifil Aseptic System and Holder unit (Millipore) connected to a vacuum pump (Boeco R-300). Store the filtered medium in a sterile glass bottle (Pyrex) and keep at 4°C until use, for a maximum of 3 weeks.

2.6 Solutions for Sample Processing and Immunocytochemistry

1. Phosphate-Buffered Saline (PBS) (1×): 20 mM phosphate buffer pH 6.8 and 150 mM NaCl).
2. 4% (*w/v*) paraformaldehyde (PF) in 1× PBS.
3. 0.1% (*w/v*) PF in 1× PBS.
4. 15% (*w/v*) gelatin (Honeywell Riedel-de Haen®) in 1× PBS.

Table 1
Composition of the KBP culture media for isolated microspore culture in barley

Macronutrients (mg/L)		**Vitamins (mg/L)**	
KNO_3	2022 mg/L	Retinol	0.01 mg/L
$CaCl_2 \times 2H_2O$	441 mg/L	Thiamine-HCl	1 mg/L
KH_2PO_4	340.2 mg/L	Nicotinic acid	1 mg/L
$MgSO_4 \times 7H_2O$	246 mg/L	Riboflavin	0.2 mg/L
NH_4NO_3	80 mg/L	Ca-Pantothenate	1 mg/L
Micronutrients (mg/L)		Folic acid	0.4 mg/L
NaFe(III)-EDTA	27.5 mg/L	Pyridoxine HCl	1 mg/L
$MnSO_4 \times 4H_2O$	11.2 mg/L	Cobalamin B12	0.02 mg/L
H_3BO_3	3.1 mg/L	Ascorbic acid	2 mg/L
$ZnSO_4 \times 7H_2O$	7.2 mg/L	Calciferol	0.01 mg/L
$NaMoO_4 \times 2H_2O$	0.12 mg/L	Biotin	0.01 mg/L
$CuSO_4 \times 5H_2O$	0.025 mg/L	Choline chloride	1 mg/L
$CoCl_2 \times 6H_2O$	0.024 mg/L	P-Aminobenzoic acid	0.02 mg/L
KI	0.17 mg/L	Myo-Inositol	100 mg/L
Amino acids, hormones, and carbohydrates (mg/L)			
L-Glutamine	438 mg/L		
BAP	0.9 mg/L		
Maltose- H_2O	90,000 mg/L		
pH 5.8			

5. Acetone solutions (30%, 50%, 70%, 90%, and 100% *v/v* in 1× PBS).
6. Historesin Plus infiltration solution (Technovit® 8.100) prepared by dissolving 60 mg Hardener I in 10 mL Basic solution.
7. Polymerization solution prepared by mixing 160 μL Hardener II with 3 mL infiltration solution.
8. 1% (*w/v*) toluidine blue (Merck) in distilled water, filtered using a 0.22 μm Minisart® Syringe Filter.
9. APTES ((3-aminopropyl) triethoxysilane) (Sigma).
10. 5% (*w/v*) bovine serum albumin (BSA) (Sigma).
11. Monoclonal anti-IAA mouse antibody (Sigma A0855), diluted 1:1000 in 1% BSA in 1× PBS.

12. Secondary antibody Alexa Fluor 488-labeled anti-mouse IgG (Molecular Probes), diluted 1:25 in 1% BSA in 1× PBS.
13. Mowiol® 4-88 (Sigma-Aldrich).
14. Eukitt (Kindler Gmbh & Co) mounting medium.
15. Nail polish.

3 Methods

3.1 Donor Plant Growth Conditions

1. Sow the seeds in trays with 4 cm Ø × 4 cm height cells, one seed per cell, using a substrate of peat, sand, and vermiculite (2:1:1 ratio) (*see* **Note 2**).
2. Keep the seeds in a cold chamber at 4°C under dark conditions for 4 weeks. Water only once with abundant water at the time of sowing and cover the seeding tray with perforated cling film (*see* **Note 3**).
3. After the vernalization period, transfer the seeds to a growth chamber set at 18°C, 80% humidity, and 16 h photoperiod, with moderate watering three times per week. Seedlings emerge in approximately 5 days.
4. After 1 month, transplant the plants into 18 cm Ø × 15 cm height pots with the same substrate, containing one or two seedlings per pot.
5. Transfer the plants to a greenhouse with a controlled temperature of 19°C and 12 h photoperiod, and water abundantly three times a week. Add a liquid NPK (20-20-20) universal fertilizer to one of the weekly waterings, following the product instructions. Spike production begins after 8–10 weeks.
6. Harvest spikes from each plant over a 2–3-month period following the initial flowering. After this period, spike production and quality decline, negatively affecting microspore reprogramming success.

3.2 Spike Selection and Sterilization

Selecting spikes with the microspores at the optimal developmental stage is crucial for the successful induction of embryogenesis.

1. Collect spikes and select them based on morphology to identify those containing microspores at the vacuolated stage, optimal for embryogenesis. For the Igri and Reinette genotypes, spikes with microspores at the vacuolated stage have an awn length of 0.1–1 cm (Fig. 2A). In contrast, in the Hop genotype, spikes containing vacuolated microspores are those with an awn length of at least 2 cm below the flag leaf ligule (Fig. 2B). The correlation between awn length and microspore developmental stage in the three cultivars has been established and is presented in Table 2.

2. Confirm the developmental stage by collecting anthers from the central spike region, squashing them in 10 μg/mL DAPI on a microscope slide. Under a fluorescence microscope with UV light, DAPI-stained DNA reveals the microspore nucleus (Fig. 2C). Select more mature spikes if the nucleus is central, or younger spikes if there is more than one nucleus.
3. Clean the collected suitable spikes before sterilization by removing the awns, the base up to the first node, the flag leaf, and the peduncle (Fig. 3A–C) (*see* **Note 4**).
4. Sterilize spikes in 0.18% sodium hypochlorite solution for 20 min with constant agitation, followed by four washes with sterile distilled water precooled to 4°C, 10 min each. To prevent damage, place spikes in perforated tubes or strainers during sterilization (Fig. 3C, D) (*see* **Note 5**).

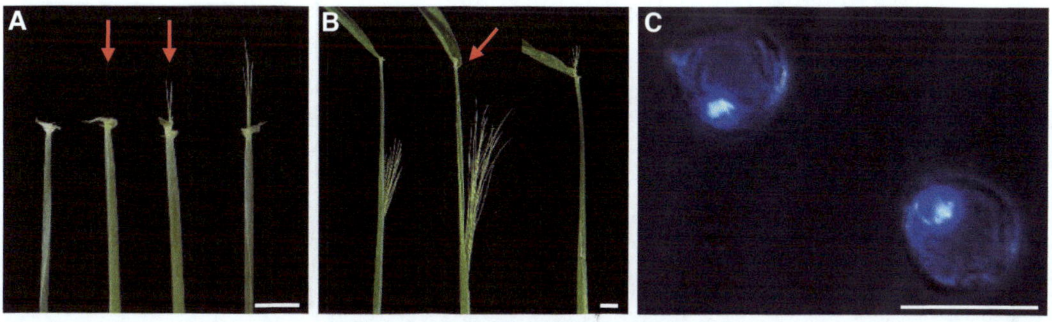

Fig. 2 Selection of the optimal microspore developmental stage for initiating microspore culture. (**A**) Immature barley spikes of cv. Igri at different developmental stages. (**B**) Immature barley spikes of cv. Hop at different developmental stages. (**C**) Vacuolated microspores stained with DAPI. Red arrows indicate the spikes with the corresponding range of awn length for the most responsive stage, containing vacuolated microspores. Bars represent 1 cm (**A, B**), 50 μm (**C**)

Table 2
Correlation between awn length and microspore developmental stage in *Hordeum vulgare* cvs. Igri, Reinette and Hop

Awn length		
Igri and Reinette	**Hop**	**Microspore developmental stage**
Not visible	>2 cm below the flag leaf ligule	Pollen mother cell
0.1–1.0 cm above the flag leaf ligule	1–2 cm below the flag leaf ligule	Vacuolated microspore
1.0–2.0 cm above the flag leaf ligule	0.1–1.0 cm above the flag leaf ligule	Bicellular pollen
>2.0 cm above the flag leaf ligule	>1.0 cm above the flag leaf ligule	Tricellular pollen

3.3 Cold Stress Treatment

1. Place sterilized spikes in 100 × 20 mm Petri dishes on moistened grade 1 filter paper (Whatman), 10–15 spikes per dish without overlapping (Fig. 3E) (*see* **Note 6**).
2. Seal the Petri dishes with Parafilm and store at 4°C for 21–24 days (Fig. 3F).

Fig. 3 Sterilization and treatment of the spikes. (**A**) Spikes collected at vacuolated microspore stage. (**B**) Opening of the sheath to extract the spike. (**C, D**) Cleaned spikes ready for sterilization in a perforated tube. (**E**) Sterilized spikes placed in a Petri dish on wet Whatman paper. (**F**) Petri dishes sealed with Parafilm and stored at 4°C

3.4 Isolated Microspore Culture

The microspore culture of barley spikes pretreated at 4°C is carried out as follows:

1. Initiate the microspore isolation in a sterile environment within a laminar flow hood (*see* **Note 7**).
2. Cut cold pretreated spikes (Fig. 4A) into approximately 1 cm fragments and place in a Waring Blender (Eberbach) with 20 mL of 0.4 M mannitol for disaggregation (*see* **Note 8**). Spikes should be processed in batches of 10–15 at a time (Fig. 4B–D).
3. Filter the homogenate through a 100 μm nylon mesh (Scharlab), supported by a funnel structure, into a beaker to remove larger tissue fragments (Fig. 4E) (*see* **Note 9**).
4. Transfer the filtrate to 50 mL Falcon tubes (Fig. 4F) and centrifuged at 100 × *g* for 10 min at 4°C using a swinging-bucket rotor (5702 RH, Eppendorf).
5. After centrifugation, discard the supernatant, and prepare a density gradient to isolate alive, vacuolated microspores. For the gradient, add 4 mL of 0.55 M maltose solution to one of the tubes combining all pellets obtained from the previous centrifugation into a single 15 mL Falcon tube. Carefully add a 1 mL layer of maltose to the previous mixture (4 mL of maltose with all the combined pellets) without mixing, followed by 1.5 mL of 0.4 M mannitol solution. It is important to ensure that the different layers remain unmixed to form a proper gradient. Centrifuge at 100 × *g* for 10 min at 4°C (*see* **Note 10**).
6. After centrifugation, the interface of the gradient contains the viable, vacuolated microspores, which are retained in the 1.5 mL of mannitol due to the buoyancy provided by their large vacuole (Fig. 4G). Transfer this fraction, often referred to as "the ring," using a Pasteur pipette to a new 15 mL Falcon tube for the subsequent washing process (Fig. 4H).
7. Wash the collected interface with 10 mL of 0.4 M mannitol and centrifuge at 100 × *g* for 10 min at 4°C to remove any remaining maltose. Then, wash the pellet with 10 mL of KBP culture medium (Table 1) to ensure a cleaner cell suspension. Centrifuge again at 100 × *g* for 10 min at 4°C and discard the supernatant. Then, resuspend the pellet in 10 mL of KBP medium, mixed well, and take a 100 μL aliquot for cell counting using a Neubauer counting chamber.
8. Calculate the final volume of KBP medium to achieve a cell density of 1.1×10^5 microspores/mL, the optimal density for barley microspore culture. The final culture medium volume is determined using the formula: $Ci \times Vi = Cf \times Vf$, where:

   Vf: Final volume of culture medium required to reach the final concentration Cf.

Fig. 4 Steps for the in vitro culture of isolated barley microspores. (**A**) Spikes pretreated at 4°C. (**B, C**) 1 cm spike fragments in the blender. (**D**) Homogenate obtained after blending. (**E, F**) Filtration of the homogenate and its collection in a 50 mL tube, respectively. (**G**) Tube with the maltose and mannitol gradient after centrifugation. The red arrow points to the interface ("the ring") containing viable isolated microspores in mannitol. (**H**) Suspension of viable isolated microspores in 10 mL of KBP media, used for cell counting in the Neubauer chamber

Cf: Optimal cell concentration (1.1×10^5 microspores/mL).

Vi: Initial volume (10 mL).

Ci: Cell concentration obtained from the count (microspores/mL) in the Neubauer counting chamber.

Ci = (total number of microspores in the eight quadrants/number of quadrants) × (1 quadrant/0.0001 cm^3) × (1 cm^3/1 mL).

9. Plate the microspores in sterile 60 × 15 mm Petri dishes with a minimum volume of 1.5 mL per dish to ensure the bottom of the plate is covered, preventing drying and allowing proper

embryo development. Seal dishes with Parafilm and incubate them in the dark at 25°C for embryogenesis induction (*see* **Note 11**).

3.5 Culture Monitoring

In order to detect the reprogramming rate of microspores in response to the stress treatment and monitor the proper development of the embryos, periodic observations of the cultures are conducted using an optical microscope.

1. After 4–6 days in culture, reprogrammed microspores underwent continuous divisions, forming multicellular structures still surrounded by the exine (a specialized microspore wall), known as proembryos. Under a stereomicroscope or a light inverted microscope, proembryos are identified by their larger size and denser cytoplasm compared to isolated microspores (Fig. 5A–D). From this stage, cells continue dividing rapidly, progressing through globular, transitional, scutellar, and finally coleoptilar stages of monocot embryo development (Fig. 5E–H). Barley cultures are quite asynchronous, with different developmental stages occurring simultaneously in the same dish. Main developmental stages that can be found in culture over time are detailed in Table 3.

After 20–30 days in culture, the embryos reach the coleoptilar stage, the most advanced embryo developmental stage. At this moment, the embryos can be germinated in vitro to produce fully developed plants.

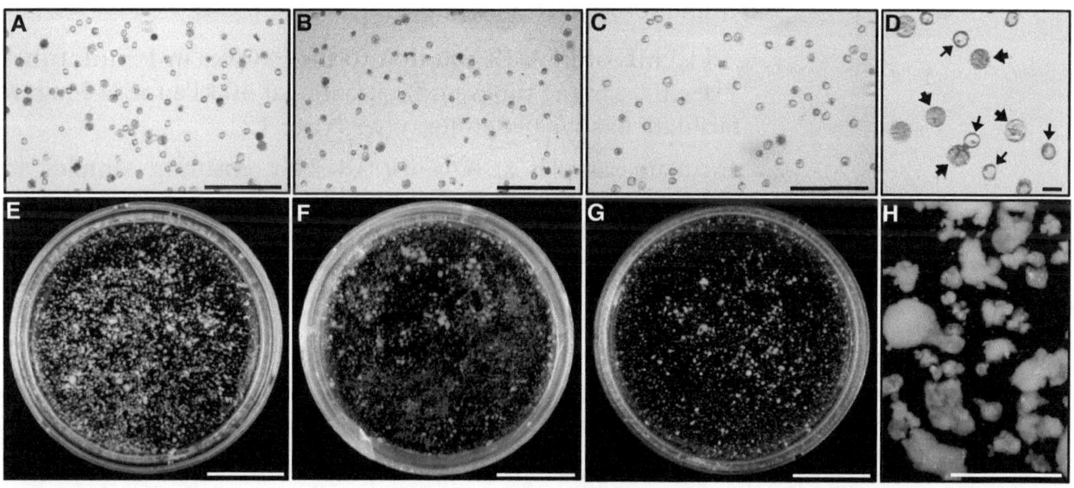

Fig. 5 Monitoring of barley cvs. Igri, Reinette, and Hop microspore culture. (**A, E**) Barley cv. Igri. (**B, F**) Barley cv. Reinette. (**C, G**) Barley cv. Hop. (**A–C**) Isolated barley microspore culture after 5 days, (**D**) detailed view at higher magnification. Thin arrows indicate nonresponsive vacuolated microspores, while thick arrows indicate proembryos (larger rounded structures with dense cytoplasm). (**E–G**) Petri dish containing microspore-derived embryos after 30 days, observed under stereomicroscope. (**H**) Detail at higher magnification of several embryos after 30 days in culture. Bars represent (**A–C**) 500 μm, (**D**) 50 μm, (**E–G**) 1 cm, (**H**) 0, 1 cm

Table 3
Developmental stages during in vitro culture of isolated barley microspores in the cultivars Igri, Reinette, and Hop

Developmental stage	Culture time (days)
Isolated vacuolated microspore	0
Proembryo	4–6
Globular embryo	7–9
Transitional scutellar embryo	10–20
Coleoptilar embryo	>20

3.6 Sample Collection for Microscopic Studies

Samples from various developmental stages of barley microspore cultures are collected for processing.

1. Transfer cultures in liquid medium from Petri dishes into 15 mL Falcon tubes.
2. Centrifuge the tubes at 620 × *g* for 5 min at 4°C using a 5702 RH swing-out rotor centrifuge (Eppendorf).
3. Remove the supernatant, leaving 1.5 mL to resuspend the pellet. Transfer it to a 1.5 mL tube and centrifuge again at 350 × *g* for 3 min at 4°C.
4. Discard the supernatant completely and proceed with fixation.

3.7 Fixation

Fixation is performed to preserve cellular structure for cytochemical and immunolocalization assays.

1. Add 1 mL of PFA 4% solution to the samples in 1.5 mL tubes. Place the sample tubes under vacuum at 40 kPa for 1–2 min to facilitate fixative penetration (*see* **Note 12**).
2. Incubate samples at 4°C for 18–24 h, ensuring samples are fully immersed.
3. After fixation, wash the samples three times in 1× PBS for 10 min each and stored in 0.1% PF in 1× PBS at 4°C to maintain fixation integrity (*see* **Note 13**).

3.8 Preembedding in Gelatin

Preembedding is performed for microspore cultures at the stages prior to coleoptilar embryo stages to facilitate handling during subsequent processing steps.

1. Centrifuge fixed samples, stored in 0.1% PF, at 1400 × *g* for 5 min at 4°C.
2. Discard the supernatant, and resuspend the pellet in 150 μL of 15% (*w/v*) gelatin (Honeywell Riedel-de Haen®) in 1× PBS preheated at 37°C.

3. Quickly, to prevent gelatin solidification, centrifuge the suspension at 1400 × *g* for 5 min at 37°C and then, cool on ice for 10 min to solidify the gelatin, preventing sample resuspension (*see* **Note 14**).
4. Store the gelatin blocks at 4°C for 24 h to ensure proper gelation, then unmold using a toothpick. Trim the excess gelatin (without the sample) from the upper layer using a single-edged razor blade (RB-009). Section the sample-containing area into approximately 3 mm fragments and store them in 0.1% PF at 4°C until further processing.

3.9 Acrylic Resin Embedding

Acrylic resins that polymerize at low temperatures are the best choice to optimize structural and antigenic preservation. Among them, Historesin Plus (Technovit® 8.100) is widely recognized for its effectiveness, delivering excellent results for this type of material. Its hydrophilic nature makes it particularly well-suited for general optical microscopy studies (*see* **Note 15**).

1. **Dehydration**
    - Wash the samples in 1× PBS on ice for 5 min, repeating three times to ensure complete removal of the fixative.
    - Dehydrate the samples sequentially in acetone solutions of increasing concentrations (30%, 50%, 70%, 90%, and 100% *v/v*) at 4°C, as outlined in Table 4.
2. **Infiltration**
    - Completely remove any residual 100% acetone and add 1 mL of infiltration solution (Historesin Plus, Technovit® 8.100) to the samples. Prepare the solution by dissolving 60 mg of Hardener I in 10 mL of Basic solution in amber glass, following the kit instructions (*see* **Note 16**).
    - Carry out the infiltration at 4°C for 48 h on a rotary shaker.
3. **Polymerization**
    - Perform polymerization using a rack adapted to size 2 gelatin capsules (Agar Scientific) placed on a transilluminator,

Table 4
Dehydration steps with acetone for acrylic resin embedding (Historesin)

Acetone concentration	Incubation time
Acetone 30%	1 h
Acetone 50%	1 h
Acetone 70%	Overnight
Acetone 90%	1 h
Acetone 100%	1 h (×3)

which facilitates sample visualization during the process (Fig. 6A). Carefully remove the sample from the tube using a toothpick and place at the bottom of the capsule (Fig. 6B).

- After positioning the sample and ensuring it remains in the desired orientation, fill the capsule with resin polymerization

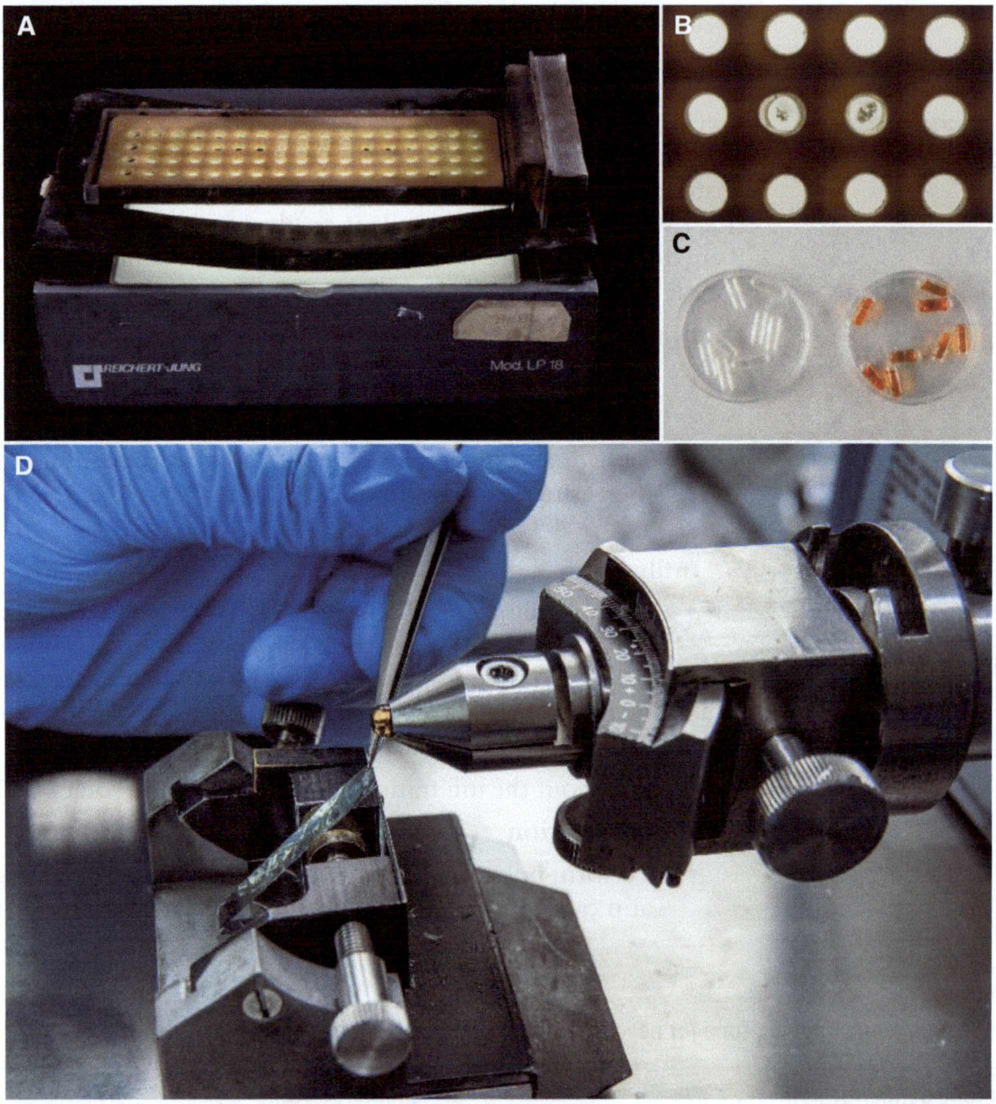

Fig. 6 Processing of culture samples for microscopic analyses: resin polymerization into capsules and semithin sectioning. (**A**) Rack adapted to gelatin capsules placed on a transilluminator. (**B**) Close-up of samples positioned inside the capsule for polymerization. (**C**) Empty gelatin capsules on the left and resin blocks of culture samples after polymerization, on the right. (**D**) Close-up view showing the positioning of the resin block and the glass knife in the ultramicrotome to obtain the sections

solution, prepared by mixing 160 μL of Hardener II with 3 mL of infiltration solution (*see* **Note 17**).

- Seal the capsules with their gelatin cups and polymerize the samples in the rack at 4°C for 24 h, protected from light (Fig. 6C) (*see* **Note 18**).

3.10 Semithin Sectioning

1. Section resin blocks at 2 μm thickness using an ultramicrotome equipped with a glass knife (Taab Superglass) preshaped using a knifemaker (Fig. 6D).
2. Stretch sections on distilled water droplets on slides, allow them to dry at room temperature (RT) for 24 h, and store until staining.

3.11 Cytochemical Staining with Toluidine Blue

1. Stain sections with 1% (*w/v*) toluidine blue in distilled water for 1–2 min, ensuring that the sections are completely covered by the staining solution (*see* **Note 19**).
2. Remove the excess stain by washing the sections with distilled water using a Pasteur pipette. Dry the sections, protected from dust, before being mounted with Eukitt.

3.12 Mounting with Eukitt

Mount the slides with Eukitt (Kindler Gmbh & Co) to enhance optical clarity and preservation.

1. Apply a generous amount of mounting medium to one side of a coverslip and place it quickly over the section. Use a toothpick to gently press the coverslip, ensuring an even distribution of Eukitt over the section and eliminating any air bubbles that could interfere with subsequent visualization.
2. Allow the mounted slides to dry for 5–15 min before microscopic analysis.

3.13 Observation with Light Microscopy

1. Observe the mounted slides under a light microscope equipped with a digital camera.
2. Perform observations under brightfield mode to assess sample quality and cellular structures. Microscopic analyses of toluidine blue-stained sections enable the monitoring of changes in the cellular organization during key developmental stages of microspore embryogenesis(Fig. 7A–D).

3.14 Immunolocalization of Auxin

Resin sections from microspore embryogenesis samples processed as described before, can also be used for immunofluorescence assays with different antibodies. Here, we describe the protocol for immunolocalization of auxin.

1. Use semithin sections (2 μm thickness) of microspore cultures mounted on multiwell slides coated with APTES

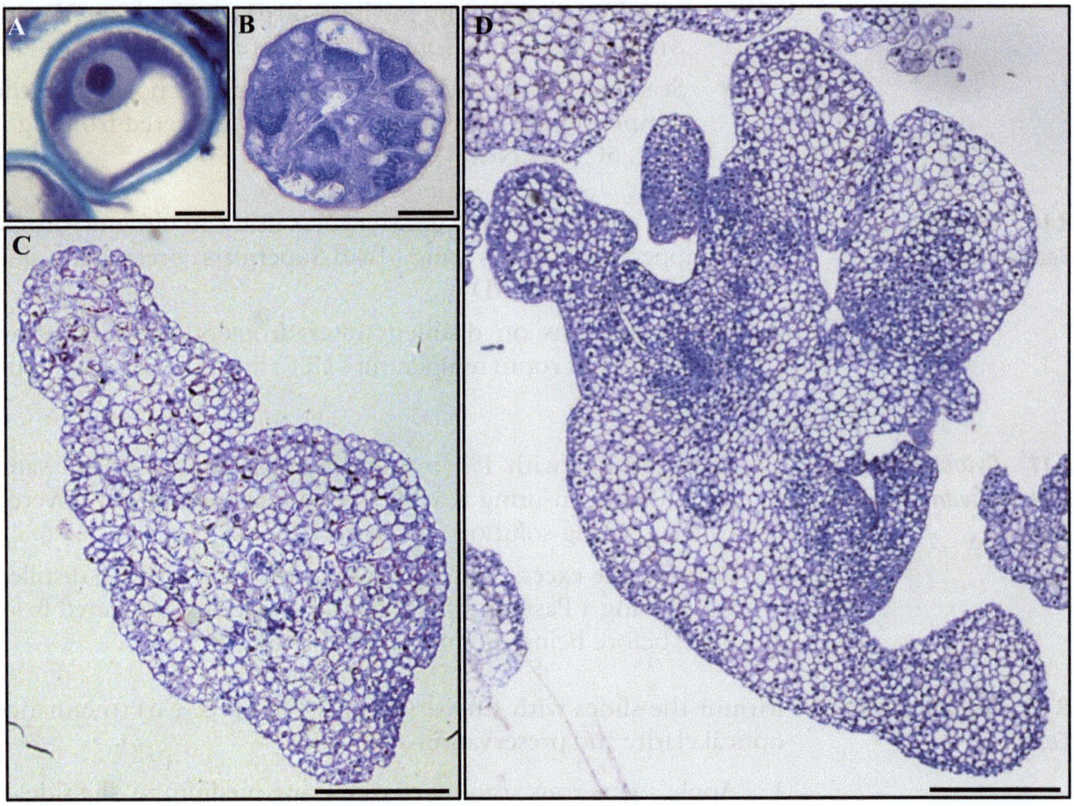

Fig. 7 Monitoring of cellular structural organization during microspore embryogenesis in barley. Micrographs of semithin resin sections stained with toluidine blue and observed under bright field microscopy. (**A**) Vacuolated microspore, at the beginning of culture. (**B**) Microspore derived proembryo. (**C**) Scutellar embryo. (**D**) Coleoptilar embryo. Bars represent (**A, B**) 10 μm, (**C, D**) 200 μm

((3-aminopropyl) triethoxysilane) to enhance section adhesion to the surface.

2. Wash the sections three times for 5 min each with 1× PBS to hydrate the samples (*see* **Note 20**).
3. Block nonspecific bindings by incubating the sections for 10 min with 5% (*w/v*) bovine serum albumin (BSA) in 1× PBS.
4. Incubate the samples with the primary monoclonal anti-IAA mouse antibody (Sigma A0855), diluted 1:1000 in 1% BSA in 1× PBS, for 1 h at RT in a humidified chamber.
5. Perform three to five washes, each for 5 min with 1× PBS to remove unbound primary antibody.
6. Incubate the sections with the secondary antibody Alexa Fluor 488-labeled anti-mouse IgG (Molecular Probes), diluted 1:25 in 1% BSA in 1× PBS, for 45 min in the dark in a humidified chamber.
7. Perform five washes, each for 5 min with 1× PBS to remove any unbound secondary antibody.

8. Incubate the sections with DAPI solution (10 μg/mL in 1× PBS) for 10 min to stain the nuclei.
9. Remove excess DAPI solution by washing twice with sterile distilled water for 5 min each.

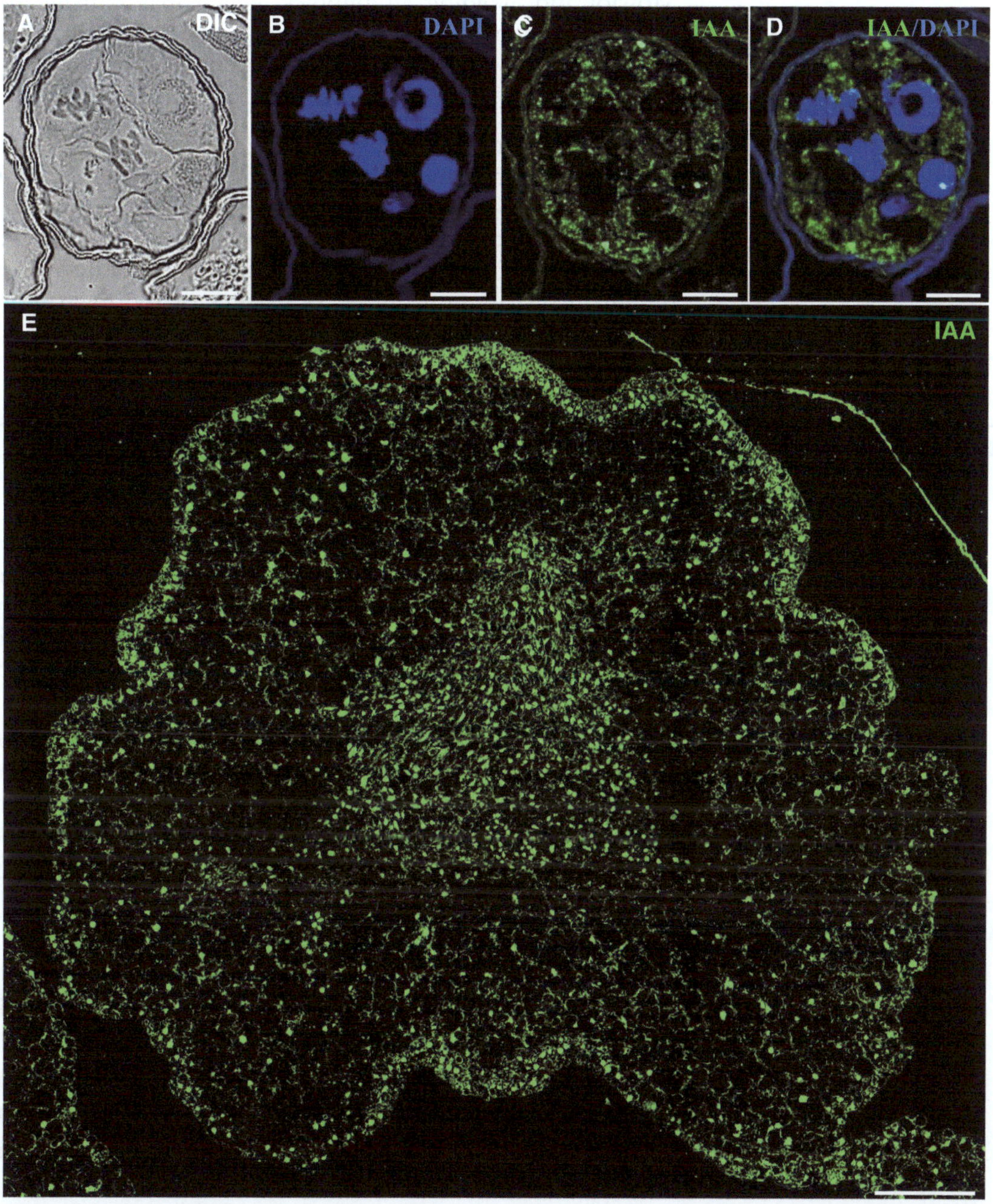

Fig. 8 Immunolocalization of endogenous auxin during microspore embryogenesis in barley. Confocal microscopy images of the immunolocalization of indole-3-acetic acid (IAA) in semithin resin sections of barley cv. Igri. The images show the IAA immunofluorescence signal (green), DAPI stained nuclei (blue) and Nomarski differential interference contrast (DIC). (**A–D**) Microspore-derived proembryo showing active mitotic cell divisions in its nuclei. (**E**) Scutellar embryo. Bars represent (**A–D**) 10 μm, (**E**) 100 μm

10. Mount the sections using Mowiol® 4-88 (Sigma-Aldrich), a water-miscible, semipermanent mounting medium that preserves fluorescence. To mount, apply a thin layer of Mowiol to the surface of a coverslip using a small stick. Carefully place the coverslip onto the section, ensuring even distribution of Mowiol and avoiding air bubbles, which may damage the sample and interfere with microscopic analysis. Seal the edges of the coverslip with nail polish to prevent air exposure.
11. Store the mounted samples at 4°C in the dark until analysis.
12. Examine the prepared samples using an epifluorescence or confocal laser scanning microscope (CLSM). Excitation with the appropriate filter or laser excitation band provide images with the emission signal from DAPI-stained nuclei (in blue) and from auxin (in green), permitting the monitoring of the auxin distribution during key stages of the microspore embryogenesis (Fig. 8A–E).

4 Notes

1. The DAPI working solution (10 μg/mL) is prepared from a commercial stock 1 mg/mL in distilled water, with 1% (*v/v*) Triton X-100 added for enhanced dye penetration. This solution can be stored at 4°C, protected from light, for several months. Due to the potential toxicity of DAPI solution, it is essential to use a lab coat and nitrile gloves to ensure protection against skin contact.
2. The vigor of the donor plants is critical for successful microsporemicrospore embryogenesis embryogenesis. During plant growth, any form of stress, such as dehydration or overwatering, extreme temperature fluctuations, pests, or intensive pesticide treatments, can negatively impact the reprogramming efficiency of microspore embryogenesis.
3. For optimal germination, development, and flowering, barley seeds require a period of vernalization.
4. You can remove these parts of the spike later, but once sterilized, handling will be more difficult.
5. Avoid overcrowding the spikes to ensure proper circulation of the sodium hypochlorite solution for optimal sterilization.
6. Ensure filter paper remains adequately moistened to prevent desiccation, without risking rotting.
7. One main cause of reduced microspore embryogenesis efficiency is cell death, so plant material and tools are kept cold, with precooled equipment at −20°C and liquids maintained at 4°C.

8. Optimal disaggregation is achieved by setting the blender to manual mode and applying three pulses of 5 s at low speed.
9. The filter is rinsed twice with an additional 10 mL of 0.4 M mannitol to collect as many cells as possible.
10. For processing more than 40 spikes, use 8 mL of maltose instead of 4 mL and divide the sample between two tubes to create two gradients, preventing saturation, which could otherwise make it difficult to separate the alive vacuolated microspores from other cells and tissues.
11. If the microspore yield is high, increase the volume per plate or use larger 100 × 20 mm plates. For culture volumes below 1.5 mL, smaller plates are recommended (35 × 12 mm).
12. 4% paraformaldehyde can be aliquoted and stored at −20°C, then thawed as needed prior to use. Due to the toxic nature of paraformaldehyde and its potential to release harmful vapors, it is recommended to use a lab coat, nitrile gloves, and safety goggles to ensure protection against contact, splashes, and inhalation hazards.
13. Storage in 0.1% PF prevents the reversal of fixation. Samples can be preserved for several weeks under these conditions without degradation.
14. Once the gelatin has solidified, 50 μL of 1× PBS is added to prevent desiccation.
15. Due to the toxic nature of the Historesin Plus and its potential to release harmful vapors, it is recommended to use a lab coat, nitrile gloves, and safety goggles to ensure protection against contact, splashes, and inhalation hazards.
16. Once prepared, the solution can be stored at 4°C for up to 1 week.
17. The polymerization solution must be prepared immediately before use, as it solidifies rapidly and becomes unsuitable for sample processing.
18. It is important to cover the capsules, as polymerization does not occur in the presence of oxygen.
19. It is recommended to filter the staining solution after preparation using a Minisart® Syringe Filter (0.22 μm, Sartorius) attached to a syringe. The solution can be stored at RT for several months.
20. It is critical to keep the sections hydrated throughout the procedure to facilitate the antigen–antibody reaction. The volume of reagents added to each well (15–20 μL) should completely cover the section.

Acknowledgments

The authors wish to thank Dr. A. Castillo and Dr. M.P. Vallés for kindly provide us with barley seeds of the Hop and Reinette varieties. Work supported by projects funded by the Spanish National Agency of Research of the Ministry of Science, Innovation and Universities (MICIU/AEI/10.13039/501100011033) and FEDER, EU (grant PID2023-146824OB-I00), Spanish National Agency of Research of the Ministry of Science, Innovation and Universities (MICIU/AEI/10.13039/501100011033), and European Union program NextGenerationEU/ PRTR (grant CPP2022-009890).

References

1. Food and Agricultural Organization of the United Nations (2024) FAOSTAT Database. http://www.fao.org/faostat
2. Cistué L, Ramos A, Castillo AM, Romagosa I (1994) Production of large number of doubled haploid plants from barley anthers pretreated with high concentrations of mannitol. Plant Cell Rep 13:709–712
3. Cistué L, Vallés MP, Echávarri B et al (2003) Barley anther culture. In: Malupszynski M, Kasha K, Foster B (eds) Doubled haploid production in crop plants, a manual. FAO/ IAEA Division, Wien, pp 29–35
4. Castillo A, Vallés M, Cistué L (2000) Comparison of anther and isolated microspore cultures in barley. Effects of culture density and regeneration medium. Euphytica 113:1–8. https://doi.org/10.1007/s002990050379
5. Olsen FL (1991) Isolation and cultivation of embryogenic microspores from barley (*Hordeum vulgare* L.). Hereditas 15(3):255–266. https://doi.org/10.1111/j.1601-5223.1992.tb00568.x
6. Li H, Devaux P (2005) Isolated microspore culture overperforms anther culture for green plant regeneration in barley (*Hordeum vulgare* L.). Acta Physiol Plant 27:611–619
7. Castillo A, Cistué L, Vallés MP, Soriano M (2009) Chromosome doubling in monocots. In: Touraev A, Forster BP, Jain SM (eds) Advances in haploid production in higher plants. Springer, Dordrecht. 10.1007/978-1-4020-8854-4_27
8. Kumlehn J, Serazetdinova L, Hensel G et al (2006) Genetic transformation of barley (*Hordeum vulgare* L.) via infection of androgenetic pollen cultures with *Agrobacterium tumefaciens*. Plant Biotechnol J 4(2):251–261. https://doi.org/10.1111/j.1467-7652.2005.00178.x
9. Rodríguez-Serrano M, Bárány I, Prem D et al (2012) NO, ROS, and cell death associated with caspase-like activity increase in stress-induced microspore embryogenesis of barley. J Exp Bot 63(5):2007–2024. https://doi.org/10.1093/jxb/err400
10. El-Tantawy AA, Solís MT, Risueño MC, Testillano PS (2014) Changes in DNA methylation levels and nuclear distribution patterns after microspore reprogramming to embryogenesis in barley. Cytogenet Genome Res 143:200–208. https://doi.org/10.1159/000365232
11. Solís MT, El-Tantawy AA, Cano V et al (2015) 5-azacytidine improves microspore embryogenesis initiation by decreasing global DNA methylation but prevents subsequent embryo development in rapeseed and barley. Front Plant Sci 6:472. https://doi.org/10.3389/fpls.2015.00472
12. Bárány I, Berenguer E, Solís MT et al (2018) Autophagy and cathepsins are activated and involved in cell death during stress-induced microspore embryogenesis in barley. J Exp Bot 69:1387–1402. https://doi.org/10.1093/jxb/erx455
13. Testillano PS (2019) Microspore embryogenesis: targeting the determinant factors of stress-induced cell reprogramming for crop improvement. J Exp Bot 70(11):2965–2978. https://doi.org/10.1093/jxb/ery464
14. Pérez-Pérez Y, El-Tantawy AA, Solís MT et al (2019a) Stress-induced microspore embryogenesis requires endogenous auxin synthesis and polar transport in barley. Front Plant Sci 10:1200. https://doi.org/10.3389/fpls.2019.01200
15. Pérez-Pérez Y, Bárány I, Berenguer E et al (2019b) Modulation of autophagy and protease activities by small bioactive compounds to reduce cell death and improve stress-induced microspore embryogenesis initiation in rapeseed and barley. Plant Signal Behav 14(2):1559577. https://doi.org/10.1080/15592324.2018.1559577
16. Ibáñez S, Carneros E, Testillano PS, Pérez-Pérez JM (2020) Advances in plant regeneration: shake, rattle and roll. Plants 9(7):1–19. https://doi.org/10.3390/plants9070897

Check for updates

Chapter 13

Production of Homozygous Maize Lines Using Maternal Haploid Inducer Lines

Vijay Chaikam, Leocadio Martinez, John Ochieng and Prasanna M. Boddupalli

Abstract

The production and use of doubled haploid (DH) lines have gained a significant momentum in maize breeding worldwide due to several advantages they offer over conventional inbred lines. The use of genetically homozygous DH lines simplifies the breeding operations, offers significant time and cost savings, and facilitates greater integration of molecular marker applications to increase selection intensity and genetic gains. The production of DH lines in maize includes several steps, including *in vivo* production of maternal haploids using appropriate haploid inducers, identification of haploid seeds from the diploid seeds, germination of putative haploid seeds, treatment of the putative haploid seedlings with antimitotic chemicals to double the chromosome number, and seed production from the putative doubled haploids. This chapter outlines the materials and methods used at different steps in maize DH line production.

Keywords Doubled haploid (DH), Maternal haploid inducers, Maternal haploids, In vivo haploid induction, Haploid identification, Chromosome doubling

1 Introduction

Maize is the most widely cultivated cereal crop in the world due to its increasing demand for an array of uses, such as food, animal feed, and commercial uses like bioethanol, industrial starch production, etc. [1]. The maize crop is well adapted to a range of geographical conditions and is, therefore, cultivated in most parts of the world, with the exception of the polar regions. However, due to the changing climate and the emergence of new biotic threats, maize production is facing changes in the tropical environments that are particularly susceptible to stress, and where millions of people depend on the crop for their food security and livelihoods [2]. To address these issues, maize breeders need to develop improved, high-yielding maize varieties that can withstand climate variabilities and extremes, and emerging diseases and pests, in an accelerated manner [2–4]. Novel

Ana Maria Castillo et al. (eds.), *Doubled Haploid Methods in Cereals*, Methods in Molecular Biology, vol. 3029,
https://doi.org/10.1007/978-1-0716-5245-9_13,

tools like molecular markers and doubled haploids (DH) facilitate the faster development of new maize inbreed lines.

The conventional production of inbred lines typically involves six to eight generations of recurrent inbreeding. This process could take a minimum of 30 months, assuming that the crop cycle is finished in 4–6 months. It demands significant time from breeders and is resource intensive [5–7]. However, the production of DH lines can be accomplished in just two crop cycles (8–12 months), thus freeing the breeder's time from the management of cumbersome selfing nurseries and allowing them to devote more attention to the critical aspects of identifying the best parental lines and hybrid combinations [5–7]. Due to their complete homozygosity, DH lines are ideal for phenotypic and molecular marker-based selection, resulting in greater selection intensity and, consequently, genetic gains [5].

DH production in maize has become cost-effective and scalable with the development of improved maternal haploid inducers with sufficiently high haploid induction rates, phenotypic marker systems that aid in effective identification of maternal haploids at the seed and early seedling stages, and efficient chromosome doubling protocols. Currently, haploid inducers with >10% haploid induction rates adapted to both temperate and tropical environments are available [8, 9]. Even though *R1-nj* anthocyanin marker-based haploid identification is widely used, it can be partially or fully inhibited by inhibitor genes [10]. Therefore, additional markers such as high oil content [11] and red root marker [12] are being integrated into the haploid inducers to complement *R1-nj*-based haploid identification. Recently, improved chromosomal doubling protocols with >25% overall success rates have become available in the public domain [13]. Together, these technical improvements have made DH line production more efficient and cost-effective.

In addition to the technical improvements, maize DH facilities have been established by a few institutions to offer DH development services. In the USA, Iowa State University offers maize DH line production services for temperate breeding programs (https://www.doubledhaploid.biotech.iastate.edu/). Similarly, CIMMYT has established maize DH facilities for tropical breeding programs in Latin America, Africa, and Asia [14]. Any public or private sector organization can access the DH development service on a cost-recovery basis. Such DH facilities are accelerating the integration of DH lines in the maize breeding programs that cannot establish sophisticated inhouse DH line production facilities.

2 Materials, Equipment, and Facilities

2.1 Haploid Induction

1. Maternal haploid inducer: Maize genotypes with the capability to induce haploids when used as pollen parents are referred to as maternal haploid inducers. These inducers need to be appropriately chosen based on their adaptation

and haploid induction rate (HIR) (*see* **Note 1**). Currently, CIMMYT's DH production pipelines use maternal haploid inducer hybrid CIM2GTAIL009 × CIM2GTAIL006. This is a vigorous hybrid with prolific pollen production in tropical environments and a high haploid induction rate (9%–13%) [9] (*see* **Note 2**). This inducer is equipped with *R1-nj* marker for haploid identification.

2. Source germplasm: CIMMYT's DH production pipelines receive source populations from CIMMYT breeders, national agricultural research systems (NARS), and private companies. Typically, these are F_1, F_2, or backcross (BC_1, BC_2) populations that are generally adapted to lowland tropics, subtropics, or tropical highlands. Occasionally, open pollinated varieties or landraces are submitted for line extraction and inbred lines are submitted for line purification.
3. Field space: For haploid induction, it is best to choose fertile fields in locations with low disease and insect pressure, a reliable source for irrigation water, and optimal growing conditions for maize. The location should also be ideal for seed production from populations belonging to different maturity groups.
4. Irrigation facility: Irrigation water should have a neutral pH and low salt content. Drip irrigation is advisable to conserve water and prevent soil erosion. Sprinkler irrigation can be used to loosen up the soil for ploughing before planting.
5. Field tags.
6. Fertilizers: Diammonium phosphate (DAP) and calcium ammonium nitrate (CAN).
7. Agrochemicals (insecticides, fungicides, and herbicides) (Table 1).
8. Plastic/glassine silk/shoot bags.
9. Water repellent paper pollination bags of size 39.5 cm in length and 15.5 cm in width.
10. Nylon mesh harvesting bags of different sizes based on the number of ears to be harvested.
11. Stapler and staple pins.
12. Paper clips.
13. Knife or cutter (to cut silks).

2.2 Haploid Identification

1. Plastic trays.
2. Mesh bags.
3. Labels.
4. Marker pens.
5. Desk lamp.

Table 1
Herbicides, insecticides, and fungicides used in the haploid induction nurseries and the treated haploid nursery (D0) at recommended dosage according to the product data sheet

Type of chemicals	Commercial name[a]	Active ingredient	Target weeds/ insects/ diseases	Stage of application
Herbicides	Primagram Gold 660 SC	S-metolachlor 290 g/L and Atrazine 370 g/L	Broad spectrum weeds	Land preparation
	Governor® 580SE	Acetochlor 340 g/L + Mesotrione 40 g/L + Atrazine 200 g/L	Broad spectrum of weeds	Land preparation and early vegetative growth
Insecticides	Gaucho 600	Imidacloprid	Fall armyworm and sucking insect pests	Seed treatment before planting
	Termidor 25 Ec	Fipronil	Termites	Before transplanting after land preparation
	Gladiator	Chlorpyrifos	Chaffer grubs and wireworms	Land preparation and as and when the insects are present
	Escort 19EC	Emamectin benzoate	Fall armyworm and stem borers	All stages of plant growth
	Pursuit 6% EC	Emamectin Benzoate 10 g/L and Lambda Cyhalothrin 50 g/L	Fall armyworm and stem borers	All stages of plant growth
	Orthene	Acephate	Fall armyworm and stem borers	All stages of plant growth
	Confidor 200 SL	Imidacloprid	Whiteflies, Aphids, Thrips	At vegetative stages
	Actara 25 WG	Thiamethoxam	Aphids, Thrips, Whiteflies, and Leaf hoppers	At flowering
	Prosper 44 EC	Profenos 40% and Cypermethrin 4% w/v	Red spider mites	At the pollination, and grain filling
Fungicides	Amistar Top 325 Sc	Azoxystrobin	Vegetative and flowering stages	Rust

(continued)

Table 1
(continued)

Type of chemicals	Commercial name[a]	Active ingredient	Target weeds/ insects/ diseases	Stage of application
	Tilt	Propiconazole	Vegetative and flowering stages	Broad spectrum fungal diseases

[a]The mention of specific brand names of commercial chemicals (including herbicides, fungicides, and pesticides) is not intended as an official endorsement of the products by CIMMYT. There may be other equal or better products available in the market for achieving the same objectives

2.3 Germination

1. Polyhouse: CIMMYT facilities use polyhouses with no temperature and humidity controls. Shade nets on the top of seedling benches are manually opened or closed to adjust the temperature and light intensity when needed.
2. Seedling trays: Styrofoam trays with 128 cavities, each with a depth of 7.3 cm and a volume of 45 mL, are used. Plastic seedling trays with similar specifications can also be used.
3. Irrigation facility.
4. Germination media: CIMMYT uses peat moss and soil mixed in equal proportions as media for germination and seedling growth in polyhouses. Red sandy soil is preferred to promote drainage and aeration in the root zone.
5. Soil sifter.
6. Labels/masking tape to mark the seedling trays.
7. Metal spatula (to extract haploid plants).

2.4 Chromosomal Doubling

1. Laboratory.
2. Chromosome doubling solution: 0.075% colchicine and 0.1% Dimethyl sulfoxide (DMSO) solution, dissolved in purified water.
3. Purified water derived from reverse osmosis and UV treatment units.
4. Personal protective equipment: People handling colchicine should be protected from head to toe by wearing disposable Tyvek suits, powered air-purifying respirator hoods, heavy duty chemical resistant gloves, and gumboots.
5. Emergency shower.
6. Chemical spill kit: This kit contains essential personal protective equipment, absorbent materials, a scoop or brush, and a waste disposable container. This kit is used to clean up the colchicine solution spilled over the laboratory surfaces.
7. Solid hazardous waste disposal bins.

8. Laboratory equipment (beakers, measuring cylinders, plastic jars, etc.).
9. Refrigerator to store chemicals.
10. Chemical safety fume hood.
11. Precision weighing balance.
12. Magnetic stirrer.
13. Domestic water pump to mix colchicine solution.
14. Shelve for seedling storage during treatment.
15. Plastic crates to move the treated plants to the field.
16. Fertilizers: Urea and Fercristal (13:40:13, N:P:K).
17. Tap water.
18. Metal sinks for washing colchicine treated plants.
19. Plastic containers (5000–10,000 L volume) to store toxic waste or solid materials.
20. Labels.

2.5 Seed Production

1. Irrigation system as described in Sect. 2.1.
2. Fertilizers: DAP, CAN, and OMEX foliar feed (24:24:18, N:P:K+trace elements).
3. Agrochemicals (insecticides, herbicides, and fungicides) (Table 1).
4. Plastic or glassine silk/shoot bags.
5. Scissors.
6. Water repellent paper pollination bags of size 32 cm length and 12 cm width.
7. Nylon mesh harvesting bags of different sizes.
8. Pollination apron.
9. Sticker labels.
10. Paper bags for seed packaging.
11. Cartons for packing and seed shipments.
12. Sticker printer.
13. Tag printer.
14. Latex gloves.
15. Dust masks to cover the nose.

3 Methods

3.1 In Vivo Haploid Induction

1. Prepare the soil in advance to a fine consistency by plowing and harrowing. Apply a preemergent herbicide such as Primagram Gold 660 SC or Governor® 580SE, as may be

needed, to control weeds at the early stages of crop growth (Table 1). Seeds can be planted in flat beds or on ridges.

2. Plant the haploid inducer staggered over several days before and after planting the source populations to synchronize flowering, especially with those that flower early or late (*see* **Note 3**). If the day on which source populations are planted is considered as day 0, haploid inducers may be planted on −7, −5, −3, 0, +3, +7, +10, and +14 days. Most of the inducer plantings can be done on day 0 together with the source populations. Staggered planting of the haploid inducer can be adjusted based on the estimated days to silking of the source populations.
3. Early and late inducer plantings (−7, −5, +7, +10, and +14 staggers) can be planted along the borders of the induction nursery, while the rest of the staggers can be planted alongside the source populations. On average, one row of inducer is planted for every two rows of source population.
4. The number of plants required in each of the source populations depends on the haploid induction rates of the inducer. At the CIMMYT's DH facilities, we use 200–300 plants from each of the source populations with a target of producing a minimum of 100 DH lines.
5. Provide proper agronomic management (irrigation, crop protection, and fertilization) to ensure good crop growth.
6. Inspect the fields daily to determine the irrigation requirements. When there is no rain, irrigate the crop every third day.
7. To enhance plant growth, apply 8–10 g of DAP per plant at planting or immediately after the seedlings have emerged. Apply 8–10 g of CAN when plants reach knee height (approximately 4–5 weeks after planting).
8. Follow regimented scouting for detecting insect and disease occurrence. If the insect/disease pressures exceed the economic thresholds, apply appropriate insecticides or fungicides at recommended concentrations described in the product data sheet (Table 1).
9. Remove tassels from the source germplasm manually as they emerge. A typical haploid induction nursery at the pollination stage, with tassels from the source populations removed, is depicted in Fig. 1A.
10. Carefully cover the ears of the source populations with silk/shoot bags before the silks emerge to avoid pollen contamination when the haploid induction nursery is not spatially or temporally isolated. When the induction nursery is spatially or temporally isolated, the ears are not covered with silk bags to facilitate open pollination (*see* **Note 4**).

Fig. 1 Maternal haploid induction in maize. (**A**) Haploid induction nursery, where the tassels from the source populations are removed and are pollinated with pollen from maternal haploid inducers. (**B**) Manual pollination in the induction nursery, where pollen is collected in pollination bags from the maternal haploid inducers and spread on the silks of the source germplasm. (**C**) Ears harvested from the source population after crossing with maternal haploid inducers

11. Carry out manual or open pollinations when the silks have fully emerged. Manual pollination is depicted in Fig. 1B.
12. After cutting the silks with scissors/cutter, manual pollinations can be carried out twice on consecutive days to ensure the ears are well-filled. The ears must be properly covered with pollination bags and stapled/clipped to prevent any outcrossing. In open pollinated nurseries, the ears are not covered with pollination bags.
13. Harvest the induced ears at physiological maturity in mesh bags. Harvesting usually happens between 120 and 150 days after planting. Typical ears harvested from a source germplasm are depicted in Fig. 1C.
14. Dry the induced ears under the sun or using seed dryers to 12% moisture content.
15. Fumigate the ears to eliminate possible infestation by insects from the field.
16. Store the induced ears or shelled and clean seed in a cold room till subjected to haploid identification.

3.2 Haploid Identification at the Seed Stage

The identification of haploid seed on the ear can be performed immediately after harvesting the ears at physiological maturity or after shelling the seed from ears (*see* **Note 5**). Figure 2A describes haploid identification on the ears.

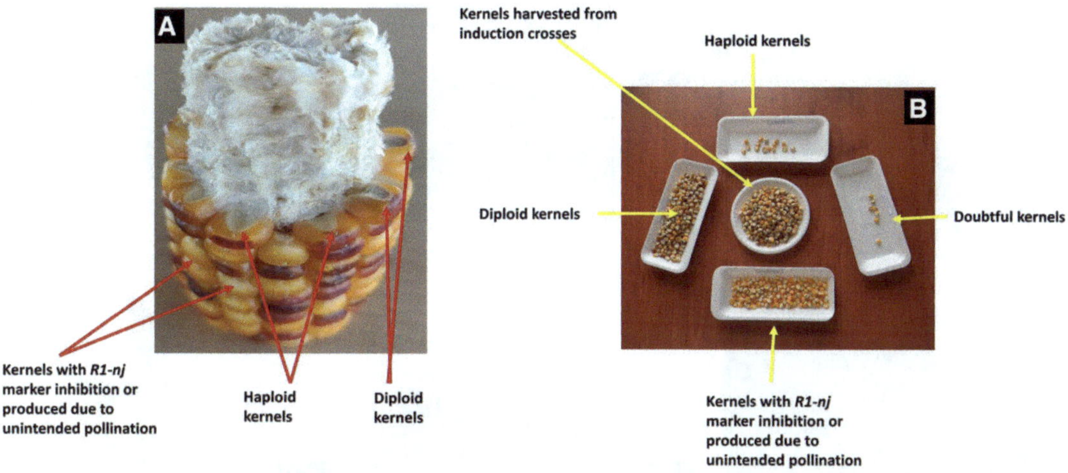

Fig. 2 Haploid identification. (**A**) Differentiating haploid kernels from the diploids based on *R1-nj* anthocyanin color marker expression on the ear. (**B**) Different classes of seeds separated during the haploid identification

1. Visually separate the seeds into the following fractions based on *R1-nj* marker expression (Fig. 2B) till the desired number of haploids are identified (*see* **Note 6**). A desk lamp with a magnifying glass can be used for easy visualization of the anthocyanin marker on the seeds. Each fraction should be placed into labeled plastic or styrofoam containers:
   a) Putative haploids with anthocyanin coloration on the endosperm but not on the embryo.
   b) Putative diploids with coloration on both the endosperm and the embryo.
   c) Anthocyanin color-inhibited (*see* **Note 7**) or contaminated seed.
   d) Doubtful seeds.

3.3 Germination of Haploids

The procedure for the germination of putative haploid seeds is depicted in Fig. 3, and in the text below.

1. Prepare the soil in advance for germination of haploids.
2. Fill the plastic or styrofoam seedling trays with the germination soil.
3. Treat the desired number of putative haploid seeds (*see* **Note 6**) with a systemic insecticide such as Gaucho 600, before planting to protect the seedlings from fall armyworm and other insect pests (Table 1).
4. Organize the seedling trays on the greenhouse benches and plant the seeds at approximately 1–1.5 cm depth with the embryo pointing downward.

Fig. 3 Germination of putative haploids and chromosomal doubling treatment

5. Irrigate the seedling trays with tap water immediately after sowing till the germination soil medium is saturated.
6. Check the seedling trays daily, once in the morning and once in the evening, and water the trays as needed.
7. Apply 750 mL per tray of an equal mix of urea and phosphorous-rich fertilizer, such as Fercristal, dissolved in water to 6–8 days old seedlings to support their growth.
8. Allow the seedlings to grow till the proper emergence of the third leaf. It may take 9–15 days depending on environmental conditions and genotype.
9. Remove any false positives (diploids) from the seedling trays, based on seedling vigor and dark green leaves, compared to the haploids, which usually have relatively poor vigor, and pale, erect leaves.
10. Remove the putative haploid seedlings (at the three-leaf stage) from seedling trays using the spatula, without damaging the roots.
11. Wash the seedlings to get rid of soil and peat moss debris.
12. Keep 50 seedlings in 500 mL plastic containers filled with water till they are subjected to chromosome doubling treatment. When the seedling sizes vary within a population, sort the seedlings according to size and bundle them into batches of 50 seedlings.
13. Label each container with the details of the population.

3.4 Chromosome Doubling Treatment

The chromosomal doubling procedure is described in the text below and shown in Fig. 3. Follow proper occupational health and safety recommendations during the process (*see* **Note 8**):

1. Transport the containers with putative haploid seedlings to the chromosome doubling lab.
2. Prepare the chromosome doubling solution in advance. If large volumes of the solution are required, a stock solution can be prepared and then diluted to the needed volume. Stock solution can be kept in the fridge at 4°C for up to a week.
3. Remove the water from the containers with seedlings and fill them with the chromosome doubling solution (approximately 300 mL) till the solution reaches 4–5 cm above the crown region of the seedlings.
4. Keep the containers undisturbed for 5 h in the lab at ambient temperature.
5. After 5 h, remove the colchicine solution from the container and rinse the seedlings with tap water three times.
6. After rinsing with tap water, place the seedlings in a solution with urea and Fercristal mixed in equal proportions for 30 min in the plastic container. Fill the container with approximately 300 mL of the fertilizer solution so that it reaches up to 4–5 cm above the crown area.

3.5 Seed Production from Putative Doubled Haploids

The procedure for producing seed from putative doubled haploid plants is described in the text below and shown in Fig. 4.

1. Prepare the field in advance for transplanting by ploughing and harrowing. The soil should be fine and deep to facilitate easy transplanting.
2. Apply appropriate preemergent herbicides such as Primagram gold 660 SC or Governor® 580SE to control weeds (Table 1).
3. Prepare ridges in the field for transplanting with 75 cm spacing between the ridges.
4. Make sure that the soil is well saturated with water before transplanting.
5. Prepare transplanting holes 7–8 cm deep and 5 cm in diameter.
6. Transplant-treated putative haploid seedlings (referred to as D_0 plants) directly into the field, preferably in the evening time when the sun is not intense. The use of shade nets (*see* **Note 9**) and plastic mulch (*see* **Note 10**) facilitates better agronomic management of D_0 plants, which are very sensitive to abiotic stresses.
7. Apply necessary insecticides within the first week after transplanting (*see* **Note 11**) to prevent any possible damage by

Fig. 4 Production of D_1 (DH) seed in the D_0 nursery of putative doubled haploid plants

cutworms, chaffer grubs, wireworms, or fall armyworms, as indicated in Table 1. Also, apply the necessary insecticides to control the sucking insect-pests like leaf hoppers, thrips, and aphids that transmit viral diseases (Table 1).

8. Irrigate the seedlings every morning and evening in the first 2 weeks to prevent moisture stress (*see* **Note 12**).

9. After the first 2 weeks, apply irrigation as and when necessary, depending on soil moisture status. When there is no rain, apply irrigation once every 2 days.

10. Control weed growth by mechanical, chemical, or manual methods as appropriate (*see* **Note 12**). Chemical weed control at early stages of crop growth can be achieved with herbicides such as Governor® 580SE (Table 1).

11. Apply 8–10 g of DAP per plant immediately after transplanting. Apply 8–10 g of CAN thrice during crop growth after 2, 4, and 6 weeks after transplanting. In addition, foliar fertilizers such as Omex foliar feed can be applied at 2-weeks intervals from transplanting to grain filling stage.

12. Scout the fields regularly for possible insect-pest or disease incidence at all stages of crop growth (*see* **Note 13**), and apply appropriate pesticides/fungicides (Table 1) to prevent any crop damage.

13. Inspect the plants regularly for ear and silk emergence, as well as anther emergence from the tassels during the flowering stage (*see* **Note 14**). Cover the ears with silk/shoot bags before the silks emerge to prevent any pollen contamination.

14. Cover the tassels with small tassel bags the day before pollination or in the morning of pollination.
15. Collect pollen either in the pollination or glassine bags from the tassel at the time of pollination.
16. Cut the top of the ears to accelerate silk emergence when there is no synchrony between male and female flowering in the same plant. This will enable the silks to be ready for pollination on the same day or the following day.
17. Self-pollinate the D_0 plants two to three times on consecutive days, depending on pollen availability, to ensure seed production.
18. Harvest the D_1 (DH) ears individually from the haploid plants at physiological maturity. The ears are generally harvested 120–150 days after germination.
19. Dry the ears under the sun or in dryers till they reach a moisture level of 12% and fumigate them to eliminate any possible incidence of post-harvest insects.
20. Process the seed from each ear separately.
21. Pack the seeds, label the seed packets, and store them in the cold room at 5°C–8°C and a 50% relative humidity till planting.

4 Notes

1. Temperate maternal haploid inducers like RWS, RWK, UH400, UH600, etc. are widely used in temperate environments [7]. However, they are poorly adapted to tropical conditions. Tropically adapted inducer lines can be more effectively used in tropical environments [9]. If inducers with alternate markers (e.g., high oil, red root) are available, using them in haploid induction can alleviate the problems associated with the inhibition of *R1-nj* marker.
2. Inducer hybrids are generally used in haploid induction as they have superior plant vigor and prolific pollen production.
3. Staggering the planting of haploid inducers is necessary to synchronize the shedding of inducer pollen with the silk emergence in source populations, especially when source populations vary in flowering time.
4. Open pollination can be cost-effective compared to manual pollination in induction nurseries if the haploid inducer is well-adapted to the target environment, produces pollen in sufficient quantities, and the haploid induction nurseries are grown in isolation (either spatially or temporally) from other

maize plantings to prevent pollen contamination. The recommended spatial isolation distance is at least 200 m, while temporal isolation should be a minimum of 4 weeks.

5. Haploid identification on ears could be faster but can result in worker fatigue. Hence identification of haploids among shelled seeds is preferred.
6. The number of putative haploid seeds to be identified and the number of putative haploid seedlings subjected to chromosome doubling varies based on the success rates obtained in the target environment. At the CIMMYT DH facilities, 1000–2000 putative haploid seeds are used to produce a minimum of 100 DH lines.
7. In some genetic backgrounds, the expression of the *R1-nj* marker could be partially or fully inhibited. When partially inhibited, more seeds may need to be sorted to identify the necessary number of haploids. When the *R1-nj* marker is fully inhibited, haploids can be identified using alternative markers like the red root marker or the high oil content marker, if the haploid inducers are equipped with these markers. If not, seeds can be grown in seedling trays and haploids can be identified based on seedling vigor, erect and pale leaves [15].
8. Colchicine is very toxic and needs to be handled with care. Proper occupational health and safety procedures need to be established and rigorously followed to avoid any exposure. Personnel handling colchicine should wear personal protective equipment from head to toe. Colchicine should be handled in a designated lab with restricted access. Colchicine powder should be weighed and dissolved in water in a chemical safe fume hood. Diluting the colchicine concentrate and proper mixing of the colchicine solution can be achieved using small domestic water pumps, thus avoiding exposure to the colchicine solution while mixing. After treatment, colchicine waste should be directed to large waste storage containers directly from the sinks and disposed when the tanks are full. The colchicine waste can be handed over to approved chemical waste disposal companies for incineration.
9. The use of shade nets is advised in environments with high temperatures to reduce ambient temperatures and to enhance the growth of D_0 plants. However, the use of shade-nets can delay the tassel and silk emergence, affect pollen production and silk viability, and can also increase the ear rots. So, the use of shade nets needs to be optimized, especially during the pollination and postpollination stages, as per the location/ environmental conditions.
10. Transplanting the putative haploid seedlings on the plastic mulching sheet can facilitate better agronomic management

by reducing the weed growth around the plants and limiting the evaporation, thereby reducing the demand for irrigation water.

11. Agronomic management in the D_0 nursery is very critical to ensure proper growth of the chromosome-doubled plants and for production of D_1 seed. In the early stages after transplanting, seedlings could be easily damaged by insect-pests like cutworm, fall armyworm, chaffer grubs etc. Other sucking pests like leaf hoppers, thrips, and aphids could transmit viral diseases, to which haploids plants could be highly vulnerable. Seedlings therefore need to be treated with an appropriate systemic insecticide to protect against these insect-pests at the early stages.
12. Haploid plants are also very sensitive to moisture stress and competition. Hence, soil moisture needs to be rigorously checked, and when needed, water should be applied preferably using drip irrigation. Weeds must be controlled at all stages of crop growth.
13. Fall armyworm and stem borers could be a major threat to the plants in the D_0 nursery throughout the crop growth cycle and need to be properly protected.
14. Most of the haploid plants that have undergone chromosome doubling treatment could be partially fertile, producing only one to a few anthers. The pollination crew needs to be well-trained to identify anthers producing pollen, collect the pollen carefully, and self-pollinate the plants in such a way that prevents pollen contamination.

Acknowledgments

The authors acknowledge the financial support received from the following projects for CIMMYT work on the maize DH technology. The projects include: (a) "A Doubled Haploid Facility for Strengthening Maize Breeding Programs in Africa" from the Gates Foundation (GF); (b) "Stress Tolerant Maize for Africa (STMA)" project funded by GF and U.S. Agency for International Development (USAID); and (c) "Accelerating Genetic Gains for Maize Improvement (AGG-Maize)" project funded by GF, USAID, and Foundation for Food and Agriculture Research (FFAR). Additional funding support came from the CGIAR Research Program on Maize (MAIZE) during 2011–2022; CGIAR Accelerating Breeding Initiative (ABI); MasAgro-Maize Project funded by the Secretariat of Agriculture, Livestock, Rural Development, Fisheries and Food (SAGARPA), Government of Mexico; and Limagrain.

References

1. Erenstein O, Jaleta M, Sonder K et al (2022) Global maize production, consumption and trade: trends and R&D implications. Food Secur 14:1295–1319
2. Prasanna BM, Nair SK, Babu R et al (2020) Increasing genetic gains in maize in stress-prone environments of the tropics. In: Kole C (ed) Genomic designing of climate-smart cereal crops, Springer Nature, Switerland, pp 97–132. https://doi.org/10.1007/978-3-319-93381-8_3
3. Miedaner T, Juroszek P (2021) Global warming and increasing maize cultivation demand comprehensive efforts in disease and insect resistance breeding in north-western Europe. Plant Pathol 70:1032–1046. https://doi.org/10.1111/ppa.13365
4. Ulfat A, Abasi F, Munir A et al (2022) How to deal with climate change in maize production. In: Liu F, Li X, Hogy P et al (eds) Sustainable crop productivity and quality under climate change. Elsevier, Amsterdam, pp 157–169. https://doi.org/10.1016/B978-0-323-85449-8.00006-3
5. Chaikam V, Molenaar W, Melchinger AE, Boddupalli PM (2019) Doubled haploid technology for line development in maize: technical advances and prospects. Theor Appl Genet 132:3227–3243. https://doi.org/10.1007/s00122-019-03433-x
6. Prasanna BM (2012) Doubled Haploid technology in maize breeding: an overview. In: Prasanna BM, Chaikam V, Mahuku G (eds) Doubled Haploid technology in maize breeding: theory and practice. CIMMYT, Mexico DF, pp 1–8. https://repository.cimmyt.org/entities/publication/436ec3cf-28cc-4948-8020-db2491d363d2
7. Röber FK, Gordillo GA, Geiger HH (2005) In vivo haploid induction in maize-performance of new inducers and significance of doubled haploid lines in hybrid breeding. Maydica 50:275–283
8. Hu H, Schrag TA, Peis R et al (2016) The genetic basis of haploid induction in maize identified with a novel genome-wide association method. Genetics 202:1267–1276. https://doi.org/10.1534/genetics.115.184234
9. Chaikam V, Nair S, Martinez L et al (2018) Marker-assisted breeding of improved maternal haploid inducers in maize for the tropical/subtropical regions. Front Plant Sci 9:385997. https://doi.org/10.3389/fpls.2018.01527
10. Chaikam V, Nair SK, Babu R et al (2015) Analysis of effectiveness of R1-nj anthocyanin marker for in vivo haploid identification in maize and molecular markers for predicting the inhibition of R1-nj expression. Theor Appl Genet 128:159–171. https://doi.org/10.1007/s00122-014-2419-3
11. Melchinger AE, Schipprack W, Friedrich Utz H, Mirdita V (2014) In vivo haploid induction in maize: identification of haploid seeds by their oil content. Crop Sci 54:1497–1504. https://doi.org/10.2135/cropsci2013.12.0851
12. Chaikam V, Martinez L, Melchinger AE et al (2016) Development and validation of red root marker-based haploid inducers in maize. Crop Sci 56:1678–1688. https://doi.org/10.2135/cropsci2015.10.0653
13. Chaikam V, Gowda M, Martinez L et al (2020) Improving the efficiency of colchicine-based chromosomal doubling of maize haploids. Plants 9:459. https://doi.org/10.3390/plants9040459
14. Chaikam V, Prasanna BM (2020) Doubled Haploid technology for rapid and efficient maize breeding. In: Gosal SS, Wani SH (eds) Accelerated plant breeding, Volume 1: cereal crops, Springer Nature, Cham, Switzerland, pp 257–292
15. Chaikam V, Lopez LA, Martinez L et al (2017) Identification of in vivo induced maternal haploids in maize using seedling traits. Euphytica 213:1–9. https://doi.org/10.1007/s10681-017-1968-3

Chapter 14

In Vitro Anther Culture of Japonica and Indica Rice (*Oryza sativa* L.)

Csaba Lantos, Árpád Székely, Tímea Szalóki, Mihály Jancsó and János Pauk

Abstract

Well-established doubled haploid (DH) plant production methods enhance the efficiency of modern plant breeding programs. In vitro anther culture is one biotechnological tool that can produce homozygous lines in one generation, while conventional breeding programs require more generations to develop genetically pure lines. In this chapter, an efficient protocol for DH rice production via anther culture is described after a cool pretreatment for 3–5 days at 10°C. We combined the beneficial effects of the "N6NDK" induction medium and the "MSNBK1" regeneration medium to obtain healthy green DH plantlets from japonica and indica rice genotypes. These plantlets can be introduced directly into research or breeding projects after a successful acclimatization in glasshouses and under field conditions.

Keywords In vitro androgenesis, Anther culture, Doubled Haploid, *Oryza sativa*, Japonica and indica rice

1 Introduction

The doubled haploid (DH) plant production techniques offer an excellent opportunity for breeders and researchers to reduce the long process of breeding, develop homogenous lines, and increase the efficiency of breeding programs. The most significant advantage of DH plant production methods is the generation of homozygous lines within approximately 1 year. DH lines can be directly used to select valuable varieties or as partners for further crosses. The homogeneity of breeding materials is an essential requirement for the registration of new varieties and hybrids.

The first anther culture-derived rice plantlets were reported by Japanese researchers more than 50 years ago [1]. Chu described the N_6 medium, which was superior to earlier media for androgenesis induction in rice [2]. This basal medium has been frequently used for androgenesis in different breeding and research programs. After the

Ana María Castillo et al. (eds.), *Doubled Haploid Methods in Cereals*, Methods in Molecular Biology, vol. 3029,
https://doi.org/10.1007/978-1-0716-5245-9_14,

optimization of culture media and conditions [3–7], androgenesis has become an efficient tool for DH plant production in rice, especially in japonica genotypes [8].

The genotype dependency of androgenesis in rice has been studied by several research groups. Efficient protocols have been reported for DH plant production in japonica genotypes, while indica genotypes are more recalcitrant to anther culture. The necrosis of anthers, the low frequency of callus production and plant regeneration, and the high proportion of albino plants limit the practical application of androgenesis in recalcitrant rice genotypes [5, 9–13].

The percentage of spontaneous doubling determines the quantity of the DH lines among the regenerated plantlets. These values range from 30%–50% in both indica and japonica species [9, 13–17].

A well-established DH plant production method has been applied in different breeding programs [14, 15, 18–20]. Recently, the selection of DH lines has led to the commercialization of more than 40 rice varieties [21].

Two methods are frequently applied for DH plant production in rice, namely (i) anther culture and (ii) isolated microspore culture. In this chapter, we describe an efficient protocol recently published for the production of DH plants by anther culture in japonica and indica rice genotypes, which shows higher efficiency in japonica (on average, 24.48 green plantlets/100 anthers) than in indica (on average, 4.7 green plantlets/100 anthers) [22, 23].

2 Materials

2.1 Plant Material

In the last 5 years, approximately 100 japonica breeding combinations (F_1–F_4) and 20 indica genotypes were tested by this anther culture protocol.

2.2 Equipment

1. Flooding technology for cultivating donor plants in the nursery.
2. Air-conditioned glasshouse set at 20°C/15°C day/night temperature for acclimatization of regenerated plantlets and seed production.
3. Olympus CK-2 inverted microscope.
4. Laminar air flow cabinet.
5. Incubators (28°C).
6. Culture growth chamber (22°C–24°C, with 16 h photoperiod and 30 μmol m^{-2} s^{-1} light intensity provided by full-spectrum white LED).
7. Autoclave for media preparation and glassware sterilization.
8. Balance.
9. pH meter.

10. Refrigerator.
11. Freezer.
12. Tissue Lyser II (Qiagen GmbH).
13. CytoFLEX flow cytometer with UV lamp (Beckman Coulter International S.A.).

2.3 Plant Growth and Laboratory Material

1. Plastic pots with 1:1:1 sand, peat, and humus mixture.
2. Selective herbicides [e.g., florpyrauxifen-benzyl and penoxsulam as a brand name of Novixid (Corteva Agriscience Ltd.)] for donor plant cultivation under conventional flooding.
3. Split nitrogen fertilizer (preferably slow-release urea or ammonium-nitrate) for donor plant cultivation under conventional flooding.
4. Volldünger complex fertilizer (N:P:K:Mg 14:7:21:1, plus 1% microelements: B, Cu, Fe, Mn, and Zn).
5. PVC bags.
6. Pipettes and pipette tips.
7. Beakers and measuring cylinders.
8. Tweezers and scissors.
9. Glass slides and coverslips.
10. Glass rod.
11. Eppendorf tubes.
12. Glass bottles and glass tubes.
13. Jars.
14. Sterile Petri dishes (60 mm Ø) for anther culture.
15. Sterile Petri dishes (90 mm Ø) for plant regeneration.
16. Sterile glass tubes (100 × 20 mm) and jars (170 × 70 mm) for rooting of in vitro green plantlets.
17. Stainless steel beads, 20 μm nylon sieves and plastic tubes (1.5 and 2 mL) for flow cytometry analyses.
18. Plastic racks (6 × 11 places) for acclimatization of plantlets in a glasshouse.

2.4 Solutions and Culture Media

1. Panicle sterilizing solution: 300 mL of 2% NaOCl in water with one drop of Tween-20.
2. Culture media preparation: Table 1 lists all media and components used for androgenesis induction, plantlet regeneration, and rooting [22, 23]. Add all the components with exception of Gelrite, L-Proline, and L-Glutamine, to 800 mL of deionized water, adjust pH (5.8), take the volume to 1 L, add Gelrite, and sterilize it in the autoclave at 121°C for 20 min. After autoclaving, add the filter-sterilized L-Proline and L-Glutamine to the induction medium.

Table 1
Compounds of media for in vitro androgenesis induction, regeneration, and rooting of rice plantlets

Compounds	"N_6NDK" induction medium (mg/L)	"MSNBK1" regeneration medium (mg/L)	"1/2MS_0" rooting medium (mg/L)
KNO_3	2830	1900	950
NH_4NO_3	–	1650	825
$CaCl_2 \times 2H_2O$	125.33	332.02	166.01
$MgSO_4 \times 7H_2O$	90.27	180.54	90.27
KH_2PO_4	400.00	170	85
$(NH_4)_2SO_4$	463.00	–	–
Na_2EDTA	37.3	37.3	37.3
$FeSO_4 \times 7H_2O$	27.8	27.8	27.8
$MnSO_4 \times 4H_2O$	4.4	16.9	8.45
$ZnSO_4 \times 7H_2O$	1.85	8.6	4.3
H_3BO_3	1.6	6.2	3.1
KI	0.8	0.83	0.415
$CuSO_4 \times 5H_2O$	–	0.025	0.0125
$CoCl_2 \times 6H_2O$	–	0.025	0.0125
$Na_2MoO_4 \times 4H_2O$	–	0.25	0.125
Glycine	–	2	1
Thiamine HCl	1	0.1	0.05
Pyridoxine HCl	0.5	0.5	0.25
Nicotinic acid	0.5	0.5	0.25
Myo-Inositol	–	100	50
L-Proline	500	–	–
L-Glutamine	500	–	–
Maltose	40,000	–	–
Sucrose	–	30,000	20,000
2,4-D	1	–	–
Kinetin	0.5	1	–
NAA	2.5	1	–
BAP	–	1	–
pH	5.8	5.8	5.8
Gelrite	2800	2800	2800

3. Galbraith buffer for flow cytometry (100 mL): 912.5 mg $MgCl_2 \cdot 6H_2O$, 887.5 mg Na-citrate, 412.5 mg MOPS, 100 µL Triton, adjust pH to 7 [24].
4. RNase solution: 1 mg/mL solution for degradation of RNA content.
5. Propidium iodide (PI) solution: 1 mg/mL solution for DNA staining.

3 Methods

3.1 Growing Conditions of Plant Material

1. Sow directly the seeds of the selected donor genotypes in the nursery using conventional European rice-growing technology (*see* **Note 1**).
2. Apply flooding technology to grow the donor plants after tillering (after 38–40 days of sowing) (*see* **Note 2**).
3. Fertilize the donor plants with 30 kg/ha of nitrogen at each tillering and panicle initiation stage (*see* **Note 2**).
4. Control weeds using selective herbicides or manually as required (*see* **Note 2**).

3.2 Pretreatment of Donor Panicles

1. Check the developmental stages of microspores under a microscope (*see* **Note 3**). Crush the anthers of donor genotypes with a glass rod in a drop of water on a glass slide and examine the developmental stages of microspores with an inverted microscope.
2. Select the tillers of donor genotypes containing anthers with early- and mid-uninucleated microspores (Fig. 1A). The position of the nucleus should be close to the microspores wall, not too far from the micropyle.
3. Place the selected tillers into Erlenmeyer flasks containing tap water and cover with PVC bags to keep high humidity.
4. Keep the donor tillers at 10°C for 3–5 days for a cool pretreatment in the dark.

3.3 Anther Culture

1. Recheck the developmental stages of microspores in pretreated materials under a microscope (*see* **Note 3**).
2. Remove the sheath leaf and sterilize the panicles containing anthers with early- and mid-uninucleated microspores in the sterilizing solution for 20 min on a shaker.

Fig. 1 Rice doubled haploid plant production by anther culture. (**A**) Uninucleate, vacuolated microspore for androgenesis induction in rice. (**B**) Microspore-derived calli developed after 4 weeks of culture. (**C**) Green plantlets on the regeneration medium. (**D**) Green rooted plantlets in individual glass tubes, and (**E**) green plantlets developed in jars. (**F**) Acclimatized plants in the glasshouse. Bars = 10 μm for (**A**); 10 mm for (**B–E**); 50 mm for (**F**)

3. Rinse the panicles three times with sterile distilled water in a laminar airflow cabinet (*see* **Note 4**).
4. Isolate the anthers manually with tweezers, and culture 100 anthers in a 60 mm Ø Petri dish containing the "N_6NDK" induction medium (Table 1).
5. Place and keep the anther cultures in an incubator at 28°C in the dark to obtain microspore-derived calli after 4 weeks of culture (*see* **Note 5**).

3.4 Plant Regeneration of Anther Culture-Derived Calli

1. Select well-developed calli (Fig. 1B) with a size of 1–2 mm and transfer them to 90 mm Ø Petri dishes containing the "MSNBK1" regeneration medium (Table 1). Green and albino plantlets regenerate within 2–4 weeks (Fig. 1C). Transfer calli to fresh regeneration medium after 2 weeks (*see* **Notes 5 and 6**).
2. Transfer well-developed green plantlets with two-three leaves individually into sterile glass tubes (Fig. 1D, E), containing ½MS_0 medium for rooting (Table 1).
3. Keep cultures at 24°C and a photoperiod of 16/8 h day/night during plantlet regeneration (approximately 2–4 weeks) and rooting (approximately 3–4 weeks).

3.5 Acclimatization of Green Plantlets

1. In the glasshouse, transplant the well-rooted green plantlets individually into 1 L plastic pots containing the 1:1:1 sand, peat, and humus mixture.
2. Cover the transplanted plantlets with a transparent PVC bag during the 4–5 days acclimatization period (*see* **Note 7**). The plants are grown in the glasshouse until they are transferred to the nursery.
3. Fertilize the acclimatized plants with a solution of 0.02%–0.03% Volldünger complex fertilizer (N:P:K:Mg 14:7:21:1, plus micronutrients) once a fortnight.
4. Maintain 20°C/15°C day/night temperature for plantlets in the glasshouse (Fig. 1F).
5. Transfer the plants to a paddy field for additional testing in May.

3.6 Flow Cytometry

Ploidy levels of the anther culture-derived plantlets can be determined by flow cytometry analysis.

1. Collect leaf samples (100 mg/plantlet) from young leaves of plantlets grown in the glasshouse.
2. To isolate the nuclei from the collected young leaves, homogenize the samples in 2 mL Eppendorf tubes containing 1 mL of Galbraith buffer and two steel beads using a TissueLyser II at 20 Hz for 2 min.

3. Purify the suspension using 20 μm sieves and place in 1.5 mL Eppendorf tubes.
4. Add 10 μL RNase solution to each sample for 60 min at room temperature (RT) to eliminate the RNA content.
5. Stain DNA with 40 μL PI solution for 30 min at RT.
6. Determine the DNA content of samples (control, haploid, diploid, tetraploid, and mixoploid) based on histograms of flow cytometry analyses (Fig. 2) (*see* **Note 8**). Use leaves from 3-week-old plants grown from seed as a control.

3.7 Growing of Anther Culture-Derived Plantlets in Nursery

1. Transplant the acclimatized anther culture-derived rice plantlets manually from the glasshouse to the nursery in April–May (above 15°C minimum) (*see* **Note 9**).
2. Water the plantlets as needed to support the development of new roots and acclimatization of the plantlets to field conditions (*see* **Note 2**).
3. Use shallow flood irrigation to avoid water-related stresses.
4. Maintain a balanced nutrient supply at every growth stage (*see* **Note 2**).
5. Apply manual weed control, if necessary.
6. Evaluate the anther culture-derived rice plants using basic selection criteria (plant habit, lodging, awn length, seed shape).
7. Harvest the panicles of the fertile DH plants separately in the full ripening stage in October.
8. Determine the percentage of spontaneous genome doubling based on seed production.

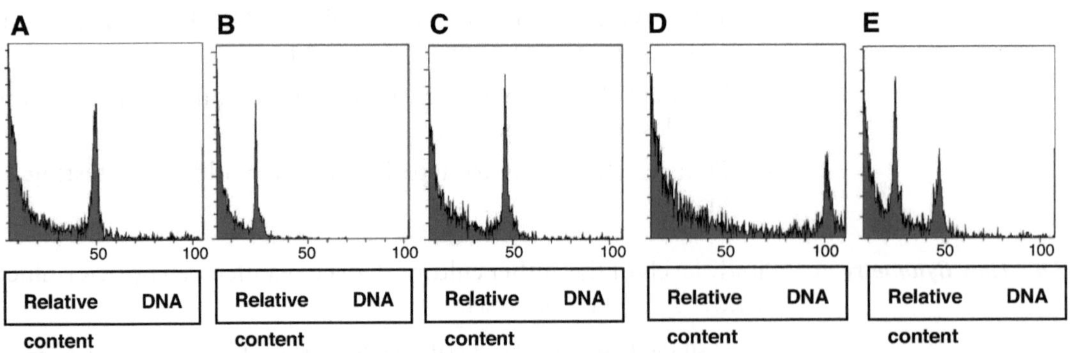

Fig. 2 Ploidy level determination of anther culture-derived plantlets based on flow cytometry analysis. Histograms of seed-derived control plants (**A**), haploid (**B**), diploid (**C**), tetraploid (**D**), and mixoploid (**E**) plantlets from anther culture

3.8 Integration of DH Lines into the Breeding Program

In the next generation, sow the seeds of each DH line in either a one-row system or a microplot, depending on the breeding system used (*see* **Note 10**).

Homogeneous DH lines can shorten breeding programs by producing homozygous plant materials in one generation.

4 Notes

1. The growing conditions of donor plants are one of the most critical factors for the efficient induction of in vitro androgenesis. The growth of healthy donor plants, tillers, and panicles under ideal conditions is important for large-scale DH plant production. In our method, field-grown donor material is preferred for DH plant production. Field-grown donor plants usually receive sufficient sunlight, and flooded conditions provide irrigation water as needed.
2. The standard flooded cultivation of rice under the temperate conditions of Hungary has the following steps:

   Tillage of the field in autumn and seedbed preparation with a combined cultivator in the spring to create a leveled soil surface and fine soil structure. Apply nitrogen (60 kg/ha), phosphorus (50 kg/ha), and potassium (50 kg/ha) fertilizers during the seedbed preparation. The amounts can be adjusted based on specific soil tests.

   Direct dry seeding between late April and mid-May (above 10°C minimum), either manually or with a plot seeder, depending on the amount of seed set available.
   Provide flush flood irrigation during the first month of the growing period to maintain optimum soil moisture content (partially saturated soil).

   Use selective herbicides such as florpyrauxifen-benzyl and penoxsulam (e.g., Novixid (Corteva Agriscience Ltd., at 2 L/ha).

   Apply 30 kg/ha of nitrogen fertilizer at each tillering and panicle initiation stages. Foliar fertilizers are also recommended, if available, at the same growth stages (nitrogen applied as a 2% urea solution, dosage can be reduced by 20%–30% compared to soil application). Avoid over-application of nitrogen when the length of the growing season is limited, especially in temperate regions.

   Apply flood irrigation from mid-June to avoid water deficiency and to prevent donor plants from high temperature fluctuations during the vegetative phase. The booting stage usually starts in mid-July but is highly dependent on its duration and the photoperiod sensitivity of the plant genotype.

3. Spikes at the appropriate microspore developmental stages must be carefully selected to induce androgenesis in rice. Early- and mid-uninucleate vacuolated microspores are optimal for efficient androgenesis induction in rice anther culture. The stage of microspores can be checked quickly and easily in a drop of tap water under an inverted microscope.
4. The preparation of anther culture methods requires in vitro manual work (isolation of anthers, transfer of calli, and green plantlets). The application of antibiotics is not necessary, but it is essential to work under sterile conditions.
5. The genotype of the donor materials influences the efficiency of rice anther culture [21–23]. The responsiveness of japonica genotypes is higher than that of indica genotypes. Using our protocols, significant numbers of green plantlets can be produced from japonica genotypes, while results for indica genotypes are lower. It is worthwhile using several genotypes when adaptating the method.
6. For the efficient production of rice DH plants, it is important to follow the schedule: collection and pretreatment of donor materials, transfer of calli, and green plantlets, etc. Significant delays in different steps can reduce the efficiency of green plantlet production.
7. In vitro plantlets are sensitive to rapid changes in humidity. The well-rooted green plantlets should be transplanted into plastic pots as quickly as possible and covered with a transparent plastic bag, which should be removed after the acclimatization period (4–5 days).
8. The ploidy level of anther culture-derived plantlets can be determined by flow cytometry analysis. Haploid, DH, tetraploid, and mixoploid plantlets can be identified based on the position of the peaks in the histograms of the samples, allowing only DH plants to be transplanted into the nursery (Fig. 2).
9. Due to the high rate of spontaneous chromosome doubling obtained with our protocol (approximately 50%), and the high toxicity of duplication agents like colchicine, all acclimatized plants are transferred to the nursery without ploidy analyses.
10. The homogeneity of DH lines can be checked visually in $DH_{1\text{-}2}$ generations in the nursery. The segregating lines (somatic tissue-derived plants, cross-pollination, seed mixtures, volunteer plants) can be identified and selected on the basis of the different phenotypes in the plot (different plant habit, plant

height, panicle shape, panicle length, panicle branches, maturation time, awning, length, and width of grain).

5 Funding

This research was funded by the Ministry for Innovation and Technology and the National Research, Development and Innovation Office, Budapest, Hungary: grant number "OTKA-FK_21-F138042," "OTKA-K_21-K138416," "TKP2020-NKA-21," and "GINOP-2.3.3-15-2016-00042."

Acknowledgments

The authors thank the conscientious work of Ferenc Markó, Krisztina Beregszászi-Kéri, and Szilvia Palaticki and all the staff of the MATE ÖVKI Rice Research Station, Szarvas, Hungary.

Conflict of Interest The authors declare that they have no conflict of interest.

References

1. Niizeki H, Oono K (1968) Induction of haploid rice plant from anther culture. Proc Jpn Acad 44:554–557
2. Chu CC, Wang CC, Sun CS et al (1975) Establishment of an efficient medium for anther culture of rice through comparative experiments on the nitrogen sources. Sci Sin 1975(18):659–668
3. Woo SC, Tung IJ (1972) Induction of rice plants from hybrid anthers of indica and japonica cross. Bot Bull Acad Sin 16:19–24
4. Heszky L, Pauk J (1975) Induction of haploid rice plants of different origin in anther culture. Il Riso 24:197–204
5. Guiderdoni E, Galinato E, Luistro J, Vergara G (1992) Anther culture of tropical japonica × indica hybrids of rice (*Oryza sativa* L.). Euphytica 62:219–224
6. Lentini Z, Reyes P, Martínez CP, Roca WM (1995) Androgenesis of highly recalcitrant rice genotypes with maltose and silver nitrate. Plant Sci 110:127–138
7. Chen CC (1977) In vitro development of plants from microspores of rice. In Vitro Cell Dev Biol Plant 13(8):484–489
8. d'Hooghvorsdt I, Ferreres I, Nogues S (2021) Anther culture and chromosome doubling in mediterranean japonica rice. In: Segui-Simarro JM (eds) Doubled haploid technology. Methods in molecular biology, vol 2287. Humana, New York, NY, pp 333–341
9. Bhattacharya A, Mikkilineni V, Verma L et al (2014) Evaluation of doubled haploid culture conditions and regeneration of an indica rice hybrid. Indian J Genet Plant Breed 74:384–386
10. Usenbekov BN, Kaykeev DT, Yhanbirbaev EA et al (2014) Doubled haploid production through culture of anthers in rice. Indian J Genet Plant Breed 74:90–92
11. Hooghvorst I, Ramos-Fuentes E, López-Cristofannini C et al (2018) Antimitotic and hormone effects on green double haploid plant production through anther culture of Mediterranean japonica rice. Plant Cell Tissue Organ Cult 134:205–215
12. Hooghvorst I, Ribas P, Nogués S (2020) Chromosome doubling of androgenic haploid plantlets of rice (*Oryza sativa* L.) using antimitotic compounds. Plant Breed 139:754–761
13. Bishnoi U, Jain R, Rohilla J et al (2000) Anther culture of recalcitrant indica × Basmati rice hybrids. Euphytica 114:93–101
14. Zapata-Arias FJ (2003) Laboratory protocol for anther culture technique in rice. In: Maluszynski M, Kasha KJ, Forster BP, Szarejko I (eds) Doubled haploid production in crop plants: a manual. Kluwer Academic Publisher, Dordrecht, The Netherlands, pp 109–116

15. Serrat X, Cardona M, Gil J et al (2013) A Mediterranean japonica rice (*Oryza sativa*) cultivar improvement through anther culture. Euphytica 195:31–44
16. Grewal D, Manito C, Bartolome V (2011) Doubled haploids generated through anther culture from crosses of elite indica and japonica cultivars and/or lines of rice: large-scale production, agronomic performance, and molecular characterization. Crop Sci 51:2544–2553
17. Lantos C, Jancsó M, Pauk J (2005) Microspore culture of small grain cereals. Acta Physiol Plant 27:631–639
18. Thomas WTB, Forster BP, Gertsson B (2003) Doubled haploids in breeding. In: Maluszynski M, Kasha KJ, Forster BP, Szarejko I (eds) Doubled haploid production in crop plants: a manual. Kluwer Academic Publisher, Dordrecht, The Netherlands, pp 337–349
19. Pauk J, Jancsó M, Simon-Kiss I (2009) Rice doubled haploids and breeding. In: Touraev A, Foster BP, Jain SM (eds) Advances in haploid production in higher plants. Springer Science, Berlin/Heidelberg, Germany, pp 189–197
20. Germanà MA (2011) Gametic embryogenesis and haploid technology as valuable support to plant breeding. Plant Cell Rep 30:839–857
21. Tripathy SK (2021) High-throughput doubled haploid production for indica rice breeding. In: Segui-Simarro JM (eds) Doubled haploid technology. Methods in molecular biology, vol 2287. Humana, New York, NY
22. Lantos C, Jancsó M, Székely Á et al (2022) Improvement of anther culture to integrate doubled haploid technology in temperate rice (*Oryza sativa* L.) breed. Plants 11(24):3446
23. Lantos C, Jancsó M, Székely Á et al (2023) Development of in vitro anther culture for doubled haploid plant production in indica rice (*Oryza sativa* L.) genotypes. Plants 12(9):1774
24. Galbraith DW, Harkins KR, Maddox JM et al (1983) Rapid flow cytometric analysis of the cell cycle in intact plant tissues. Science 220:1049–1051

Check for updates

Chapter 15

Indica Rice Anther and Shed Microspore Culture

Samira Tajedini, Mehran E. Shariatpanahi, Abdelbagi Mukhtar Ali Ghanim, Maryam Tavakoli and Sakineh Farhadi-Tooli

Abstract

During the last few decades, the induced mutation and selection procedure has been widely used as an efficient method for the rapid production of biotic stress-resilient plants in a short period of time. The efficiency of this method has been shown to be further enhanced when combined with the doubled haploid (DH) technology. DH technology facilitates the rapid and efficient identification of recessive and dominant mutants in the first generation at either haploid or DH levels. Additionally, it also fixes the mutation without the need for further selection, which is commonly used in the conventional mutation breeding. To be practically feasible in breeding programs, DH technology needs to be well optimized in the desired species to obtain sufficient numbers of stable DH plants in a wide range of genotypes. The protocol described here represents an efficient procedure for DH production via both anther and shed-microspore culture (SMC) techniques in *O. sativa* L.

Keywords Rice, Androgenesis, Anther culture, Shed-microspore culture, Haploid embryogenesis

1 Introduction

Rice (*Oryza sativa* L.) is one of the most important cereal crops and is considered a staple food for 60% of the world's population [1]. Anticipating a global population exceeding nine billion by 2050, the demand for rice is projected to outstrip supply [2]. This steeply increasing human population demands an increase in rice production. In this context, doubled haploid (DH) technology, a biotechnology-based breeding method, could play an important role in rice improvement [3]. DH induction is the fastest route to produce pure lines and accelerate breeding of self-pollinated crops [4–6]. The use of DH increases the efficiency of selection and facilitates the immediate fixation of mutant traits in mutation breeding programs [7, 8]. Since there would be no segregation, stable mutants can be obtained in one generation after irradiation, significantly reducing the breeding cycles for the production of breeding materials [9]. DH lines are

Ana Maria Castillo et al. (eds.), *Doubled Haploid Methods in Cereals*, Methods in Molecular Biology, vol. 3029,
https://doi.org/10.1007/978-1-0716-5245-9_15,

also useful for genetic studies, such as QTL mapping, gene transformation, reverse breeding, genomics, and gene identification [7, 8]. Androgenesis, the process in which male gametophytes are induced to follow the sporophytic pathway by the application of a stress treatment and culture [10], is the most widely used in vitro method for producing DH plants [11]. Anther culture and isolated microspore culture are the two main methods of androgenesis. Additionally, a new two-layer culture method called shed-microspore culture (SMC) has been developed [12]. Anther culture has been reported as the preferred and proven efficient method for DH production in rice [1, 13]. Here, we describe protocols for DH production in rice using both anther and SMC procedures, in the context of the integration of this technology into a Coordinated Research Project (CRP) to enhance the efficiency of mutation breeding for *Striga* resistance in cereals. The protocol was validated on the first generation of irradiated material (M1) in two laboratories, including ABRII-Iran and PBGL-IAEA-Austria.

2 Materials

2.1 Plant Material

The gamma irradiated (at 100 Gy) rice cvs. including Wakara, Sahel 305, Remar, Kosti, and Umgar provided by the Plant Breeding and Genetics Laboratory (PBGL) of the Joint Food and Agriculture Organization/International Atomic Energy Agency (FAO/IAEA)-Division of Nuclear Techniques in Food and Agriculture, along with the Iranian Hashemi, were used. The rice cultivar Taipei, an internationally known responsive genotype to androgenesis, also provided by the FAO/IAEA, was also used to validate the established protocol.

2.2 Equipment

1. Epifluorescent microscope with a 358 nm excitation filter and a 461 nm emission filter for DAPI visualization, and a 490 nm excitation filter and a 525 nm emission filter for FDA visualization.
2. Stereomicroscope.
3. Inverted microscope with 200× magnification.
4. Flow Cytometer: PA-I Ploidy Analyzer (Partec-Sysmex) with a 488 nm Argon laser for excitation of Propidium Iodide (PI) fluorochrome.
5. Autoclave.
6. Incubators with controlled temperatures set at 25°C, maintaining a constant humidity (≥95% relative humidity-RH-) without light.
7. Growth chamber with controlled temperature set at 24 ± 2°C, a light intensity of 3000 lux provided by LED lamps, and 16 h light.

8. Growth chamber set at 30°C, a light intensity of 3000 lux provided by LED lamps and 24 h light.
9. Refrigerated centrifuge.
10. Vacuum filtration System.
11. Laminar flow cabinet.
12. pH meter
13. Refrigerator
14. Freezer.
15. Self-contained Gammacell 220 Cobalt-60 research irradiator at FAO/IAEA.

2.3 Greenhouse Materials

1. Greenhouse with controlled conditions at 25°C with a 16 h photoperiod and supplemented with incandescent lamps that provide 20,000 lux light intensity.
2. Plastic pots (3400 cm^3, 18 cm diameter).
3. Substrate: Peat: farm soil (1:2, *w:w*).
4. Fertilizers: NPK (20:20:20, 3000 mg/L), micro fertilizers (2000 mg/L), humic acid (2000 mg/L), amino acid (1 mL/L), calcium nitrate (2000 mg/L), magnesium chelate (2000 mg/L).

2.4 Laboratory Materials

1. Mortar and pestle.
2. Microscope slides and coverslips.
3. Pasteur and automatic pipettes.
4. Eppendorf microtubes.
5. Petri dishes (60 × 15 mm, 80 × 15 mm, 100 × 15 mm).
6. Parafilm, scissors.
7. Forceps.
8. Glass bottles, beakers, Erlenmeyer flasks, measuring cylinders.
9. Magnetic stirrer, gyratory shakers.
10. Scalpel, blades, forceps.
11. Sterile 0.22 μm pore size membrane filters.
12. Whatman filter paper.
13. Nylon mesh for flow cytometry (30 μm, 50 μm).
14. Aluminum foil.

2.5 Solutions and Chemicals

1. CyStain UV Precise P kit (Sysmex), store at 4°C.
2. Fluorescein diacetate (FDA) (5 mg/mL): Dissolve 50 mg of FDA in 10 mL of acetone, and store at 4°C.
3. Ethanol (50% and 70%).

4. Ethanol: acetic acid (3:1, *v:v*).
5. Tween 20.
6. Hypochlorite solution (1% active chlorine in water).
7. Sterile distilled water.
8. Chronic-formaldehyde (1:1, *v:v*): 10% Chromic acid mixed with 10% formaldehyde.
9. Sodium hydroxide (NaOH) 1 N.
10. Unsaturated α-bromonaphthalene stock solution: Add 1 mL of pure α-bromonaphthalene to 100 mL of pure ethanol and store in the dark at room temperature (RT). Prepare the working solution by adding 0.2 mL of the stock solution to 100 mL of distilled water. This solution should always be freshly prepared.
11. Glacial acetic acid (45%).
12. Hematoxylin (4%): Mix 4 g of hematoxylin powder with 100 mL of 45% acetic acid, then add 1 g of NH_4 $FE(SO_4)_2$. Store for a week at RT in the light, stirring several times, then stored at 4°C.
13. Dimethyl sulfoxide (DMSO).
14. Colchicine solution (0.1% *w:v*): Dissolve 1.0 g of colchicine in 20 mL DMSO. Make up to 1 L with distilled water, adding

Table 1
Yoshida's culture solution

	Component	Concentration mg/L	Stock solution g/L (800×)	Amount taken for 4 L (mL)
A	NH_4NO_3	114.25	91.4	5
B	$NaH_2PO_4{\cdot}2H_2O$	50.375	40.3	5
C	K_2SO_4	89.25	71.4	5
D	$CaCl_2$	110.75	88.6	5
E	$MgSO_4{\cdot}7H_2O$	405.00	324	5
F	$MnCl_2{\cdot}4H_2O$	1.875	1.5	5
	$(NH_4)6Mo_7O_{24}{\cdot}4H_2O$	0.0925	0.074	
	H_3BO_3	1.1675	0.934	
	$ZnSO_4{\cdot}7H_2O$	0.04375	0.035	
	$CuSO_4{\cdot}5H_2O$	0.03875	0.031	
	$FeCl_3{\cdot}6H_2O$	9.625	7.7	
	Citric acid (monohydrate)	14.875	11.9	
pH		5		

three drops of Tween 20 to enhance colchicine penetration and store at RT.

15. CyStain PI Absolute kit (Sysmex), store at 4°C.
16. The composition of Yoshida´s nutritive solution, used for rice plants' hydroponic growth, is detailed in Table 1 [14]. Prepare six stock solutions (800×): five of them contain each macronutrient, and the sixth one contains micronutrients, iron source, and citric acid. To prepare 4 L of nutritive solution, take 5 mL of each macronutrient stock solution. Adjust pH to 5 with nitric acid. Add 5 mL of the sixth stock solution and ensure that pH is 5.
17. B5 powder.

2.6 Culture Media

The composition of the induction, pre-regeneration, regeneration, and rooting media used for anther culture is shown in Table 2.

1. A modified N6 medium is used for callus induction [15]. The following stock solutions are prepared:
   - Macronutrient stock solution (10×): dissolve each compound separately in 100 mL of distilled water. Then, gently mix all the compounds together and bring to a total volume of 1 L.
   - Micronutrients stock solution (100×): add each compound separately to a beaker containing 100 mL of distilled water and stir until completely dissolved before adding the following one. Take up the volume to 200 mL.
   - Vitamin stock solution (100×): add each vitamin to a beaker containing 100 mL of distilled water and stir until completely dissolved before adding the following one. Take up the volume to 200 mL.
   - NAA (1 mg/1 mL): dissolve 10 mg of NAA in NaOH, take up the volume to 10 mL with distilled water, filter sterilize, aliquot in sterile tubes, and store it at 4°C.
   - 2-4-D (1 mg/1 mL): dissolve 10 mg of 2,4-D in KOH, take up the volume to 10 mL with distilled water, filter sterilize, aliquot in sterile tubes, and store it at 4°C.
   - Kinetin (1 mg/1 mL): dissolve 10 mg of kinetin in the minimum amount of NaOH, bring the volume to 10 mL with distilled water, filter sterilize, aliquot into sterile tubes, and store at −20°C.

   To prepare 1 L of medium:

   - Add 700 mL of distilled water to an Erlemmeyer flask.
   - Add 100 mL of macronutrients stock solution, 10 mL each of micronutrients and vitamins stock solutions, and 1 mL, 2 mL, and 2.5 mL of Kinetin, 2,4-D and NAA stock solutions, respectively.

- Weigh the corresponding amounts of FeNa2EDTA and FeSO4·7H2O, myoinositol, casein hydrolysate, and sucrose (Table 2), separately and add them to the medium.
- Take the volume to 1 L.
- Adjust the pH to 5.8.
- Add 8 g of agar.
- Autoclave the medium at 121°C for 15 min.
- Pour 6 mL of medium in 60 × 15 mm Petri dishes.

2. The pre-regeneration, regeneration, and rooting medium is a modified Murashige and Skoog medium [16] (Table 2). The following stock solutions are prepared:
   - MS1 stock solution (50×): 95 g/L KNO_3 + 82.5 g/L NH_4NO_3.
   - MS2 stock solution (100×): 37 g/L $MgSO_4{\cdot}H_2O$ + 2.23 g/L $MnSO_4H_2O$+0.86 g/L $ZnSO_4$·7H2O+0.0025 g/L $CuSO_4{\cdot}5H_2O$.
   - MS3 stock solution (100×): 44 g/L $CaCl_2{\cdot}2H_2O$ + 0.083 g/L KI + 0.0025 g/L $CoCl_2{\cdot}6H_2O$.
   - MS4 stock solution (100×):17 g/L KH_2PO_4 + 0.62 g/L H_3BO_3 + 0.025 g/L $Na_2MoO_4{\cdot}2H_2O$.
   - Vitamins stock solution (200×): 10 mg/100 mL nicotinic acid, 10 mg/100 mL pyridoxine HCl, 2 mg/100 mL thiamine HCl, 40 mg/100 mL glycine.
   - NAA (1 mg/mL), Kinetin (1 mg/mL) are prepared as described previously.

   To prepare 1 L of medium:
   - Add 700 mL of distilled water to an Erlemmeyer flask.
   - Add 20 mL of MS1 stock solution, 10 mL each of MS2, MS3, and MS4 stock solutions, and 5 mL of vitamins stock solution. For pre-regeneration medium, add 2 mL of Kinetin and 5 mL of NAA stock solutions. For regeneration medium, add 2 mL of Kinetin and 1 mL of NAA stock solutions (Table 2).
   - Weigh and add separately the corresponding amounts of $FeNa_2EDTA$, $FeSO_4{\cdot}7H_2O$, myoinositol, and carbon sources to the medium, depending on the type of medium (pre-regeneration, regeneration, and rooting) (Table 2).
   - Make up the volume to 1 L.
   - Adjust the pH to 5.8.
   - Add the corresponding amounts of Gelrite (pre-regeneration, regeneration, or rooting) (Table 2).
   - Autoclave the medium at 121°C for 15 min.
   - Pour 15 mL of medium in 80 × 15 mm and 100 × 15 mm Petri dishes.

Table 2

Composition of modified N6 medium for callus induction and MS pre-regeneration, shoot regeneration, and rooting media used in rice anther culture

Compounds	Callus induction	Pre-regeneration	Regeneration	Rooting
	N6-1	MS-PREG	MS-REG	MS-Root
Macronutrientes (mg/L)				
KNO_3	2.830	1.900	1.900	1.900
$(NH_4)_2SO_4$	463	–	–	–
KH_2PO_4	170	170	170	170
$CaCl_2 \cdot 2H_2O$	440	440	440	440
$MgSO_4 \cdot 7H_2O$	185	370	370	370
NH_4NO_3	–	1.650	1.650	1.650
Micronutrientes (mg/L)				
$MnSO_4 \cdot 4H_2O$	4.4	22.3	22.3	22.3
$ZnSO_4 \cdot 7H_2O$	1.5	8.6	8.6	8.6
H_3BO_3	1.6	6.2	6.2	6.2
KI	0.83	0.83	0.83	0.83
$Na_2MoO_4 \cdot 2H_2O$	0.25	0.25	0.25	0.25
$CuSO_4 \cdot 5H_2O$	0.025	0.025	0.025	0.025
$CoCl_2 \cdot 6H_2O$	0.025	0.025	0.025	0.025
Vitamins (mg/L)				
Glycine	2	2	2	–
Thiamine HCl	2.5	–	–	–
Pyridoxine HCl	2.5	5	1	–
Nicotinic acid	2.5	100	100	100
Growth regulator and other organic compounds (mg/L)				
Kinetin	1	2	2	–
2,4-D	2	–	–	–
NAA	2.5	5	1	–
Myo-inositol	100	100	100	100
Casein Hydrolysate	500	–	–	–
Iron source (mg/L)				
EDTA, disodium salt	37.285	37.285	37.285	37.285
$FeSO_4 \cdot 7H_2O$	27.85	27.85	27.85	27.85

(continued)

Table 2
(continued)

Compounds	Callus induction	Pre-regeneration	Regeneration	Rooting
	N6-1	MS-PREG	MS-REG	MS-Root
Carbon source (mg/L)				
Sucrose	40,000	–	30,000	3000
Maltose	–	30,000	–	
Sorbitol	–	20,000	–	
Other compounds (mg/L)				
Agar	8000	–	–	–
Gelrite	–	3000	3000	3000
pH	5.8	5.8	5.8	5.8

3. The composition of the double-layered modified B5 medium [17] used for SMC is shown in Table 3.

   To prepare 1 L of B5 media:

   - For both the upper and lower layer B5 medium, add 3.17 g of B5 macro and micro nutrients powder in an Erlenmeyer flask containing 700 mL of water.
   - For B5 lower layer medium, add 90 g maltose and 10 g active charcoal, adjust pH at 5.8, make up to 1 L, and add 7 g Agar. Autoclave the medium at 121°C for 15 min and pour 4 mL of medium in 60 × 15 mm Petri dishes.
   - For B5 upper layer medium, add 1,256 mg glutamine, 10 mg Arabic gun and 30 g maltose, adjust pH at 5.8, make up the volume to 1 L, and sterilize by filtration through a 0.22 μm PES filter.

All the macronutrients and micronutrients stock solutions can be made in advance and stored in a refrigerator. Vitamins and growth regulators stock solutions are kept in plastic tubes in a refrigerator. Solid media are stored at RT, whereas liquid media can be stored in the refrigerator for a few days (*see* **Note 1**).

3 Methods

3.1 Growth of the Donor Plants

1. Irradiate seeds at low doses, usually between 100 and 200 Gy, in order to produce M0 plants (*see* **Notes 2 and 3**).
2. Germinate the irradiated seeds in Petri dishes (100 × 15 mm) containing water and filter paper, in a growth chamber with

Table 3
Composition of double-layered modified B5 medium for shed microspore culture

Compounds	B5 lower layer (mg/L)	B5 upper layer (mg/L)
Macronutrients (mg/L)		
$CaCl_2$	113.23	113.23
KNO_3	2500	2500
$MgSO_4$	121.56	121.56
NaH_2PO_4	130.44	130.44
$(NH_4)_2SO_4$	134	134
Micronutrients (mg/L)		
$CoCl_2 .6H_2O$	0.025	0.025
$CuSO_4 .5H_2O$	0.025	0.025
H_3BO_3	3	3
KI	0.75	0.75
$MnSO_4.4H_2O$	10	10
$Na_2MoO_4{\cdot}2H_2O$	0.25	0.25
$ZnSO_4 .7H_2O$	2	2
Vitamins (mg/L)		
Myo-Inositol	100	100
Nicotinic acid	1	1
Pyridoxine HCl	1	1
Thiamine HCl	10	10
Other compounds (mg/L)		
EDTA disodium salt dehydrate		37.300
$FeSO_4.7H_2O$		27.800
Glutamine		1256
Arabic gum		10
Maltose	90,000	30,000
Active charcoal	10,000	
Agar	7000	
pH	5.8	5.8

controlled temperature set at 24±2°C, with 3000 lux light intensity provided by LED lamps for 16h (Fig. 1a, b). About 7 days after germination, transplant the seedlings into boxes or pots.

3. Keep the plants in the greenhouse, at a constant day/night temperature of 25°C with a 16h photoperiod and 20,000 lux.

Fig. 1 Growth of rice mother donor plants. (**a**) Irradiated seeds of selected genotypes/varieties/F1 hybrids plated in Petri dishes. (**b**) Germinated seeds after 7 days. (**c, d**) Plantlets transferred and grown in the greenhouse (**e**) Panicles harvested and wrapped in moistened filter paper. (**f**) Panicles placed in aluminum foil for cold-shock pre-treatment at 4°C for 8–10 days

Irrigate donor plants regularly with tap water once every 5 days. Additionally, apply Yoshida´s solution to the donor plants weekly (Table 1), and regularly inspect to ensure they are well watered (Fig. 1c, d).

3.2 Panicle Harvest

1. Collect panicles from the primary tillers when the distance between the flag leaf and the lower (penultimate) leaf is approximately 8–12 cm (Fig. 2a). This distance varies depending on the genotype.
2. Excise anthers from the central part of the panicles to determine the correlation of the distance described previously and the developmental stage of the microspores by DAPI staining (*see* Sect. 3.2.1). Viability of microspores is also studied by FDA staining, (*see* Sect. 3.2.2).
3. Further collection of panicles can be done from secondary tillers containing microspores at late uninucleate stages, according to the DAPI studies (*see* **Notes 4 and 5**).

3.2.1 Determination of the Developmental Stage of the Microspores

1. Use mortar and pestle to release microspores from the rice florets with mannitol 0.3 M solution. Filter the solution through a 50 μm mesh and centrifuge the suspension to obtain the microspores. Fix 50 μL of the cell suspension in ethanol: acetic acid (3:1, *v:v*) for 15–20 min.
2. Wash twice in 50% ethanol, centrifuge for 3 min at 367 × *g* and 1 min at 15,493 × *g*.

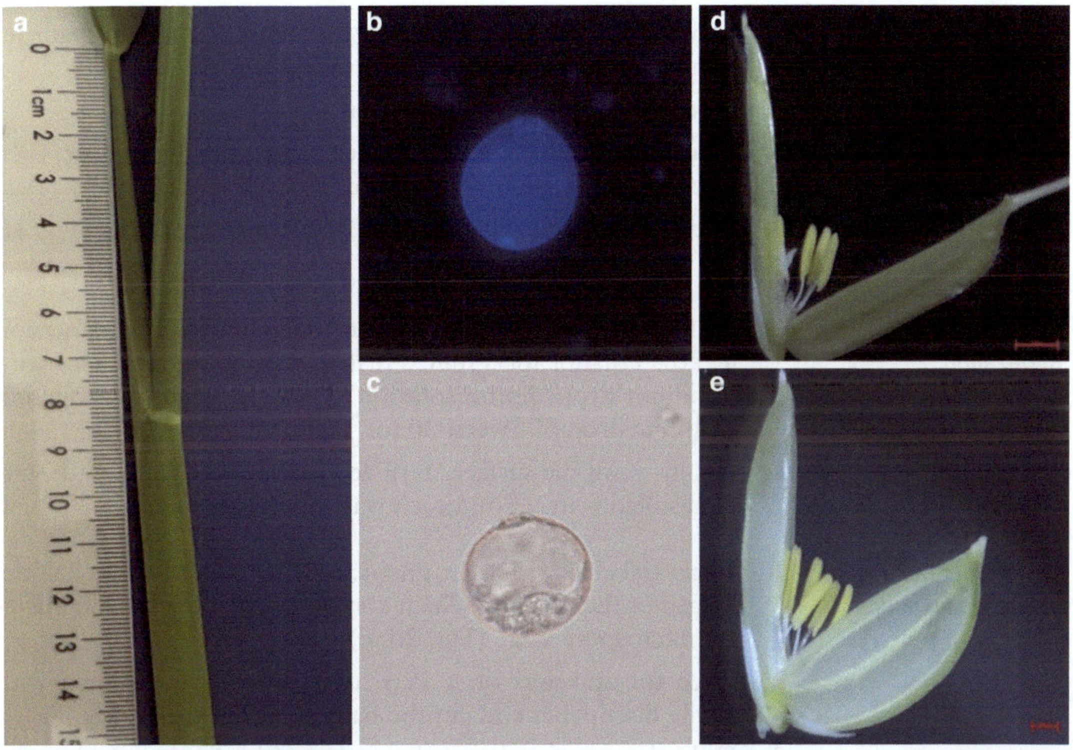

Fig. 2 Determination of the microspore developmental stage. (**a**) Panicle at the appropriate developmental stage (the ruler indicates the distance between the flag and the penultimate leaf). (**b, c**) Microspore at the late uninucleate stage, stained with DAPI and observed under brightfield, respectively. (**d, e**) Florets containing microspores at the late uninucleate stage (the anther length should not exceed half the floret length)

3. Remove supernatant and add 40 μL DAPI solution from CyStain UV Precise P kit (Sysmex).
4. Examine the microspore suspension under an epifluorescence microscope using UV-light and a 350 nm excitation filter and a 460 nm emission filter (Figs. 2b, c and 3) [18].

3.2.2 Determination of Microspore Viability

1. Use a mortar and pestle to isolate microspores from excised anthers used for SMC culture or microspore suspension extracted from upper layer of B5 medium (Fig. 4a–c).
2. Centrifuge at 367 × *g* for 2 min at RT (Fig. 4d).
3. Remove the supernatant, take 100 μL of microspore suspension into an Eppendorf tube, add 5–10 μL of FDA stock solution (5 mg/mL) to the suspension (Fig. 4e, f).
4. Gently mix the sample and incubate it in the dark at RT for 5–10 min (Fig. 4g).
5. Place a small drop of the stained suspension onto a microscope slide and cover with a coverslip (Fig. 4h).
6. Observe the slides under an epifluorescence microscope using UV-light and a 490 nm excitation filter and a 525 nm emission filter (Fig. 4i).

3.3 Cool Stress Treatment

1. Remove the outer leaf sheaths and ensure that the panicles remain wrapped within the inner leaf sheath.
2. Wash the panicles with tap water.
3. Wrap the panicles with moistened Whatman filter paper, then place them in a plastic bag or aluminum foil (Fig. 1e).
4. Apply a cold-shock at 4°C for 8–10 days to the panicles (Fig. 1f) (*see* **Note 6**).

3.4 Panicle Sterilization

1. Surface-sterilize the cold pre-treated panicles by immersion in 70% ethanol for 1 min. Further sterilize the panicles with a sodium hypochlorite solution (1% active chlorine) together with one drop of Tween 20 for 10 min.
2. Finally, wash the surface-sterilized panicles 3-4 times with sterile distilled water, 5 min each wash (*see* **Note 7**).

3.5 Anther Culture and Callus Induction

1. Select spikelets with light-green anthers no longer than half the length of the spikelet, which corresponds to the late uninucleate microspore stage (Fig. 2d, e).
2. Place the spikelets on a Petri dish (60 × 15 mm), lined with sterile filter paper and cut the base of each spikelet (Fig. 5a).
3. Hold the tip of each spikelet and cut with forceps. Isolate the anthers with the forceps and gently tap on the edge of the Petri dish containing modified N6 medium for callus induction

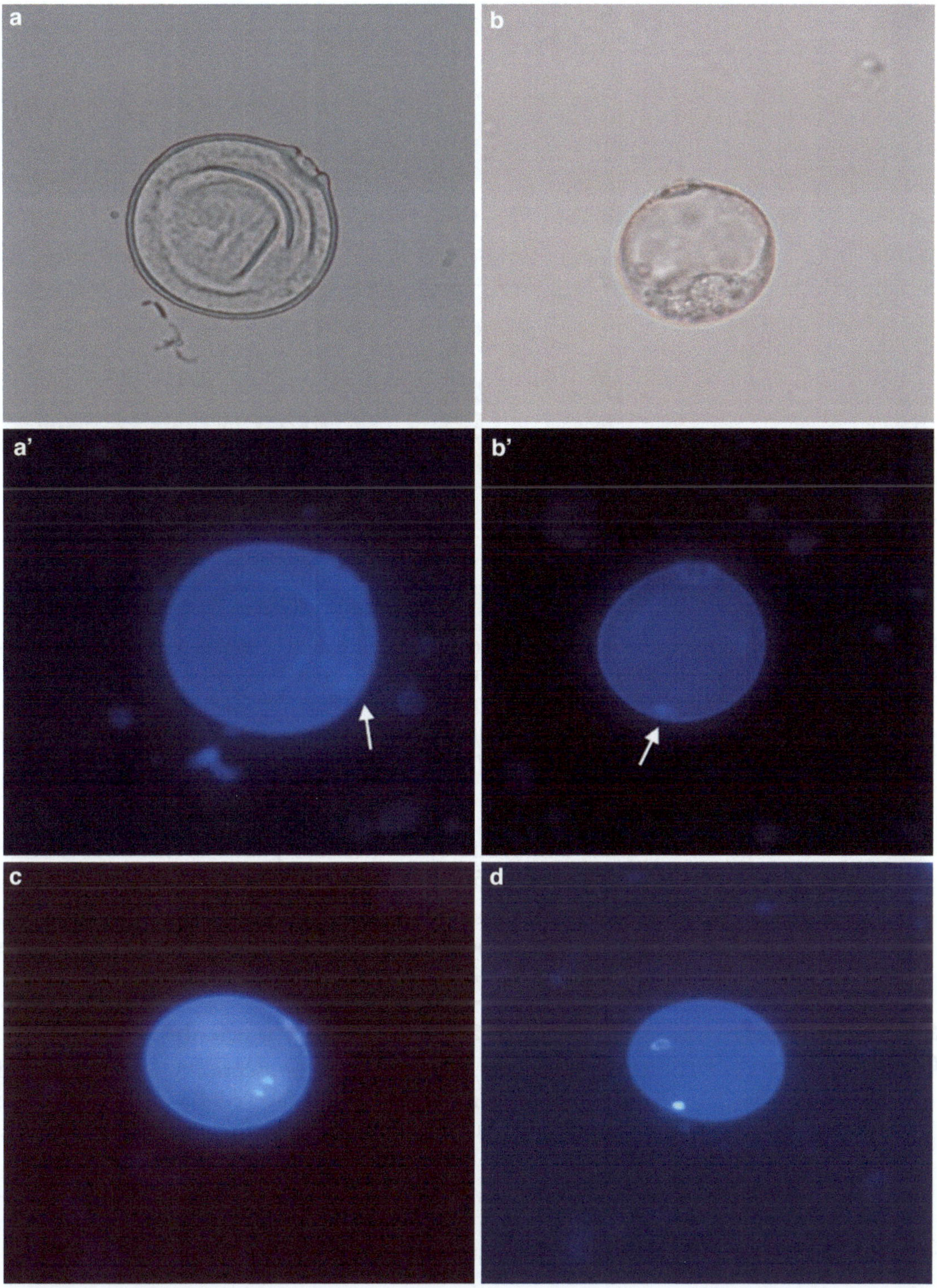

Fig. 3 DAPI staining for determination of the developmental stage of the microspores. (**a, a′**) Microspore at the early uninucleate stage. (**b, b′**) Microspore at the late uninucleate stage. (**c**) Microspore at the early bi-nucleate stage. (**d**) Microspore at the late bi-nucleate stage

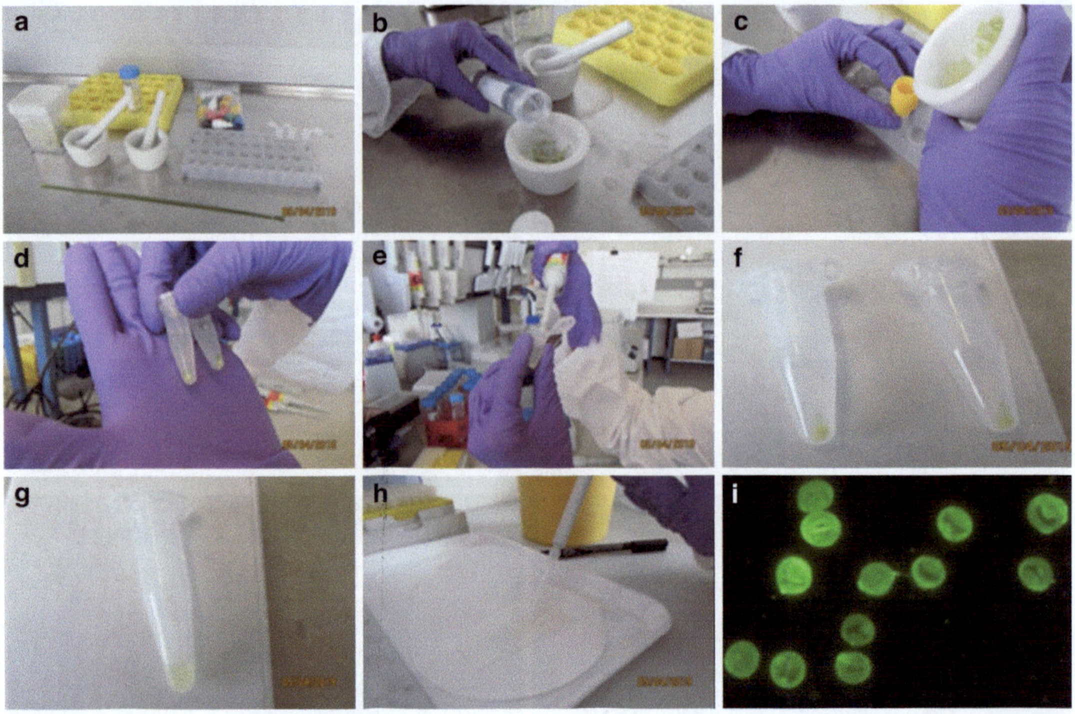

Fig. 4 Microspore isolation procedure and determination of microspore viability. (**a**) Materials needed for microspore isolation under sterile conditions. (**b**) Release of microspores from rice florets with a mortar and pestle. (**c**) Filtration of the microspore suspension through a 50 μm mesh. (**d**) Centrifugation of the microspore suspension. (**e**) Discarding the supernatant. (**f**) Microspore pellet (**g**) Addition of fluorescein diacetate (FDA). (**h**) Slide preparation. (**i**) Visualization of viable microspores stained by FDA under an epifluorescence microscope

(Table 2). About 60 anthers are plated in each 60 × 15 mm Petri dish (Fig. 5b, c).

4. Seal the Petri dishes with Parafilm, wrap them with aluminum foil, and keep the cultures in the dark at 25°C. Observe callus formation after 4 weeks (Fig. 5d–f).

3.6 Shed Microspore Culture (SMC)

1. Follow Steps 1–2 of anther culture as described previously.
2. Dissect anthers from the panicles aseptically and transferred them onto the modified double-layered B5 culture medium (Table 3; Fig. 6a). The anthers will dehisce after absorbing the liquid medium (Fig. 6b–d).
3. Seal Petri dishes (60 × 15 mm) with Parafilm and wrap in aluminum foil. Incubate them at 25°C in the dark for 4 weeks to obtain embryo structures (Fig. 6e).

3.7 Plant Regeneration to Obtain Haploid Plants

1. Transfer the calli generated from either anther culture or SMC to the MS pre-regeneration medium (Table 2; Fig. 7a, b). The origin of the calli can be identified by flow cytometry analysis.

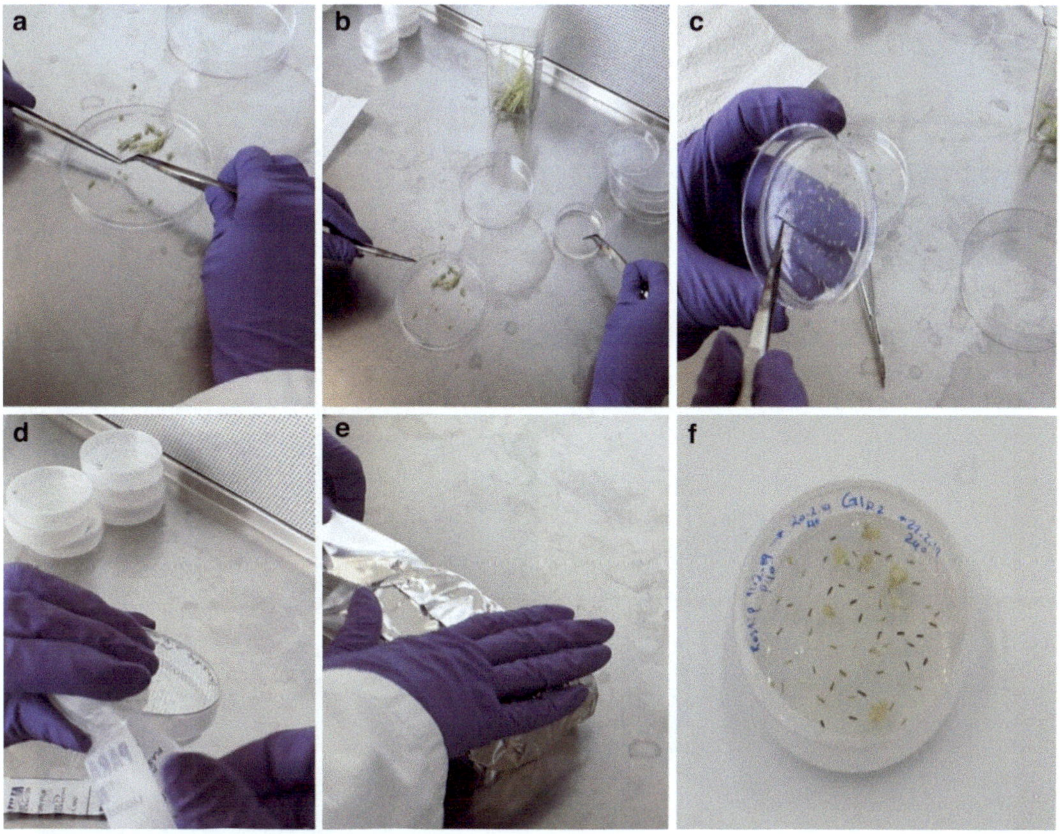

Fig. 5 Anther culture and callus induction. (**a**) Cutting the base of each spikelet. (**b**) Plating anthers on callus-induction medium. (**c**) Plating 60 anthers in each 60 × 15 mm Petri dish. (**d, e**) Sealing Petri dishes with parafilm and wrapping with aluminum foil. (**f**) Induced calli after 4 weeks of culture

2. Incubate the cultures for 4 weeks under continuous light at 30°C, a light intensity of 3000 lux provided by LED lamps until green spots appear (Fig. 7c).
3. Transfer the proliferated calli to the MS regeneration medium (Table 2; Fig. 7d) and incubate the cultures under the same growing conditions.
4. Subculture every 10 days for shoot regeneration in the same medium (Fig. 7e).
5. Transfer the regenerated shoots to the MS rooting medium at 30°C, a light intensity of 3000 lux provided by LED lamps and 24 h light (Table 2; Fig. 7f) (*see* **Notes 8–10**).

3.8 Acclimatization

1. After 7–14 days in the rooting medium, transfer the plantlets with well-developed roots to Yoshida's culture solution for 1–2 months (Table 1) at 25°C under 60% RH, with a 16 h photoperiod, supplemented with incandescent lamps that provide

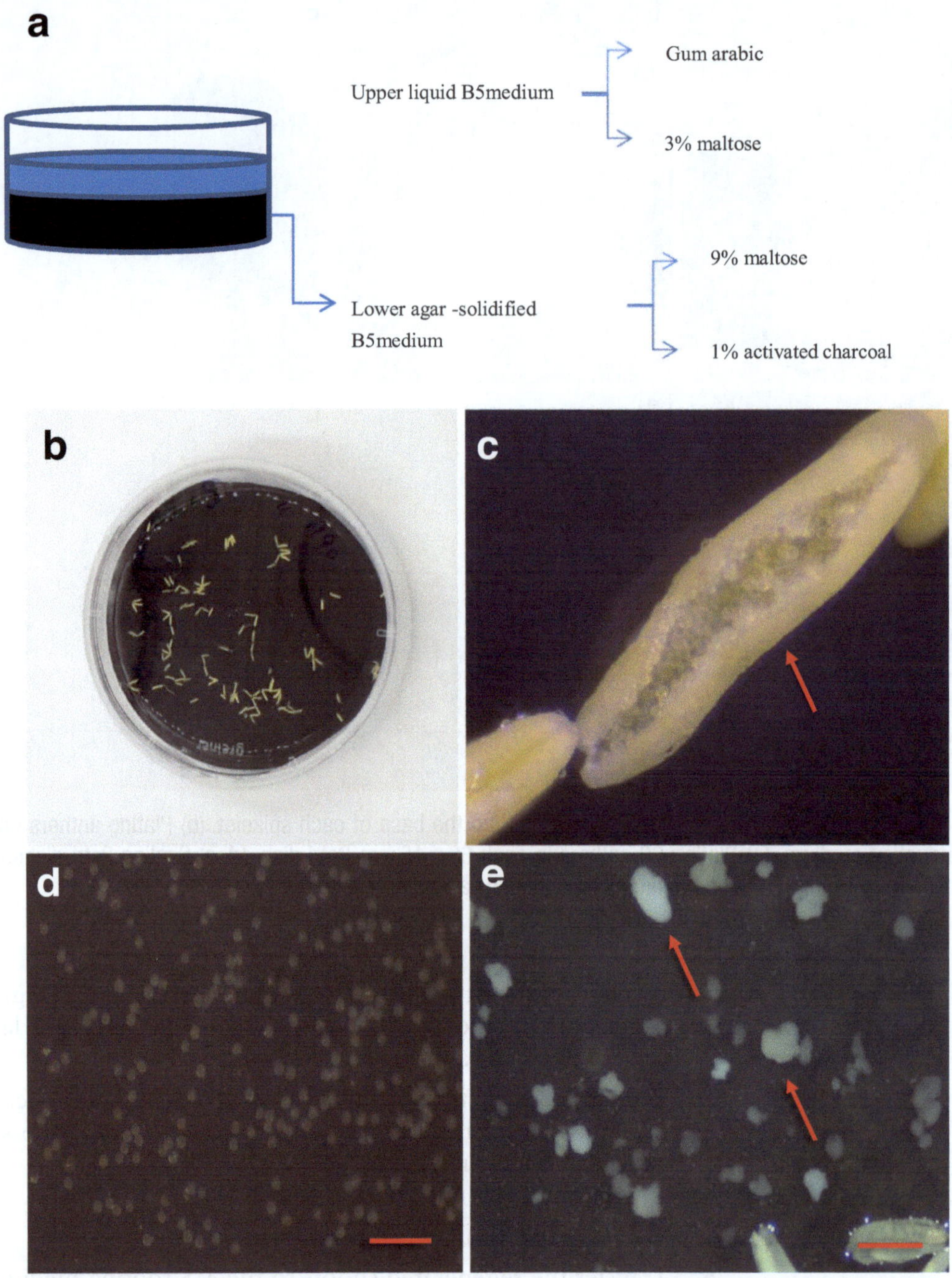

Fig. 6 Shed microspore culture (SMC) of rice mutant cultivars. (**a**) Schematic illustration of the double-layered culture system. (**b**) Anthers in the double-layered B5 medium. (**c**) Splitting of floating rice anthers in the upper liquid medium after 7 days of culture. (**d**) Microspores shed in the liquid induction medium after 7–10 days of culture (scale bar = 1000 μm). (**e**) Emergence of embryo structures from shed microspores after 4 weeks of culture (scale bar = 0.5 mm)

Fig. 7 Plant regeneration and haploid plant production. (**a**) Embryogenic calli from anther culture of Indica rice formed after 4 weeks of culture. (**b**) Transfer of embryogenic calli to MS pre-regeneration medium. (**c**) Callus showing green spots in regeneration medium. (**d, e**) Regenerated green shoots. (**f**) Plantlets in MS rooting medium

20,000 lux light intensity covered with plastic glass in a greenhouse (Fig. 8a).

2. Transfer the plantlets with well-developed roots to pots filled with peat: farm soil (1:2, *w:w*). Cover the plants with a plastic glass and maintain with 60% RH for acclimatization in the greenhouse (25°C).
3. Remove the plastic cover gradually and allow the plantlets to continue growing in the greenhouse until they reach the 5-6-leaf stage (Fig. 8b).
4. Check the ploidy level of the regenerated plants using an appropriate and available procedure, e.g., chromosome counting, flow cytometry or morphological observations, before colchicine treatment (Fig. 8c).

3.9 Ploidy Analysis by Flow Cytometry

1. Take leaf discs (0.5 cm^2 in diameter) from the third upper young leaf of the stem of both diploid control and the putative regenerated haploid plantlets.
2. Finely chop the samples with a clean and sharp razor blade in 2–3 mL of ice-cold nuclei isolation buffer from Cystain PI kit inside a Petri dish.
3. Gently pipette the chopped mixture a few times with a clean Pasteur pipette in order to accelerate the release of nuclei.

Fig. 8 Hardening haploid plants. (**a**) Haploid-plantlets with vigorous roots in Yoshida's solution. (**b**) Transplanting regenerated plants into the soil, acclimatization, and maintaining them in the glasshouse at 25°C with 16 h photoperiod. (**c**) Haploid plantlets before colchicine treatment. (**d**) DH (left) and haploid (right) panicles at seed setting stage

4. Filter the crude extract through a 50 μm nylon mesh into test tubes.
5. Add 200 μL of PI staining buffer, 12 μL of PI, and 6 μL of RNase per sample to the filtered suspension. Nuclei can be kept in this solution for 5 min to 48 h at 4°C.
6. Filter the sample for a second time through a 30 μm nylon mesh.
7. Analyze the samples by flow cytometry (Fig. 9a, b) [19] (*see* **Note 11**).

3.10 Ploidy Analysis by Chromosome Counting

1. Take 1.5–2 cm of the root tip of the regenerated plantlets and treat them with α-bromonaphthalene at 4°C for 4 h.
2. Rinse the treated root tips in distilled water and fix them in chromic oxide (10%): formaldehyde (10%) (1:1, *v:v*) solution for 48 h.

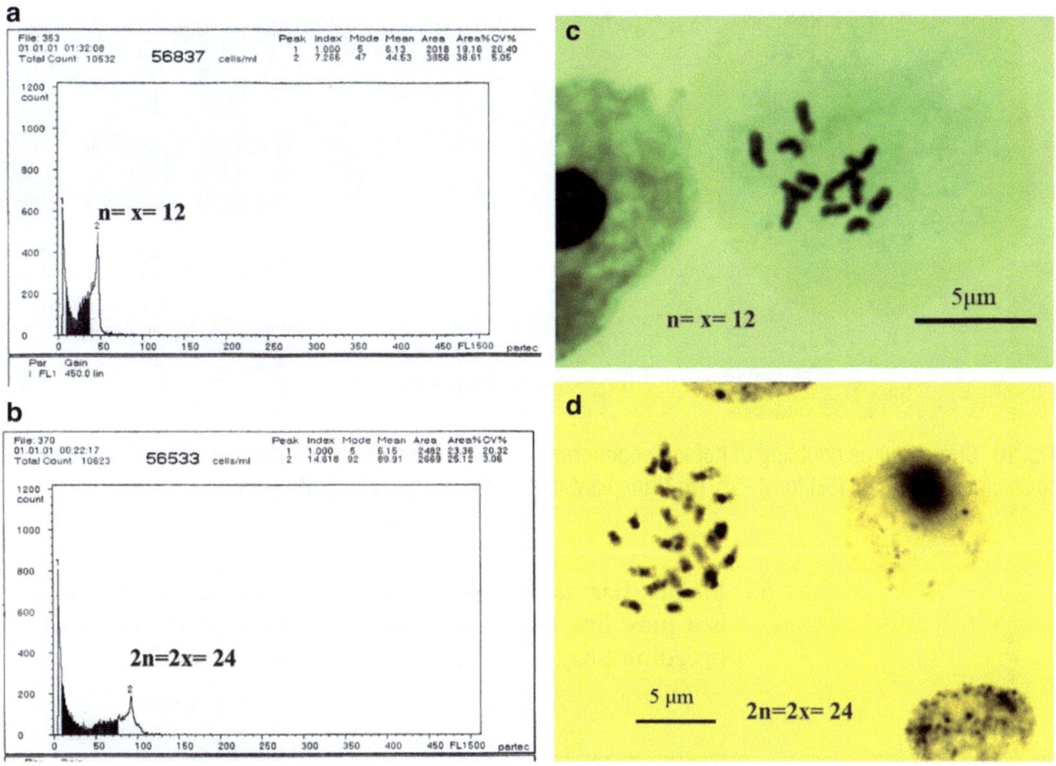

Fig. 9 Ploidy determination by flow cytometry and chromosome counting of regenerated haploid plantlets and control diploid plants of *Oryza sativa* L. (**a, c**) Haploid plantlets. (**b, d**) Diploid plants (donor plants)

3. Keep the samples in 70% ethanol at 4°C for 24 h.
4. Hydrolyze root tips in 1 N NaOH for 10 min at 60°C.
5. Rinse and transfer samples to hematoxylin (4%) and incubate them at 30°C for 16 h for staining.
6. At the final step, transfer each root tip onto a microscope slide, add a drop of glacial acetic acid (45%), and then squash. Count chromosomes of three to four scattered cells (Fig. 9c, d).

3.11 Chromosome Doubling of Haploid Plants and Seed Production from DH Plants

1. Cut the roots and the upper parts of the plants at approximately 4 cm and 15 cm from the plant base, respectively.
2. Completely immerse the roots of the plants in colchicine solution (0.1%) for 4–5 h in the greenhouse [20]. Make sure that enough amount of solution is prepared in order to immerse the plants up to 5 cm above the base of their stems (Fig. 10a).
3. Wash the roots of the colchicine-treated plants carefully under the running tap water for 2.5 h.
4. The washed plantlets are then transferred to pots filled with peat:farm soil (1:2, *w:w*) (Fig. 10b).

Fig. 10 Chromosome doubling of haploid regenerants to restore fertility. (**a**) Immersion of the plant roots in the colchicine solution (0.1%) for 4–5 h. (**b**) Transplantation of the treated plant into pots

5. Finally, harvest the seeds of the DH plants. Each DH plant is a pure line (with 100% homozygosity) that can be used in breeding programs (Fig. 8d) (*see* **Notes 12 and 13**).

4 Notes

1. Liquid media should be checked for contamination by keeping the freshly-made liquid media at RT for 2–3 days.
2. M1/M2 or M3/M4 seeds can be used to integrate the haploidy system. Practically, plants of any generation from M1 to Mn can be used to produce DH lines, but for accelerated breeding, the earlier the better.
3. Mutant plants are used as donor material for DH production. When the M1 generation is used to produce haploids, any recessive mutations are easily identified phenotypically as soon as the chromosome doubling is performed.
4. Select anthers containing majority (more than 80%) of microspores at the late-uninucleate stage, as it is the most responsive stage to anther/SMC. The panicle should be at the proper developmental stage. The distance between the flag and the penultimate leaf is correlated with the microspore developmental stage.
5. In order to induce embryogenesis in microspores, certain changes must occur in the cytoplasm and nucleus, which are possible at a certain time. In many species, the most appropriate stage is uninucleate or only early binucleate (just after the first mitosis of the pollen grain). The detection of this stage is

one of the key steps, which is done by staining the extracted microspores with DAPI.

6. Cold pretreatment of panicles is critical for blocking gametophytic development and reducing the number of albino regenerated plants in anther and SMC systems. A 10-day cold pre-treatment at 4°C is optimal for callus/embryo induction in all rice cultivars tested. Cold shock pre-treatment of rice panicles delays the senescence of anther wall tissue, providing sufficient time for microspore development [1].
7. Perform all tissue culture processes in the laminar flow cabinet that has been previously disinfected with 70% ethanol.
8. In anther culture, callus induction and regeneration were achieved in all cultivars tested. However, Hashemi and Taipei cultivars were the most responsive ones. The highest percentage of calli/100 anthers was 31.66 ± 7.63%. The frequency of regeneration of haploid plants/100 calli was up to 45 ± 5%.
9. In SMC, embryo induction and regeneration were observed in all cultivars tested. The highest percentage of regenerated embryos/100 anthers was produced in cultivars Taipie (25 ± 5%) and Hashemi (17.66±2.51%). The highest percentage of regeneration of haploid plants/100 embryos was 25.1 ± 5.3%.
10. Three technologies, including anther culture, SMC, and isolated microspore culture, were compared. Results showed that anther and SMC culture methods responded better than isolated microspore culture.
11. In flow cytometry analysis, it is necessary to use a diploid plant as control to set the position of the diploid peak on the x-axis before analyzing the samples.
12. The frequency of doubled haploid (DH) plants produced from 100 calli/embryos was up to 26 ± 1.9%.
13. Workable and reproducible DH protocols, via androgenesis (anther and SMC), applicable for the production of mutant DH populations in rice, were developed.

Acknowledgments

This work has been supported by the International Atomic Energy Agency (IAEA) CRP code: D25005 and contract number: 20742. The authors are thankful to Florian Goessnitzer, Mahnaz Oroojloo, and Behzad Ahnadi for their kind assistance in the lab and greenhouse.

References

1. Sarao NK, Gosal SS (2018) In vitro androgenesis for accelerated breeding in rice. In: Gosal S, Wani S (eds) Biotechnologies of crop improvement, vol 1. Springer, Cham
2. Haiyong G, Shihu L, Junliang Z (2022) Novel sequencing and genomic technologies revolutionized rice genomic study and breeding. Agronomy 12(1):218–218
3. Ruwan Sharma A (2018) Morphology response of growth regulators on callus of Japonica rice through anther culture. Clin Biotechnol Microbiol 2:424–442
4. Thomas WTB, Forster BP, Gertsson B (2003) Doubled haploids in breeding. In: Maluszynski M, Kasha KJ, Forster BP, Szarejko I (eds) Doubled haploid production in crop plants: a manual. Kluwer, Dordrecht, pp 337–349
5. Germanà MA (2011) Gametic embryogenesis and haploid technology as valuable support to plant breeding. Plant Cell Rep 30:839–857
6. Niazian M, Shariatpanahi ME (2020) In vitro-based doubled haploid production: recent improvements. Euphytica 216:69
7. Shariatpanahi ME, Ahmadi B (2016) Isolated microspore culture and its applications in plant breeding and genetics. In: Anis M, Ahmad N (eds) Plant tissue culture: propagation, conservation and crop improvement. Springer, Singapore, pp 487–507
8. Tajedini S, Fakheri B, Niazian M et al (2023) Efficient microspore embryogenesis and haploid induction in mutant indica rice (*Oryza sativa* L.) cultivars. J Plant Growth Regul 42:2345–2359
9. Jankuloski L, Szarejko I, Pauk J (2018) Haploids and doubled haploids in mutation breeding. In: Spencer-Lopes MM, Forster BP, Jankuloski L (eds) Manual on mutation breeding (Third Edition). Food and Agriculture Organization of the United Nations, International Atomic Energy Agency, Vienna, pp 230–248
10. Makowska K, Kaluzniak M, Oleszczuk S et al (2017) Arabinogalactan proteins improve plant regeneration in barley (*Hordeum vulgare* L.) anther culture. Plant Cell Tissue Organ Cult 131:247–257
11. Kasha KJ (2005) Chromosome doubling and recovery of doubledhaploid plants. In: Don Palmer CE, Keller WA, Kasha KJ (eds) Haploids in crop improvement II. Springer, Berlin, pp 123–152
12. Supena EDJ, Suharsono S, Jacobsen E, Custers JBM (2006) Successful development of a shed-microspore culture protocol for doubled haploid production in Indonesian hot pepper (*Capsicum annuum* L.). Plant Cell Rep 25:1–10
13. Khush GS, Virmani SS (1996) Haploids in plant breeding. In: Jain SM, Sopory SK, Veilleux RE (eds) In vitro haploid production in higher plants, vol 1. Springer Netherlands, Dordrecht, pp 11–33
14. Yoshida S, Forno DA, Cock JH, Gomez KA (1976) Laboratory manual for physiological studies of rice, 3rd edn. International Rice Research Institutes, Manila, Philip pines, p 61
15. Chu CC, Wang CC, Sun CS et al (1975) Establishment of an efficient medium for anther culture of rice through comparative experiments on the nitrogen sources. Sci Sin 18: 659–668
16. Murashige T, Skoog F (1962) A revised medium for rapid growth and bio assays with tobacco tissue cultures. Physiol Plant 15:473–497. https://doi.org/10.1111/j.1399-3054.1962.tb08052.x
17. Gamborg OL, Miller RA, Ohima K (1968) Nutrient requirement of suspension cultures of soybean root cells. Exp Cell Res 50: 151–158
18. Vergne P, Delvallee I, Dumas C (1987) Rapid assessment of microspore and pollen development stage in wheat and maize using DAPI and membrane permeabilization. Stain Technol 62:299–304
19. Dpooležel J, Binarová P, Lcretti S (1989) Analysis of nuclear DNA content in plant cells by flow cytometry. Plant Biol 31:113–120
20. Zapata-Arias (2003) Laboratory protocol for anther culture technique in rice. In: Maluszynski M, Kasha KJ, Forster BP, Szarejko I (eds) Doubled haploid production in crop plants. A manual. Kluwer Academic Publishers, Dordrecht, The Netherlands, pp 109–116

Chapter 16

Isolated Microspore Cultures of Triticale (× *Triticosecale* Wittm.)

Iwona Żur, Ewa Dubas and Monika Krzewska

Abstract

Microspore embryogenesis (ME) is the shortest route to total homozygosity, as it results in the production of doubled haploid (DH) plants. However, the huge theoretical potential of this method is only rarely realized in practice, as there are no effective protocols available for a large number of crop species. In addition, the effectiveness of DH plant production is genotype dependent and highly variable not only between different species, but also between different cultivars, lines, and even individuals. The majority of modern triticale breeding materials are moderately responsive to ME induction. Therefore, procedures are still being developed and optimized for the requirements of specific genotypes. Based on the hypothesis of the important role of redox homeostasis in effective ME induction, we describe here a protocol modified by the application of reduced glutathione (GSH) as a factor that maintains microspore viability and promotes proper embryo-like structure (ELS) development in triticale.

Keywords In vitro culture, Microspore embryogenesis, Stress pretreatment, Reduced glutathione

1 Introduction

Triticale (× *Triticosecale* sp. Wittmack) is the first man-made hybrid cereal developed to combine the beneficial traits of its parents, wheat (*Triticum* spp. L.), and rye (*Secale cereale* L.). In theory, it should have the characteristics of wheat for high-quality food production and the tolerance of rye to a wide range of adverse environmental conditions [1]. Although triticale has many advantageous traits, it has not met all the expectations of breeders and farmers. At the present time, changing climatic conditions, the growing demand for food and new energy resources, and the principles of modern agriculture have increased interest in this cereal and created the need for its further improvement. Modern triticale breeding follows both traditional and more advanced breeding approaches, including DH technology [2]. Since the first reports in the early 1970s [3], various techniques have been investigated, optimized, and incorporated into breeding programs. Techniques involving in vitro culture of anthers and isolated microspores are the most popular [4].

Ana Maria Castillo et al. (eds.), *Doubled Haploid Methods in Cereals*, Methods in Molecular Biology, vol. 3029,
https://doi.org/10.1007/978-1-0716-5245-9_16,

However, many genotypes of high breeding value have been found to be moderately or poorly responsive.

Suspensions of isolated microspores are not only a potent source of DHs, but also a very convenient system for optimizing culture protocols. This is a laborious and time-consuming task, as many environmental and intracellular factors interact to determine the effectiveness of ME induction. The ability to continuously monitor cell viability, morphology, and development allows immediate selection of beneficial factors and elimination of those that induce negative effects. Isolated haploid microspores are also advantageous targets for mutation or transformation, due to the possibility to produce DH plants that are homozygous for the mutated gene or transgene [5, 6].

The isolated microspore culture protocol presented below involves an additional tiller pretreatment with GSH, which enhances the elimination of reactive oxygen species (ROS) generated in response to the stress used for ME induction [7]. The application of GSH not only reduces microspore mortality after transfer to in vitro culture, but also promotes the development of ELS, probably by altering the intracellular redox balance [8, 9]. GSH is primarily beneficial for donor plants with low stress tolerance, where the stress used for microspore reprogramming leads to a severe reduction in cell viability. However, increasing evidence suggests that a certain genotype-specific level of ROS is required for microspore reprogramming [8–10]. ROS act as signaling molecules that trigger changes in molecular control and cellular metabolism necessary for microspore survival and the initiation of an alternative developmental program. Excessive ROS elimination could lead to the failure of microspore reprogramming and embryogenesis induction.

The procedure was tested on seven winter triticale DH lines and its effectiveness was highest on DH lines with moderate responsiveness to ME induction [7]. Depending on the genotype, the physiological conditions of the donor plants and the ME-inducing stress treatment applied, the effect of GSH tiller pretreatment could be different and some further modifications of this procedure may be required, as was the case for rye [11].

2 Materials

2.1 Plant Material and Materials for Donor Plant Growth and Development

1. Seeds of triticale (× *Triticosecale* Wittm.).
2. Petri dishes (90 × 15 mm) lined with filter paper moistened with tap water.
3. Magenta vessels containing perlite presoaked with Hoagland's salt solution (Table 1).
4. Pots containing a mixture of "Eco-Ziem Universal soil" (universal substrate, pH 5.5–7) and sand (3/1; v/v).

Table 1
Composition of media used in isolated microspore cultures of triticale

Composition	Hoagland salts[a]	190-2 ind[b]	190-2 reg[c]	MS root[d]
Macroelements [mg/L]				
KNO_3	660	1000	1000	1900
KH_2PO_4	–	300	300	170
$(NH_4)_2SO_4$	–	200	200	–
$NH_4H_2PO_4$	120	–	–	–
NH_4NO_3	–	–	–	1650
$CaCl_2$	–	–	–	332.2
$Ca(NO_3)_2{\cdot}4H_2O$	940	100	100	–
$MgSO_4{\cdot}7H_2O$	520	200	200	–
$MgSO_4$	–	–	–	180.7
KCl	–	40	40	–
Iron source [mg/L]				
Na_2EDTA	37.3	37.3	37.3	37.3
$FeSO_4{\cdot}7H_2O$	27.8	27.8	27.8	27.8
Microelements [mg/L]				
$MnSO_4{\cdot}4H_2O$	–	8.0	8.0	22.3
$MnSO_4{\cdot}H_2O$	3.4	–	–	–
$ZnSO_4{\cdot}7H_2O$	0.2	3.0	3.0	8.6
H_3BO_3	2.8	3.0	3.0	6.2
KI	–	0.5	0.5	0.83
$Na_2MoO_4{\cdot}2H_2O$	0.025	–	–	0.25
$CoCl_2{\cdot}6H_2O$	–	–	–	0.025
$CuSO_4{\cdot}5H_2O$	0.1	–	–	0.025
Vitamins [mg/L]				
Myo-inositol	–	100	100	100
Nicotinic acid	–	0.5	0.5	0.5
Pyridoxine·HCl	–	0.5	0.5	0.5
Thiamine·HCl	–	1.0	1.0	0.1
Amino acids [mg/L]				
Glycine	–	2.0	2.0	2.0
Glutamine	–	438	100	–
Carbohydrates [mg/L]				
Maltose	–	63,000	–	–

(continued)

Table 1
(continued)

Composition	Hoagland salts[a]	190-2 ind[b]	190-2 reg[c]	MS root[d]
Sucrose	–	–	30,000	20,000
Growth regulators [mg/L]				
NAA	–	–	0.5	–
KIN	–	–	0.5	–
Gelling agent [mg/L]				
Agar	–	–	6000	6000
pH	6.7	5.8	6.0	5.8

[a]Hoagland salts according to [12]
[b]190-2 induction medium according to [13] modified by [14]
[c]190-2 regeneration medium according to [13] modified by [12]
[d]MS root—Murashige and Skoog basal medium (M5519, Merck KGaA) used for regenerants rooting according to [15]

2.2 Equipment

1. Greenhouse operating at 20°C with a 16/8 h (day/night) photoperiod (6.00 a.m.–10.00 p.m.) provided by high pressure sodium lamps (400 W; Philips SON-T AGRO, Belgium).
2. Refrigerated counters operating at 4°C with an 8/16 h (day/night) photoperiod (6.00 a.m.–2.00 p.m.) provided by high pressure sodium lamps (400 W; Philips SON-T AGRO, Belgium).
3. Analytical balance.
4. Fluorescence and inverted light microscopes equipped with a digital camera (e.g., Zyla 4.2; Andor) and imaging analysis software (e.g., NIS-Elements AR 4.00 Imaging Analysis). Fluorescence microscope equipped with filters for DAPI (EX 365/DM 400/BA 420 EF) and FDA (EX 470–490/DM 510 BA/515 EF) staining.
5. Refrigerator-freezer.
6. Autoclave.
7. Microwave.
8. pH meter.
9. Laminar flow cabinet.
10. Waring laboratory variable speed control blender with stainless steel container (100 mL).
11. Centrifuge equipped with swing-bucked rotor and adapter that accommodate 10 mL tubes, preferably refrigerated.
12. Cell culture incubator operating at temperature 26°C.
13. Cell culture chamber operating at temperature of 26°C with a light intensity 80-100 μmol $m^{-2} s^{-1}$ and a 16/8 h (day/night) photoperiod.

2.3 Laboratory Materials

1. Two autoclavable pipettes (100–1000 μL, 1–5 mL) with sterile pipette tips.
2. Neubauer counting chamber.
3. 1 L beakers.
4. 250 mL Erlenmeyer flasks.
5. 1 L autoclavable bottles.
6. Measuring cylinders and glass rods.
7. 15 mL glass tubes.
8. 150 mm Pasteur pipettes with rubber bulbs.
9. Microscope slides and cover slips.
10. Glass jars.
11. 15 and 50 mL Falcon polypropylene tubes.
12. 2 mL Eppendorf tubes.
13. Polypropylene syringe.
14. 0.22 μm pore size syringe filters.
15. Polystyrene sterile Petri dishes (60 × 15 mm, 90 × 15 mm).
16. Autoclavable plant cell culture vessels (e.g., Magenta vessels).
17. Sterile stainless steel tea strainer with about 1 mm opening size.
18. Sterile cell dissociation sieve with 74 μm opening size (200 mesh; CD-1, Sigma-Aldrich).
19. Two micro-dissecting stainless steel forceps length c.a. 120 mm.
20. Stainless steel forceps length c.a. 200 mm.
21. Stainless steel scissors.
22. Razor blades.
23. Parafilm.
24. Plastic bags, clip seal plastic bags.
25. Aluminum foil.

2.4 Solutions for Plant Fertilization, Microspore Cytological Analyses, Tiller Sterilization, Tiller Pretreatment, Microspore Isolation, and Chromosome Doubling

1. The composition of the donor plant fertilizer (Hoagland's salt solution) is presented in Table 1. The solution is prepared in a volume of 200 L and stored in a barrel at room temperature. Weigh and dissolve 132 g KNO_3, 24 g $NH_4H_2PO_4$, 188 g $Ca(NO_3)_2{\cdot}4H_2O$, 104 g $MgSO_4{\cdot}7H_2O$ separately each in 1 L of deionized water. Weigh and dissolve 7.5 g NaEDTA together with 5.6 g $FeSO_4{\cdot}7H_2O$ in 0.5 L of deionized water. Weigh and dissolve together 680 mg $MnSO_4$, 40 mg of $ZnSO_4{\cdot}7H_2O$, 560 mg of H_3BO_3, 20 mg $CuSO_4{\cdot}5H_2O$, and 5 mg $Na_2MoO_4{\cdot}2H_2O$ in 0.5 L of deionized water. Mix together all the solutions in a barrel and make up the volume to 200 L with tap water.
2. Sodium phosphate buffer (PBS): 0.14 M NaCl, 2.7 M KCl, 1.5 mM KH_2PO_4, 6.5 mM $Na_2HPO_4{\cdot}2H_2O$. Add about

800 mL deionized water to 1 L glass beaker. Weigh 8 g of NaCl, 1.96 g of $Na_2HPO_4 \cdot 2H_2O$, 0.2 g of KCl, 0.24 g of KH_2PO_4, and add to the beaker. Adjust the pH to 7.4 using 1 N HCl. Add deionized water to bring the total volume to 1 L. Store at room temperature.

3. 4′6-Diamidino-2-phenylindole·2HCl (DAPI, ×10) stock solution: $2.86 \cdot 10^{-5}$ M DAPI in PBS. Dissolve 0.1 mg DAPI in 1–2 drops of DMSO and add PBS to 10 mL. Store at −20°C.
4. DAPI working solution ($2.86 \cdot 10^{-6}$ M): add 10 μL of DAPI stock-solution to 9.9 mL PBS and 90 μL of Triton-X100. Store in the refrigerator at 0°C–4°C protected from light.
5. Fluorescein diacetate (FDA, 100×) stock solution: 0.5% FDA in acetone. Dissolve 10 mg of FDA in 2 mL of acetone. Store at −20°C.
6. FDA working solution: add 10 μL of FDA stock-solution to 990 μL induction medium. Store at 0°C–4°C.
7. 0.3 mM reduced glutathione (GSH) in Hoagland salts solution: dissolve 92.2 mg GSH in 800 mL of Hoagland salts solution and make up to 1 L (*see* **Note 1**).
8. 20% solution of commercial bleach (containing 4.5% sodium hypochlorite) in deionized water.
9. 0.3 M mannitol: dissolve 54.65 g of mannitol in 800 mL of deionized water and make up to 1 L. Autoclave for 20 min at 120°C.
10. 21% maltose: dissolve 21 g of maltose in 80 mL of deionized water and make up to 100 mL. Autoclave for 20 min at 120°C.
11. 1 N NaOH for pH adjustment.
12. Colchicine solution: 0.1% colchicine, 4% DMSO, 0.025% GA_3, Tween 20 (*see* **Note 2**).

Dissolve 1 g of colchicine in 500 mL of deionized water, add 40 mL of DMSO, 1 drop of Tween 20 and 25 mg of GA3, make up to 1 L with deionized water and mix with a glass rod.

2.5 Culture Media

The composition of culture media used for ME induction, plant regeneration, and regenerants rooting are presented in Table 1.

190-2 induction and 190-2 regeneration media are prepared from stock solutions, which are stored at −20°C.

(a) 190-2 induction medium:

- Stock solution A (100×): dissolve 2 g of $(NH_4)_2SO_4$ and 2 g of $MgSO_4 \cdot 7H_2O$ in 80 mL of deionized water, make up to 100 mL and pour 10 mL into each of ten 15 mL Falcon tubes.

- Stock solution B (100×): dissolve 10 g of KNO_3, 3 g of KH_2PO_4, 1 g of $Ca(NO_3)_2{\cdot}4H_2O$ and 0.4 g of KCl in 80 mL of deionized water, make up to 100 mL and pour 10 mL into each of ten 15 mL Falcon tubes.
- Stock solution C (1000×): dissolve 80 mg of $MnSO_4{\cdot}H_2O$, 30 mg of $ZnSO_4{\cdot}7H_2O$, 30 mg of H_3BO_3 and 5 mg of KI in 8 mL of deionized water, make up to 10 mL and pour 1 mL into each of ten 2 mL Eppendorf tubes.
- Stock solution D (1000×): dissolve 373 mg of $Na_2FeEDTA$ and 278 mg of $FeSO_4{\cdot}7H_2O$ in 8 mL of deionized water, make up to 10 mL and pour 1 mL into each of ten 2 mL Eppendorf tubes.
- Stock solution E (1000×): dissolve 5 mg of nicotinic acid, 5 mg of pyridoxine HCl and 10 mg of thiamine HCl in 8 mL of deionized water, make up to 10 mL and pour 1 mL into each of ten 2 mL Eppendorf tubes.
- Stock solution F (100×): dissolve 4.38 g of glutamine, 20 mg of glycine, and 1 g of myo-inositol in 80 mL of deionized water, make up to 100 mL and pour 10 mL into each of ten 15 mL Falcon tubes.
- One tube of each stock solution (A–F) is used to prepare 1 L of 190-2 induction medium. Add 63 g of maltose and make up the volume to 1 L with deionized water. Adjust the pH to 5.8 with 1 N NaOH. Store into 50 mL clip sealed plastic bags at −20°C. The medium is defrosted and sterilized by syringe filter sterilization (0.22 μm pore size) just before use.

(b) 190-2 regeneration medium:

- Stock solutions A–E are identical to the A–E stock solutions for 190-2 induction medium.
- Stock solution F (100×): dissolve 1 g of glutamine, 20 mg of glycine, and 1 g of myo-inositol in 80 mL of deionized water, make up to 100 mL and pour 10 mL into each of ten 15 mL Falcon tubes.
- NAA stock solution (1000×, 0.5 mg/mL): dissolve 5 mg NAA in a few drops of 1 N NaOH and make up to 10 mL with deionized water. Distribute the solution into 2 mL Eppendorf tubes (1 mL per tube) and store at −20°C.
- Kinetin stock solution (1000×, 0.5 mg/L): dissolve 5 mg kinetin in a few drops of 1 N NaOH and make up to 10 mL with deionized water. Distribute the solution into 2 mL Eppendorf tubes (1 mL per tube) and store at −20°C.
- Mix stock solutions A, B, C, D, and F with 30 g sucrose in about 600 mL deionized water. Adjust the pH to 6.0 with

dilute alkali (1 N NaOH). Dissolve 6 g of agar in 300 mL of deionized water by heating in a microwave, add it and make up to 1 L with deionized water. Autoclave the solution for 20 min at 120°C. Sterilize vitamins (stock solution E) and growth regulators using syringe filters (0.22 μm pore size) and add to the medium after it has cooled to approximately 40°C (*see* **Note 3**).

(c) Half-strength Murashige-Skoog medium according to [15]: Dissolve 2.2 g (half of the standard amount recommended by the producer) of Murashige and Skoog Basal Medium (M5519, Merck KGaA) in 600 mL of deionized water. Adjust the pH to 5.8 with dilute alkali (1 N NaOH). Dissolve 6 g of Gum agar in 300 mL of deionized water by heating in a microwave, add to the solution and make up to 1 L with deionized water. Autoclave the solution for 20 min at 120°C.

3 Methods

3.1 Donor Plant Growth Conditions

3.1.1. Triticale seeds are germinated on 90 × 15 mm Petri dishes lined with filter paper moistened with tap water for 2 days in the dark at room temperature.

3.1.2. Germinating seeds of winter genotypes are placed in Magenta vessels containing perlite presoaked with Hoagland's salt solution and vernalized for 7–8 weeks at 4°C and an 8/16 h (day/night) photoperiod.

3.1.3. Seedlings of vernalized winter genotypes and germinating seeds of spring genotypes are planted in pots containing a mixture of soil and sand (3/1; v/v) and grown in a greenhouse at about 20°C and a 16/8 h (day/night) photoperiod (Fig. 1A). Supplementary lighting (high pressure sodium lamps) is used to extend the photoperiod or to provide additional light in unfavorable weather conditions. Plants are fertilized with Hoagland's salt solution twice a week (*see* **Note 4**).

3.2 Tillers Pretreatment

3.2.1. Tillers are collected when the majority of microspores are at the mid to late uninucleate stage of development (*see* **Note 5**).

3.2.2. The stage of microspore development is assessed by DAPI staining. A few anthers collected by forceps from the central part of a spike are placed on a glass slide in a drop of DAPI working solution, chopped with a razor blade, covered with a cover slip and examined under a fluorescence microscope at $\lambda_{Ex} = 365$ nm and $\lambda_{Em} = 420$ nm (blue fluorescence).

Fig. 1 Isolated microspore cultures of triticale (× *Triticosecale* sp. Wittm.). (**A**) Donor plant. (**B**) Optimum development stage of the spike to start the low temperature tiller pretreatment. (**C**) Low temperature tiller pretreatment. (**D**) Sterilization procedure. (**E**) Spike fragmentation before blending. (**F**) Viable microspores deposited at the mannitol/maltose interface after density gradient centrifugation. (**G**) Microspore suspension at day 8 of in vitro culture. (**H**) Multicellular embryo-like structures after 4 weeks of in vitro culture. (**I**) Green and albino regenerants after 4 weeks on regeneration medium. (**J**) Green regenerants on rooting medium. (**K**) Green regenerants on perlite for further acclimatization to out-of-*vitro* conditions. (**L**) Microspore-derived plants during chromosome doubling procedure. (**M**) Microspore-derived plants after transfer to soil

3.2.3. Microspore viability is tested based on the reaction with FDA. A few anthers are placed on a glass slide in a drop of FDA working solution, chopped with a razor blade, covered with a cover slip and examined under a fluorescence microscope at $\lambda_{Ex}=465$ nm, $\lambda_{Em}=515$ nm (green fluorescence).

3.2.4. Tillers containing microspores at the mid to late uninucleate stage of development are defoliated (except for the flag leaf), placed in plastic bags, wrapped in aluminum foil, and stored in glass jars containing Hoagland's salt solution at 4°C in the dark for 3 weeks (Fig. 1B, C).

3.2.5. In genotypes with low microspore viability, Hoagland's salt solution is supplemented with 0.3 mM GSH (G6013, obtained from Merck KGaA, Germany). GSH can be added at the beginning of the low temperature pretreatment, or the tillers can be transferred to the solution containing GSH for the last 4 days before microspore isolation [7] (*see* **Note 6**).

3.3 Microspore Isolation and Culture

3.3.1. Pretreated tillers are sprayed with 96% ethanol in a laminar flow hood. The stem is cutoff, the spikes are pulled out of the leaf sheath and placed in sterile Erlenmeyer flasks (5–8 for one flask). Spikes are covered with a 0.9% sodium hypochlorite solution for 15 min and then rinsed 4–5 times with sterile deionized water (*see* **Note 7**; Fig. 1D).

3.3.2. A few sterilized spikes are used as ovary donors. Ovaries are isolated with forceps and transferred to Petri dishes with modified 190-2 medium.

3.3.3. Sterilized spikes are cut into approximately 2 cm segments (Fig. 1E) with sterile scissors and blended in 0.3 M mannitol using a Waring blender (Fisher Scientific Inc.) set to mode of 2–3 (10 s) pulses at minimum rotational speed (*see* **Note 8**).

3.3.4. The resulting homogenate is filtered first through a sterile stainless steel tea strainer with about 1 mm opening size and then through 74 μm metal sieve. It is transferred to 10 mL sterile glass tubes and pelleted by centrifugation (100 × *g*, 4–6 min, at 4°C).

3.3.5. The supernatant is removed, and the microspores are resuspended in 1–2 mL of 0.3 M mannitol and gently placed on a 21% maltose solution for density gradient centrifugation (80 × *g*, 4–6 min, at 4°C) (*see* **Note 9**).

3.3.6. Viable microspores deposited at the mannitol/maltose interface (Fig. 1F) are collected with a sterile Pasteur pipette, transferred to a sterile 10 mL glass tube. For washing, microspores are resuspended in 5–6 mL of 0.3 M mannitol, and centrifuged (100 × *g*, 4–6 min, at 4°C).

3.3.7. The supernatant is removed, and the pelleted microspores are resuspended in 1 mL of modified 190-2 medium.

3.3.8. The total number of microspores collected is estimated using a hemocytometer (e.g., Neubauer counting chamber). By adding the required volume of modified 190-2 induction medium, the culture density is adjusted to approximately 80,000 microspores per mL.

3.3.9. The microspore suspension is transferred to 35×10 mm Petri dishes and cocultured with immature ovaries (10 ovaries per Petri dish containing 1.5 mL of cell suspension). Cultures are incubated in the dark at 26°C.

3.3.10. Microspore development can be monitored using a light inverted microscope. Observations are typically made after 1, 2, and 4 weeks of culture (Fig. 1G).

3.4 Plant Regeneration

3.4.1. Starting from the sixth week of culture, ELS of approximately 2 mm in size (Fig. 1H) are transferred with forceps to 90×15 mm Petri dishes containing 0.6% agar solidified 190-2 regeneration medium. Cultures are maintained at 26°C, with a 16/8 h (day/night) photoperiod in dim light (80–100 µmol $m^{-2} s^{-1}$). The transfer is repeated once or twice at 2-week intervals, depending on the number of ELS (*see* **Note 10**).

3.4.2. Green regenerants are transferred to Magenta vessels on half strength MS medium (MS root) for growth and rooting (Fig. 1I, J).

3.4.3. Poorly-rooted plants are transferred to Magenta vessels with perlite presoaked with half strength Hoagland's nutrient solution (Fig. 1K).

3.4.4. Well-rooted plants are transferred to pots containing a mixture of soil and sand (3/1; v/v) and grown in a greenhouse at about 20°C with a photoperiod of 16/8 h (day/night).

3.5 Genome Doubling by Colchicine Treatment

3.5.1. Green plants at the tillering stage (with minimum 3 tillers) are removed from the soil. The roots are washed in tap water.

3.5.2. The plants are placed in holes cut in plastic box lids and fixed with cotton wool (Fig. 1L). The boxes are filled with tap water to cover the plant shoot base and kept at 16°C–18°C until the following morning.

3.5.3. Plant leaves are cut to a length of about 5 cm above the shoot base and the water is replaced with colchicine solution. The plants are left for 7.5 h at about 25°C under a light intensity of about 1000 µmol $m^{-2}s^{-1}$. The colchicine solution must be aerated throughout the treatment (*see* **Note 11**).

3.5.4. The roots are washed several times in tap water, then cut to 5 cm, and washed in running water for approximately 24 h.

3.5.5. The plants are placed in soil, and allowed to recover for 3–4 days at about 10°C. For the first 24 h, the plants should be covered with plastic bags or cling film to reduce transpiration.

3.5.6. The plants are vernalized at 4°C for 8–9 weeks and then transferred to growth conditions as described above for donor plants (Fig. 1M).

4 Notes

1. The solution of GSH in Hoagland's salts is usually prepared fresh before use.
2. Colchicine is highly toxic, carcinogenic, and teratogenic. All producer's safety precautions must be strictly followed.
3. The agar should be heated until completely dissolved, taking care not to let it boil over. It is then added to the medium solution and mixed with a glass rod. Even distribution of the agar in the medium allows the solution to be poured into 300 mL Erlenmeyer flasks (exactly 250 mL each), facilitating subsequent pouring into Petri dishes. The exact volume of medium (250 mL) is necessary to add the correct amount of vitamins and growth regulators (1/4 the volume of the stock solution). To distribute the vitamins and growth regulators more evenly in the medium, magnetic bars should be placed inside the Erlenmeyer flasks before autoclaving. Before pouring the medium into Petri dishes, the bottles should be placed on a magnetic stirrer for 1–2 min.
4. Donor plants can be grown either in the field or in a greenhouse, but greenhouse conditions result in higher microspore viability.
5. Microspores after asymmetric division are generally no longer competent for reprogramming. The optimal late unicellular stage of microspore development is characterized by a large central vacuole and a nucleus located close to the cell wall opposite the operculum, embedded in a thin layer of cytoplasm. This stage usually corresponds to the appearance of awns (1–2 cm outside the upper leaf base).
6. A short GSH tiller pretreatment (3–8 days) is generally more effective in stimulating ELS development. However, some genotypes respond better when GSH is applied at the beginning of low temperature treatment and maintained throughout the entire 3 weeks.
7. Long awns characteristic of some genotypes should be trimmed before sterilization. In the case of contamination,

the spikes can by additionally treated with 0.05% mercuric chloride for 1 min before treatment with 0.9% sodium hypochlorite.

8. The stainless-steel container should be autoclaved earlier, covered with a double layer of aluminum foil. Depending on the size of the spikes, 5–8 spikes fit into a 100 mL Waring blender container. 0.3 M mannitol must be added to cover fragmented spikes but not more than 2/3 of the container volume to avoid splashing during blending.
9. Approximately 4 mL of 21% maltose solution should be transferred to a 10 mL sterile glass tube. Gently layer the microspore suspension using a sterile Pasteur pipette. This part of the procedure must be carried out with great care, as disturbing the equilibrium between these two solutions cannot be corrected.
10. Slowly developing ELS usually do not regenerate into plants, so their transfer should not be prolonged beyond 8 weeks of culture.
11. Specially prepared plastic boxes with holes cut in the lids are used to hold a single plant. Plastic hoses are inserted to allow aeration of the colchicine solution. This method achieves high genome doubling efficiency with minimal plant mortality.

References

1. Mergoum M et al (2009) Triticale: a "New" Crop with Old Challenges. In: Carena MJ (eds) Cereals. Handbook of Plant Breeding, vol 3. Springer, New York, NY, pp 267–290. https://doi.org/10.1007/978-0-387-72297-9_9
2. Mergoum M, Sapkota S, ElDoliefy AEA et al (2019) Triticale (× *Triticosecale* Wittmack) breeding. In: Al-Khayri J, Jain S, Johnson D (eds) Advances in plant breeding strategies: cereals. Springer, Cham, pp 405–451
3. Wang YY, Sun CS, Wang CC et al (1973) The induction of the pollen plantlets of Triticale and *Capsicum annuum* from anther culture. Sci Sin 16:147–151
4. González JM, Jouve N (2000) Improvement of anther culture media for haploid production in triticale. Cereal Res Commun 28:65–72
5. Szarejko I (2003) Doubled haploid mutant production. In: Maluszynski M, Kasha KJ, Forster BP, Szarejko I (eds) Doubled haploid production in crop plants. A manual. Springer, Dordrecht, pp 351–362
6. Shim YS, Kasha KJ (2003) Barley microspore transformation protocol by biolistic gun. In: Maluszynski M, Kasha KJ, Forster BP, Szarejko I (eds) Doubled haploid production in crop plants. A manual. Springer, Dordrecht, pp 363–366
7. Żur I, Dubas E, Krzewska M et al (2019) Glutathione provides antioxidative defence and promotes microspore-derived embryo development in isolated microspore cultures of triticale (× *Triticosecale* Wittm.). Plant Cell Rep 38:195–209
8. Żur I, Dubas E, Krzewska M et al (2021) Triticale and barley microspore embryogenesis induction requires both reactive oxygen species generation and efficient system of antioxidative defence. Plant Cell, Tiss Org Cult 145:347–366
9. Żur I, Kopeć P, Surówka E et al (2021) Impact of ascorbate—glutathione cycle components on the effectiveness of embryogenesis induction in isolated microspore cultures of barley and triticale. Antioxidants 10:1254
10. Żur I, Dubas E, Krzewska M et al (2014) Antioxidant activity and ROS tolerance in triticale (× *Triticosecale* Wittm.) anthers affect the efficiency of microspore embryogenesis. Plant Cell Tiss Organ Cult 119:79–94
11. Zieliński K, Krzewska M, Żur I et al (2020) The effect of glutathione and mannitol on androgenesis in anther and isolated microspore

cultures of rye (*Secale cereale* L.). Plant Cell Tiss Org Cult 140:577–592
12. Wędzony M (2003) Protocol for anther culture in hexaploid triticale (×*Triticosecale* Wittm.). In: Maluszynski M, Kasha KJ, Forster BP, Szarejko I (eds) Doubled haploid production in crop plants. A manual. Springer, Dordrecht, pp 123–128
13. Zhuang J, Xu J (1983) Increasing differentiation frequencies in wheat pollen callus. In: Hu H, Vega M (eds) Cell and tissue culture techniques for cereal crop improvement. Science Press, Beijing, pp 431–432
14. Pauk J, Mihály R, Monostori T et al (2003) Protocol of triticale (×*Triticosecale* Wittmack) microspore culture. In: Maluszynski M, Kasha KJ, Forster BP, Szarejko I (eds) Doubled haploid production in crop plants. A manual. Springer, Dordrecht, pp 129–134
15. Murashige T, Skoog F (1962) A revised medium for rapid growth and bio-assays with tobacco tissue cultures. Physiol Plant 15:473–497

Chapter 17

Efficient Protocol for Doubled Haploid Production in Rye *(Secale cereale L.)*

Ewa Dubas, Monika Krzewska and Iwona Żur

Abstract

The cultivation of rye (*Secale cereale* L.) for functional foods is economically important due to its high-dietary fiber content and health-related bioactive compounds. Breeders are seeking modern biotechnology techniques to accelerate traditional breeding in this predominantly cross-pollinated species. The development of efficient protocols for double haploid (DH) production in rye breeding faces challenges. Despite five decades of research, no universally efficient DH production protocol has emerged, largely due to genotypic variability and recalcitrance to in vitro regeneration. The protocols presented here incorporate antioxidants, such as reduced glutathione (GSH) and selenium (SS), into the tiller low temperature pretreatment procedure, as well as arabinogalactan proteins (AGPs) and phytagel into the induction medium. All these modifications aim to reduce the generation of reactive oxygen species (ROS). As a result, microspores are more likely to survive stress pretreatment, the isolation process and transfer to in vitro culture better. A higher population of viable microspores could result in an increase in the number of reprogrammed, embryogenic microspores and microspore-derived embryo-like structures (ELS). The protocols presented have been tested on 15 F1 rye breeding lines, with the greatest effectiveness in lines with moderate and high responsiveness to androgenesis induction, also known as microspore embryogenesis (ME) induction.

Keywords In vitro culture, Mannitol, Microspore embryogenesis, Reduced glutathione, Rye, Selenium, Stress pretreatment

1 Introduction

Rye (*Secale cereale* L.) is of considerable economic importance to produce functional foods. This cereal contains a high amount of dietary fiber and a wide spectrum of health-related bioactive compounds [1]. Homozygous double haploid (DH) rye lines produced by the androgenesis method are useful in genetic mapping studies and help traditional cereal breeding to halve the valuable time needed to produce commercially new varieties with fixed rare alleles and traits adapted to market needs [2]. Unfortunately, the growing demand for rye breeding has been accompanied by difficulties in establishing an efficient

Ana María Castillo et al. (eds.), *Doubled Haploid Methods in Cereals*, Methods in Molecular Biology, vol. 3029, https://doi.org/10.1007/978-1-0716-5245-9_17,

protocol for DH production due to its recalcitrance to haploidization procedures [3–5]. Thus, despite 50 years of modern research on haploidization in this predominantly cross-pollinated species, no universally efficient DH production protocol has been developed. The main reasons for this are high genotypic variability, low plant regeneration capacity, and a high proportion of albino plants [3, 6].

Low temperature tiller stress pretreatment (4°C) is commonly used in both rye anther and microspore culture methods [7]. Low temperatures stimulate the generation of reactive oxygen species (ROS), which play a dual role by inducing damaging oxidative stress and acting as signaling molecules that trigger microspore embryogenesis (ME) [8, 9]. The balance between ROS production and scavenging is the critical factor in determining the effectiveness of ME induction. Mannitol (MAN), an osmotic agent, is required to protect cells under conditions of oxidative stress through a complex response that contributes to the rearrangement of the microtubular cytoskeleton in embryogenic microspores [8]. The potential of microspores to adapt to oxidative stress also results from the efficiency of the endogenous antioxidant system, which is composed of enzymatic and nonenzymatic elements. This system can be supported by exogenously applied reduced form of glutathione (GSH), an antioxidant involved in ROS scavenging. Our studies have shown its beneficial effect on triticale microspore survival and ELS development [9, 10]. Based on these results, the pretreatment procedure for rye tillers was modified to include an initial mannitol treatment, followed by a GSH treatment without mannitol (MAN_GSH). Thus, 21 days of tiller incubation at 4°C with 0.7 M MAN for the first 14 days, followed by a transfer to 0.3 mM GSH, reduced the level of oxidative stress, increased microspore viability and stimulated ELS formation, when compared to the cold treatment alone [11, 12].

Similarly, preliminary results of a collaborative research indicate that selenium enhances ROS scavenging, reduces levels of oxidative stress, and alleviates recalcitrance to ME induction. Selenium is a modulator of glutathione peroxidase (GPX) activity and alleviates oxidative stress through various antioxidant pathways. The application of a pretreatment with selenium, in the form of sodium selenate (SS), during the last 7 days of the pretreatment at 4°C was found to increase microspore viability and induce ELS development [13].

The rye anther and isolated microspore culture methods are generally based on medium 190-2 [14], with subsequent modifications in both the induction and regeneration media. The addition of arabinogalactan proteins (AGPs) and phytagel (as a gelling agent) into the induction medium was based on literature data [15], confirming that they also stimulate the formation of rye microspore-derived ELS [12]. In addition, this colchicine-free protocol has been shown to result in a spontaneous doubling of the number of chromosomes in 55% of cases [11].

The stress pretreatment procedure, combining low temperature, MAN, and GSH was tested on 15 rye breeding F1 lines, and its effectiveness was the highest on lines with moderate and high responsiveness to ME induction [11]. A similar approach, combining low temperature with MAN and selenium, has also shown promising results in anther and microspore cultures for many recalcitrant winter wheat genotypes [16].

2 Materials

2.1 Plant Material

Seeds from winter rye (*Secale cereale* L.) F1 lines of various origins and pollination systems have been used. These included lines that are cross-pollinating, cross-pollinating with a self-compatibility gene (which allows for self-fertilisation in cross-pollinating plants), and self-pollinating lines (*see* **Note 1**).

2.2 Equipment

1. Greenhouse or growth chamber with controlled growth conditions, with the temperature set at 25/17°C (day/night), light intensity at 800 µmol m^{-2} s^{-1}, and 16 h/8 h light/dark. In both environments, the light intensity is maintained using high-pressure sodium lamps (400 W; Philips SON-T AGRO, Belgium) between 6–8 a.m. and 6–10 p.m., and on cloudy days in the greenhouse.
2. Vernalization growth chamber set at 2°C, light intensity at 100 µmol m^{-2} s^{-1} provided by high-pressure sodium lamps (400 W; Philips SON-T AGRO, Belgium), and 8 h/16 h light/dark.
3. In vitro chambers with temperature controlled at 26°C (day/night), light intensity at 30 and 80–100 µmol m^{-2} s^{-1} provided by lamps 140 W/33 RS DOUBLE FLUX TLM (Philips, Poland), and 16 h/8 h light/dark.
4. Incubators with controlled temperature set at 26°C.
5. Refrigerator.
6. Fume hood.
7. Laminar flow cabinet.
8. Autoclave.
9. Distiller.
10. Stereomicroscope.
11. Light and fluorescent inverted microscopes equipped with digital camera and imaging analysis software. Fluorescence microscope Nikon Eclipse-E600 equipped with a differential interference contrast (DIC) system. Images are captured with a digital CCD camera (Zyla 4.2; Andor) and processed with imaging analysis software (NIS-Elements AR 4.00 Imaging

Analysis, Microsoft PowerPoint and Corel PhotoPaint X5). Fluorescence is examined under filters EX 365/DM 400/BA 420 EF (DAPI); EX 470–490/DM 510 BA/515 EF (FDA).

12. Microwave.
13. Refrigerated centrifuge with swinging rotors.
14. Gyratory shakers.
15. Flow cytometer (CyFlow Space, Sysmex Partec) equipped with a UV-led diode 365 nm and FCM FloMAx software.
16. Analytical balances.
17. pH meter.
18. Hotplate.
19. Peristaltic pump.

2.3 Plant Growth and Laboratory Materials

1. Pots (3 L).
2. Perlite.
3. Substrate: mixture of soil and sand (3:1 *v/v*).
4. Autoclavable and automatic pipettes with sterile tips (100–1000 μL, 1–5 mL).
5. Beakers.
6. Measuring cylinders.
7. Glass bottles for media and pretreatment solutions.
8. Erlenmeyer flasks for media and pretreatment solutions.
9. Laboratory glass crystallizer.
10. Sterile conical 50 mL plastic tubes (Falcon).
11. Sterile conical 15 mL glass tubes.
12. Eppendorf tubes (1.5 and 2.0 mL).
13. Sterile vacuum filter.
14. Sterile filters (Ø 0.22 μm).
15. Nylon mesh (Ø 42 μm).
16. Glass Petri dishes (Ø 9 cm × 2.5 cm).
17. Polystyrene Petri dishes of various diameters (Ø 9 cm × 1.4 cm, Ø 5.5 cm × 1.2 cm, Ø 3.5 cm × 1.2 cm).
18. Magenta boxes.
19. Parafilm.
20. Scalpel and blades.
21. Stainless steel forceps for micro-dissection (length c.a. 120 mm).
22. Pasteur pipettes.
23. Metal sieves (Ø 5 cm; pore size Ø 1 mm and Ø 74 μm).
24. Dissecting needles.

25. Sterile plastic (1–50 mL) centrifuge tubes.
26. Aluminium and polyethylene food-foils.
27. Neubauer chamber (hemocytometer) for cell counting.
28. Scale.
29. Glass slides.
30. Cover slips.

2.4 Solutions for Plant Fertilisation, Microspore Cytological Analyses, Microspore Isolation, Sterilization, Ploidy Analyses, and Chromosome Doubling

1. Composition of donor plant fertilisation solution is presented in Table 1. To prepare a total volume of 200 L, it is first necessary to dissolve 132 g of KNO_3, 24 g of $NH_4H_2PO_4$, 188 g of $Ca(NO_3)_2 \cdot 4H_2O$, and 104 g of $MgSO_4 \cdot 7H_2O$ separately, each in 1 L of distilled water. Then, dissolve 7.5 g of NaEDTA and 5.6 g of $FeSO_4 \cdot 7H_2O$ in 0.5 L of distilled water. Next, weigh 680 mg of $MnSO_4$, 40 mg of $ZnSO_4 \cdot 7H_2O$, 560 mg of H_3BO_3, 20 mg of $CuSO_4 \cdot 5H_2O$, and 5 mg of $Na_2MoO_4 \cdot 2H_2O$, before dissolving them in 0.5 L of distilled water. Combine all the solutions in a barrel and make up to 200 L with tap water. The solution is stored at ambient temperature.
2. Fluorescein diacetate (FDA) stock solution (100×, 5 mg/mL): Dissolve 10 mg of FDA powder in 2 mL of acetone and mix thoroughly. Store at −20°C.
3. FDA working solution (0.05 mg/mL): Take 10 μL of the stock solution and transfer to a 1.5 mL Eppendorf tube containing 990 μL of fresh liquid medium. Store at 4°C for further use.
4. 4′6-diamidino-2-phenylindole*2HCl (DAPI) stock solution (100×, 1 mg/mL): Dissolve 0.01 g of DAPI powder in 50 μL of dimethyl sulfoxide (DMSO) and make up to 10 mL with sodium phosphate buffer (PBS: 0.14 M NaCl, 2.7 M KCl, 1.5 mM KH_2PO_4, 6.5 mM Na_2HPO_4 $2H_2O$, pH 7.4). Store at −20°C.
5. DAPI working solution (10 μg/mL): Take a 15 mL Falcon working solution tube. Dilute 10 μL of DAPI stock solution to 900 μL with PBS and add 90 μL of Triton-X100. Store at 4°C and protect from light.
6. 0.7 M mannitol (MAN): Dissolve 127.52 g of MAN in 700 mL of MilliQ water and make up to 1 L (*see* **Note 2**).
7. 0.3 mM reduced glutathione (GSH): Dissolve 92.2 mg of GSH in 800 mL of MilliQ water and make up to 1 L (*see* **Note 3**). The solution must be prepared immediately before use.
8. Sodium selenate (SS; Na_2SeO_4) stock solution (1000×, 50 mM): Dissolve 94 mg of SS in 8 mL MilliQ water and make up the volume to 10 mL. Store in 1.5 mL Eppendorf tubes at −20°C.

Table 1
Composition of donor plant fertilisation solution and culture media for anther and microspore culture

Components	Hoagland salts[a]	190-2_ind[b]	190-2_reg[c]	1/2 MS[d]
Macronutrients [mg/L]				
KNO_3	660	1000	1000	950
KH_2PO_4	–	300	300	85
$(NH_4)_2SO_4$	–	200	200	–
$NH_4H_2PO_4$	120	–	–	–
NH_4NO_3	–	–	–	825
$CaCl_2$	–	–	–	166
$Ca(NO_3)_2 \times 4H_2O$	940	100	100	–
$MgSO_4 \times 7H_2O$	520	200	200	–
$MgSO_4$	–	–	–	90.35
KCl	–	40	40	–
Iron source [mg/L]				
Na_2EDTA	37.3	37.3	37.3	18.63
$FeSO_4 \times 7H_2O$	27.8	27.8	27.8	13.9
Micronutrients [mg/L]				
$MnSO_4 \times 4H_2O$	–	8.0	8.0	11.5
$MnSO_4$	3.4	–	–	–
$ZnSO_4 \times 7H_2O$	0.2	3.0	3.0	4.3
H_3BO_3	2.8	3.0	3.0	3.1
KI	–	0.5	0.5	0.415
$Na_2MoO_4 \times 2H_2O$	0.025	–	–	0.125
$CoCl_2 \times 6H_2O$	–	–	–	0.0125
$CuSO_4 \times 5H_2O$	0.1	–	2.5	0.0125
Organic components [mg/L]				
myo-Inositol	–	100	100	50
Nicotinic acid	–	0.5	0.5	0.25
Pyridoxine HCl	–	0.5	0.5	0.25
Thiamine HCl	–	1.0	1.0	0.05
Glycine	–	2.0	2.0	1
Glutamine	–	200 or 438	100	–
Arabic Gum (AGP)	–	50	100	–
Other compounds [mg/L]				
Maltose	–	63,000 or 90,000	–	–

(continued)

Table 1
continued

Components	Hoagland salts[a]	190-2_ind[b]	190-2_reg[c]	1/2 MS[d]
Sucrose	–	–	30,000	20,000
Agar	–	–	6000	6000
Phytagel	–	0 or 3000	–	–
Plant Growth Regulators [mg/L]				
BAP	–	0.9	–	–
Dicamba	–	1.0	–	–
Picloram	–	1.0	–	–
NAA	–	–	0.5	–
KIN	–	–	0.5	–
Antibiotics [mg/L]				
Cefotaxime	–	1.0	–	–
pH	6.7	5.9	6.0	6.0

[a]Hoagland salts according to [17] used as donor plant fertiliser
[b]190-2 induction medium according to [14] modified by [11] used for ME induction
[c]190-2 regeneration medium according to [17] modified by [11] used for plant regeneration
[d]MS, half-strength Murashige and Skoog-based medium (MS medium) used for regenerants rooting according to [18]

9. SS working solution (50 µM): Take 0.1 mL of the stock solution and dilute to a total volume of 100 mL with tap water immediately before use (*see* **Note 4**).
10. Ethanol 96% and 70% (*v/v*).
11. 0.9% (*w/v*) sodium hypochlorite sterilizing solution: Take 0.9 g of sodium hypochlorite and dissolve it in 100 mL of distilled water immediately before use.
12. 21% (*w/v*) maltose solution: Dissolve 10.5 g of maltose in 40 mL of distilled water in a volumetric flask and make up the volume to 50 mL with water. Mix well to ensure homogeneity. Autoclave the solution for 20 min at 121°C.
13. 0.3 M MAN solution: To prepare 1 L, dissolve 54.65 g of MAN in 800 mL of distilled water in a volumetric flask. When the MAN is dissolved, make up the volume to 1 L with distilled water. Mix well to ensure homogeneity. Autoclave the solution in 1 L bottles (*see* **Note 5**).
14. Otto I buffer (0.1 M citric acid) Tween 20, pH 2–3: Dissolve 0.96 g of citric acid powder in 45 mL of Milli-Q water and add 0.25 g of Tween 20 and water to the flask. Once the Tween 20 is dissolved, adjust the pH with 5 M sodium hydroxide (NaOH) to 2.5–3. Then make up the volume to 50 mL with water in an Erlenmeyer flask and mix again to ensure uniformity (*see* **Note 6**).

15. Otto II buffer (0.4 M $Na_2HPO_4{\cdot}12H_2O$, 2 mg/mL mercaptoethanol, 2 μg/mL DAPI, pH 8–9): Dissolve 7.16 g of sodium phosphate, dibasic powder in 30 mL of Milli-Q water in an Erlenmeyer flask. When dissolved, add 100 mg of β-mercaptoethanol and 100 μg of DAPI. Gently vortex the solution until all components are dissolved and make the volume to exactly 50 mL with water (pH approximately 8–9) (*see* **Note 7**).

2.5 Culture Media

The composition of culture media for rye anther and microspore culture is presented in Table 1. Stock solutions of minerals, vitamins, aminoacids, growth regulators, and antibiotic for nutrient media are prepared using distilled water and stored at −20°C.

1. Mineral salts stock solutions for 190-2 induction (190-2_ind) and 190-2 regeneration (190-2_reg): macronutrients, divided in solution A (25×) ($(NH_4)_2SO_4$, $MgSO_4{\cdot}7H_2O$) and solution B (25×) (KNO_3, $Ca(NO_3)_2{\cdot}4H_2O$, KH_2PO_4, KCl); micronutrients (1000×); iron source (×1000).
2. The vitamin stock solution (1000×) for 190-2 induction (190-2_ind) and 190-2 regeneration (190-2_reg): Dispense 1 mL of the vitamin stock solution into 1.5 mL Eppendorf tubes. Label each tube accordingly and store at −20°C.
3. The amino acid stock and myoinositol stock solution (100×): Weigh and dissolve the corresponding amount of glutamine, glycine, and myoinositol for 1 L of each medium (Table 1) in 8 mL of water. Take the volume to 10 mL and place into 10 mL Falcon tubes.
4. $4CuSO_4{\cdot}5\,H_2O$ stock solution (1000×, 2.5 mg/mL) for 190-2-reg medium: To prepare 10 mL dissolve 25 mg of $CuSO_4{\cdot}5\,H_2O$ powder in 10 mL of deionized water. Stir gently to ensure complete dissolution. Transfer 1 mL aliquots into 10 individual 1.5 mL Eppendorf tubes, label them, and store at −20°C.
5. Plant growth regulator (PGR) stock solutions (*see* **Note 8**):
   - *Benzylaminopurine* (BAP) stock solution (1000×, 0.9 mg/mL): Dissolve 9 mg of BAP powder in 50 μL of 1 N NaOH, then adjust the volume to 10 mL with water and mix thoroughly.
   - *Kinetin* stock solution (1000×, 0.5 mg/mL): Dissolve 5 mg of Kinetin powder in 50 μ: of 1 N NaOH, and adjust the volume to 10 mL with water.
   - *Dicamba* (3,6-dichloro-2-methoxybenzoic acid) stock solution (1000×, 1 mg/mL): Dissolve 10 mg of Dicamba powder in 8 mL of hot water. Once dissolved, adjust the volume to 10 mL with water and mix thoroughly.

- *NAA* (1-Naphthaleneacetic acid) stock solution (1000×, 0.5 mg/mL): Dissolve 5 mg of NAA powder in 50 µL of 1 N NaOH, then take the volume to 10 mL with water.
- *Picloram* (4-amino-3,5,6-trichloropicolinic acid) stock solution (1000×, 1 mg/mL): Dissolve 10 mg of Picloram powder in 50 µL of 1 N NaOH and take the volume to 10 mL with water.

Filter each PGR stock solution through sterile Ø 0.22 µm filter, aliquot 1 mL into sterile 1.5 mL Eppendorf tubes and store at −20°C. Use 1 mL of this stock solution per 1 L of medium.

6. Arabic Gum (AGP) stock solution (2000×, 0.1 g/mL): Dissolve 5 g of AGP powder in 40 mL of water, mix thoroughly, and adjust the volume to 50 mL. Filter the stock solution through a sterile filter (Ø 0.22 µm) into sterile 15 mL Falcon tubes and store at −20°C. Use 0.5 mL or 1 mL for 1 L of 190-2 induction or regeneration medium, respectively.
7. Cefotaxime stock solution (3000×, 3 mg/mL): Dissolve 30 mg of cefotaxime powder in 8 mL of water, mix thoroughly, and adjust the volume to 10 mL. Filter the stock solution through a sterile filter (Ø 0.22 µm) into sterile 1.5 mL Eppendorf tubes and store at −20°C. Use 0.333 mL of stock solution for 1 L sterile induction medium (1 mg/L).
8. To prepare 1 L of liquid 190-2 induction medium: In 500 mL of deionized water, add 40 mL of macronutrients stock solutions A and B, 1 mL of micronutrients stock solution, 1 mL of iron stock solution, 1 mL of vitamins stock solution, 0.5 mL of AGP stock solution, and 10 mL of amino acids stock solution (final concentration of glutamine is 438 mg/L), 0.333 mL of cefotaxime stock solution and 1 mL of each BAP, Dicamba and Picloram stock solutions, ensuring it dissolves completely. Add 63 g of maltose mixing thoroughly. Adjust the total volume to 1 L with additional water, check and adjust the pH as necessary and mix well. Sterilize the medium using a filter with a Ø 20 µm pore diameter and store at 4°C.
9. To prepare 1 L of solidified 190-2 induction: Add all stock solutions as in liquid medium (except for vitamins, AGP, growth regulators, and cefotaxime), and 90 g maltose to 500 mL of deionized water. Take the volume to 1 L and add 3 g of phytagel and stir thoroughly to remove any lumps before autoclaving. When the phytagel has dissolved and the solution is homogeneous, pour into one bottle and autoclave at 121°C for 20 min (*see* **Note 9**). Once the medium has cooled to approximately 40°C, add 1 mL of each sterilized vitamin, BAP, Dicamba, and Picloram stock solutions, 0.5 mL of AGP and 0.333 mL of cefotaxime stock solutions. The final concentration of glutamine is 200 mg/L. After preparing, pour the medium into 5.5 cm Ø Petri dishes.

10. To prepare 1 L of 190-2 regeneration medium: In 500 mL of deionized water, add 40 mL of macronutrients stock solutions A and B, 1 mL of micronutrients stock solution, 1 mL of iron source stock solution, 1 mL of $CuSO_4{\cdot}5\,H_2O$ stock solution, and 30 g of sucrose. Adjust the pH to 6.0 with diluted 1 N NaOH. Add 6 g agar and mix the ingredients thoroughly and adjust the total volume to 1 L with additional water. When the agar has dissolved and the solution is homogeneous, pour into one bottle and autoclave at 121°C for 20 min. After autoclaving, allow the medium to cool to about 40°C before adding 1 mL of each of sterilized vitamins, NAA and KIN, 1 mL of AGP stock solutions and 10 mL of amino acids stock solution (final concentration of glutamine is 100 mg/L), ensuring it dissolves completely. After preparing, pour the medium into 9 cm Ø Petri dishes.
11. To prepare 1 L of 1/2 MS medium, first add 500 mL of distilled or deionized water to a sterile container, then dissolve 2.2 g of MS basal medium powder (Sigma-M5519) and 20 g of sucrose, stirring well. Next, adjust the pH 6.0 with diluted 1 N NaOH. Add 6 g agar and mix thoroughly, making up to 1 L with additional water. When the agar has dissolved and the solution is homogeneous, pour into one bottle. Autoclave the bottle at 121°C for 20 min. After autoclaving, allow the medium to cool to approximately 40°C before using. After preparing, pour the medium into Magenta boxes.

3 Methods

3.1 Donor Plants and Growth Conditions

3.1.1. Seedling development: For seedling development of winter rye (*Secale cereale* L.), sow seeds in perlite presoaked with Hoagland's salt solution and transfer them to an in vitro growth chamber at 25°C with 80–100 $\mu mol\,m^{-2}\,s^{-1}$ and 16 h/8 h light/dark (*see* **Note 10**).

3.1.2. Vernalization: Two- to three-day-old seedlings grown in an in vitro growth chamber at 25°C are then vernalized for 9 weeks at 2°C with a light intensity of 100 $\mu mol\,m^{-2}\,s^{-1}$ and 8 h/16 h light/night (Fig. 1A).

3.1.3. Planting and growth conditions: Vernalized seedlings (about 20 per line) are planted individually in 3 L pots containing peat and sand (3:1), and grown in the greenhouse (April/May) at 25°C/17°C (day/night) with 800 $\mu mol\,m^{-2}\,s^{-1}$, and 16 h/8 h light/dark (Fig. 1B, C). Plants are fertilised with 1 L of Hoagland's solution per pot twice a week from the beginning of the shooting/budding period (*see* **Note 11**).

3.2 Tiller Optimal Stage Assessment and Microspore Viability Determination

Assessment of the optimal microspore developmental stage and determination of microspore viability are performed using DAPI and FDA staining, respectively, followed by microscopic analysis. Prepare independent slides for each type of staining with at least three replicates per analysis.

3.2.1. DAPI staining for microspore developmental stage assessment: First, isolate anthers from different flowers and place at least four anthers per slide in a drop of sterile water. Using a razor blade or scalpel, carefully cut the anthers into small fragments to release the microspores into the water. After releasing the microspores, apply one drop of 10 μg/mL DAPI to each slide and cover with a cover slip. Allow the dye to interact with the microspores for 15 min in the dark and examine the samples under a fluorescence microscope using filters EX 365/DM 400/BA 420 EF.

3.2.2. FDA staining for microspore viability determination: Follow the same procedure for anther isolation and microspore release as described in Sect. 3.2.1. Then, apply one drop of 0.05 mg/mL FDA to each slide and cover with a cover slip. Incubate for 15 min to allow the dye to interact with the microspores. Examine the samples under a fluorescence microscope using filters EX 470–490/DM 510 BA/515 EF.

3.3 Tillers Collection, Pretreatment Procedure, and Sterilization

3.3.1. Tillers are ready for harvesting 6–10 weeks after potting. Harvest tillers whven the distance between the base of the flag leaf and the previous leaf is 10–14 cm, the length of the awns protruding above the base of the flag leaf is 3–4 cm, and the majority of microspores are at the late uninucleate stage (Fig. 1C).

3.3.2. Leave the stem with the ear after removing all the leaves except the flag leaf and label the spikes.

3.3.3. Pretreatment: Collect tillers daily and randomly select for the proposed pretreatments. Wrap the selected tillers in plastic bags and store for 3 weeks at 4°C in the dark, placing them in jars containing the respective pretreatment solutions for both anther (AC) and isolated microspore cultures (MC) (Fig. 1D).

3.3.4. For the pretreatment procedure with MAN_GSH, treat tillers with 0.7 M mannitol for the first 14 days and then transfer to 0.3 mM GSH for the last 7 days. Use approximately 12 spikes per line per each isolation. When preparing material for one isolation, set aside approximately four additional spikes for ovary dissection.

Fig. 1 Procedure for microspore isolation in rye (*Secale cereale* L.). (**A**) Seedlings after 9 weeks of vernalization. (**B**) Seedlings planted in 3 L pots, grown to maturity in the greenhouse. (**C**) Plants showing tillers at the appropriate developmental stage for microspore collection. (**D**) Tillers under stress pretreatment at 4°C before

3.3.5. For the pretreatment procedure with sodium selenate (SS; Na_2SeO_4), treat tillers with tap water for the first 14 days and then transfer to a 50 μM SS solution for the last 7 days. Use approximately 12 spikes per line per each isolation. Set aside approximately 4 additional spikes for ovary dissection.

3.3.6. For sterilization of the tillers after stress treatment (Fig. 1E), first spray the spikes with 96% (*v/v*) ethanol under sterile conditions in the laminar flow cabinet (Fig. 1F). Remove the spikes from the leaf sheath with forceps, trim the awns with scissors, and then place five to six spikes per 150 mL glass Erlenmeyer flask. Next, immerse the spikes in 70% ethanol for 1 min and pour it off. Then, immerse the spikes in a 0.9 % (*w/v*) sodium hypochlorite sterilizing solution for 15 min. Gently shake the flasks by hand several times during sterilization. Decant the bleach solution and rinse the spikes thoroughly with sterile deionized water. Perform four to five rinses: the first two should last 3 min each and the subsequent ones should last 10 min, with vigorous agitation of the flasks.

3.4 Microspore Isolation and Culture

3.4.1. Harvest approximately 40 tillers from each line (20 plants per line) over a 3-week period for microspore culture. Divide the harvested tillers into groups of 12 tillers per pretreatment for each single isolation of microspores. For each isolation, set aside 4 additional tillers specifically for ovary dissection.

3.4.2. After pretreatment, cut sterile spikes into approximately 2 cm long segments and blend them in 0.3 M mannitol using a Waring blender (Fisher Scientific Inc.; Fig. 1G, H).

3.4.3. Filter the resulting slurry through a double layer of metal sieves: the first Ø 1 mm and the second Ø 74 μm (in a glass crystallizer/Magenta) (Fig. 1I). Transfer the homogenate to the sterile conical 50 mL Falcons tubes and pellet the microspores by centrifugation (100 × *g*, 5 min) (Fig. 1J).

3.4.4. Remove the supernatant and transfer the microspore pellet to 15 mL conical glass tubes.

Fig. 1 (continued) microspore isolation. **(E)** Tiller ready for microspore isolation. **(F)** The workspace and materials required for microspore isolation in a laminar flow cabinet arranged for sterile microspores isolation. **(G)** Sterile spikes are cut into approximately 2 cm segments in a Waring blender. **(H)** Spike segments prior to mixing with mannitol solution using a Waring blender. **(I)** The resulting slurry filtered through metal sieves to collect microspores. **(J)** Microspores transferred to 50 mL conical flasks and centrifuged to pellet. **(K)** Glass tubes containing maltose phase. **(L)** The microspore pellet is transferred to 15 mL glass tubes containing maltose. **(M)** Viable microspores are deposited at the interface of mannitol and maltose after centrifugation (arrow). **(N)** Viable microspores collected into tubes containing 4 mL of mannitol

3.4.5. Resuspend the microspores in approximately 0.5 mL 0.3 M MAN and gently layer onto a 21% (*w/v*) maltose solution (Fig. 1K, L) for density gradient centrifugation (80 × *g*, 5 min). Gently collect viable microspores settling at the interface between the mannitol and maltose using sterile Pasteur pipettes (Figs. 1M, N and 2C).

3.4.6. Transfer the collected viable microspores to new conical 15 mL glass tubes, wash them in 0.3 M MAN, and pellet them again (100 × *g*, 5 min).

3.4.7. Resuspend the microspores in 1 mL of liquid 190-2 induction medium and observe under the microscope (Fig. 2).

3.4.8. Collect a sample of microspores for cell nucleus phenotyping by chromatin staining with DAPI. Add one drop of freshly isolated microspore suspension to a microscope slide. Stain microspores with a drop of 10 μg/mL DAPI (*v/v* 1/1), cover, and incubate for 15 min prior to microscopy (Fig. 2D, E).

3.4.9. Determine the viability of the microspores on the day of isolation using FDA staining. Prepare microscope slides with one drop of microspore suspension. Stain microspores with a drop of 0.05 mM FDA (*v/v* 1/1), cover, and incubate for 15 min prior to microscopy (Fig. 2B, G, H).

3.4.10. Calculate the number of viable microspores in the suspension and adjust the density for culture using a Neubauer chamber (hemocytometer). First, dilute the microspore suspension and load a 10–20 μL of the diluted suspension into the chamber grid. Count the microspores in the four large corner squares under the microscope and calculate the average number of microspores per square. Use the formula: *Cell density (cells/mL) = (average count per square × dilution factor)/volume of each square (0.0001 mL)*. This gives the number of microspores per mL of suspension. Adjust the dilution as necessary to obtain a final density of approximately 70×10^3 microspores per mL (*see* **Note 12**).

3.4.11. Dissect ovaries at the same time as microspore isolation, ensuring they come from separate tillers than those used for microspore isolation. Use ten ovaries per 1 mL of microspore suspension for coculture.

3.4.12. Transfer the microspore suspension to 15 × 60 mm Petri dishes containing the 190-2 induction medium (Table 1) (Fig. 2F) (*see* **Note 12**). Incubate cultures at 26°C in the dark for 42 days.

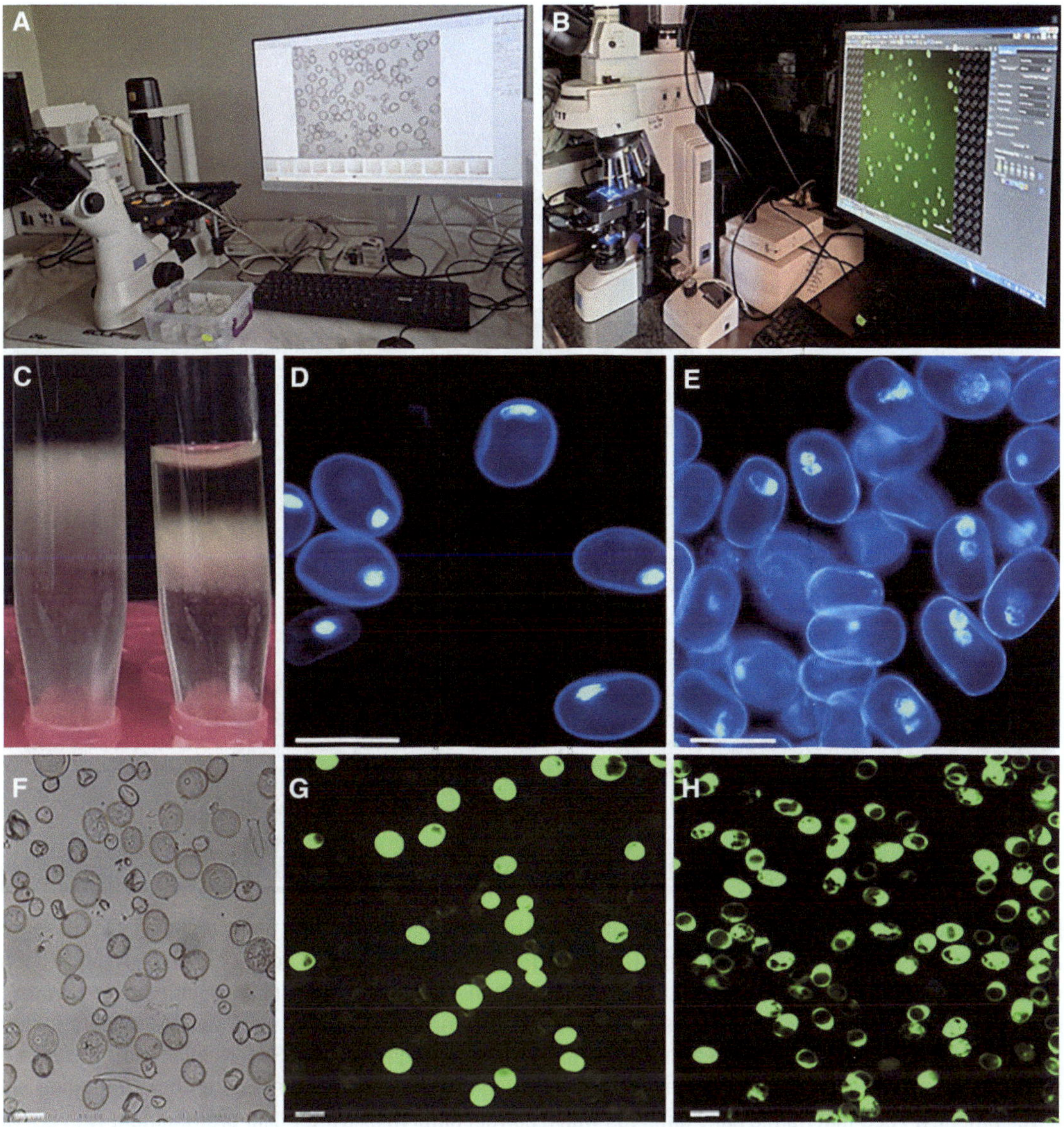

Fig. 2 Cytological analysis of the role of mannitol with GSH or selenium treatments on microspore viability and symmetric division in rye (*Secale cereale* L.). (**A**) Inverted microscope equipped with a digital camera and image acquisition program. (**B**) Fluorescent microscope equipped with a digital camera and image acquisition program. (**C**) Effect of pretreatment with mannitol and glutathione (MAN_GSH; left) and selenium (SS; right) on the efficiency of viable microspores isolation. Difference in thickness of the band of settled viable microspores collected at the mannitol/maltose interface after density gradient centrifugation. (**D**) Uninucleated microspores after treatment with MAN_GSH. (**E**) Uninucleated microspores and a high percentage of symmetric divisions marking the onset of embryogenesis after tillers pretreatment with SS. (**F**) Isolated microspores in vitro on 190-2 induction medium on the day of isolation. (**G**) Microspore viability after tiller pretreatment with MAN_GSH. (**H**) High rate of microspore viability after tiller pretreatment with SS, suggesting an effect that supports cell survival during isolation procedure. (**D, E**) Nuclei counterstained with DAPI (intense blue); (**G, H**) Microspores stained with FDA (intense green fluoresce of vital cells). (**D–H**) Bar = 20 μm

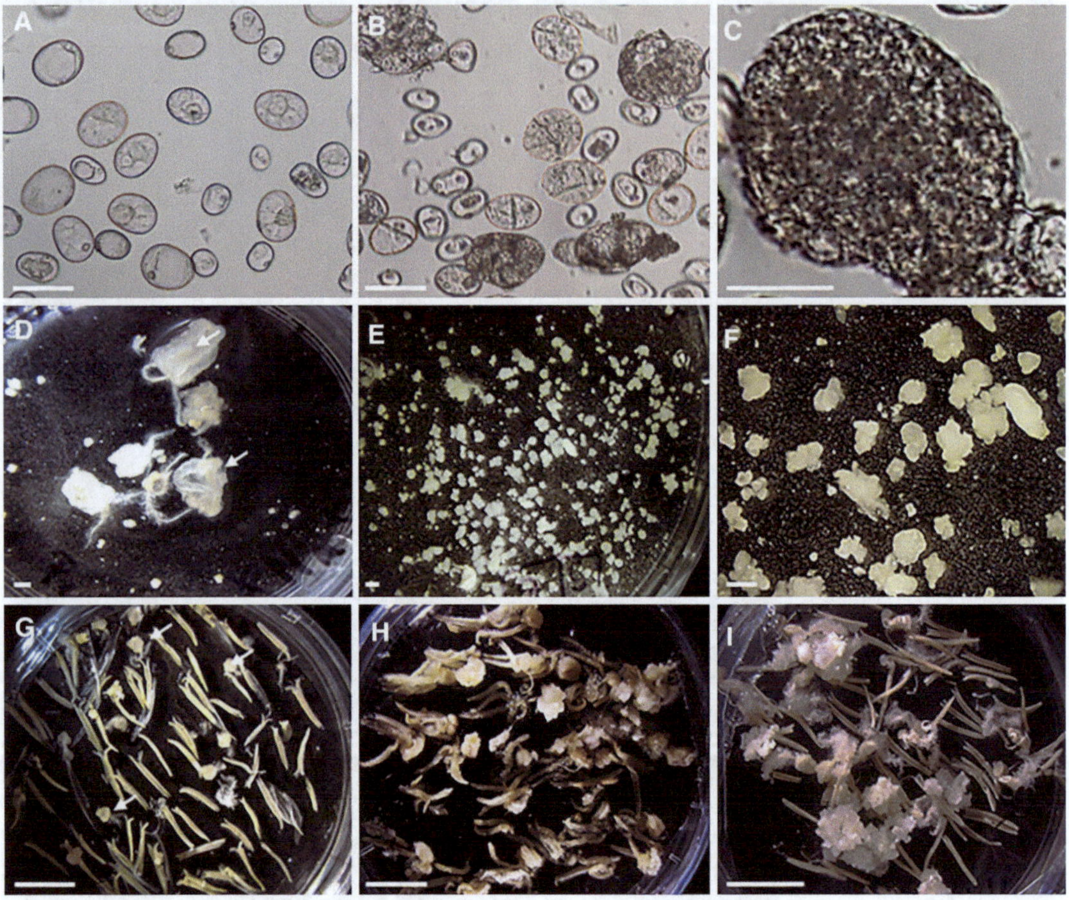

Fig. 3 Effect of the mannitol with glutathione (MAN_GSH) or selenium (SS) treatments on the in vitro development of embryo-like structures (ELS) in rye (*Secale cereale* L.) microspore and anther cultures. (**A**) Microspore culture on the induction medium immediately after isolation. (**B**) Early multicellular structures after 2 weeks of culture. (**C**) ELS visible after 3 weeks of culture. (**D**) ELS formation after 6 weeks of microspore culture and ovary coculture after tiller pretreatment with MAN_GSH. (**E**) High rate of ELSs formation after 6 weeks of microspores culture after tiller pretreatment with SS, suggesting a positive effect on embryo formation and development. (**F**) ELS that mimic early embryos with a correct embryo axis after tiller pretreatment with SS. (**G**) Anther cultures on solid induction medium immediately after anther isolation. (**H**) ELS formation after 6 weeks of anther culture from tillers pretreated with MAN_GSH. (**I**) High rate of ELS formation after 6 weeks of anther culture from tillers pretreated with SS, suggesting a positive effect on embryo formation and development. Arrows mark ovaries in isolated microspores and anther culture. (**A–B**) Bar = 20 μm; (**C–I**) Bar = 50 μm

3.4.13. For microscopic analyses, a minimum of three replicates is required per line and per pretreatment procedure. Collect a sample of microspores on the day of isolation (Fig. 2A–H) resuspended in 1 mL of medium prior to division for the in vitro experiment.

3.4.14. Observe the microspore suspensions under a microscope at weekly intervals: on the day of isolation (Fig. 3A), then after 14 and 21 days (Fig. 3B, C), and again after 4 and 6 weeks of in vitro culture (Fig. 3D–F).

3.5 Anther Culture

3.5.1. Carefully dissect anthers from spikes using fine forceps under sterile conditions in a laminar flow cabinet.

3.5.2. Place 100 anthers on 55 mm Ø Petri dishes, containing 2.5 mL of modified 190-2 induction medium (see Table 1 for modifications [14]) (Fig. 3G).

3.5.3. Additionally, randomly distribute ten ovaries in each dish to condition the induction medium.

3.5.4. Seal the Petri dishes tightly with strips of thin polyethylene food foil to prevent culture desiccation (*see* **Note 13**).

3.5.5. Incubate the cultures in the dark at 26 ± 1°C to promote embryogenic development. After 6 weeks of culture, ELS measuring ≥5 mm become visible. Transfer the ELS to fresh medium at 3-week intervals, ensuring optimal growth conditions for the developing structures (Fig. 3H, I).

3.6 Regeneration Phase of In Vitro Culture

3.6.1. Starting from the sixth week of culture, determine the number of ELS using an inverted light microscope.

3.6.2. Transfer the ELS and calli with a diameter of approximately 5 mm or larger, onto the 190-2 regeneration medium (Table 1) (30 ELS per 90 mm Ø Petri dish) (Fig. 4A).

3.6.3. Conduct the regeneration phase at 26°C under light conditions (Fig. 4A–E) initially at about 30 μmol $m^{-2} s^{-1}$ for the first week, then increase to 80–100 μmol $m^{-2} s^{-1}$ with a photoperiod of 16 h.

3.6.4. After 4 weeks of regeneration, transfer green regenerated plants to Magenta boxes with 1/2 MS half-strength medium to improve rooting (Fig. 4f).

3.6.5. After 2 weeks of rooting, transfer the plants to Magenta or plastic boxes containing perlite soaked in 1/2 Hoagland's nutrient solution to facilitate adaptation to ex vitro growth conditions in in vitro chambers with temperature controlled at 26°C (day/night), light intensity at 30 and 80–100 μmol $m^{-2} s^{-1}$ (Fig. 4G).

3.6.6. Subsequently, transfer the plants into pots filled with a soil-peat mixture and grow them in a greenhouse until the fourth leaf is fully formed.

3.6.7. Vernalize the seedlings and grow the plants following the same procedure as outlined in Sect. 3.1 (Fig. 4H).

3.7 Ploidy Level Determination

The ploidy level is determined starting from the fourth week of growth of the regenerated plants in the greenhouse.

Fig. 4 Regeneration of embryo-like structures (ELS) in rye (*Secale cereale* L.). (**A, B**) ELS immediately after transfer from induction medium to 190-2 regeneration medium. (**B**) Visible islands of ELS transferred to the surface of a solid medium in a drop of liquid medium. (**C**) ELS from microspore cultures showing the correct shoot and root axis after 1 week of culture on 190-2_reg medium. (**D**) Plant regeneration from anther culture showing a correct shoot and root development after 3 weeks of culture on 190-2_reg medium. (**E**) Green plant regeneration from anther culture after 3 weeks of culture on 190-2_reg medium. (**F**) Rooting of the regenerated green plants in Magenta boxes containing 1/2 MS medium. (**G**) Adaptation to semi ex vitro conditions. (**H**) Control plant and doubled haploid (DH) plants (left and right, respectively) at the tillering stage growing in soil

3.7.1. For flow cytometry analysis collect at least three pieces of fresh leaf tissue of approximately 1 cm^2 from each green plant in a glass Petri dish placed on ice. Use the same amount of fresh leaf tissue from a diploid plant ($2n=2x=14$) as an external reference standard.

3.7.2. Prepare a suspension of isolated nuclei, for e.g., in Otto I solution (*v/v*) to a total volume of 1000 μL. Chop the tissues finely with a razor blade and filter the suspension through a 42 μm Ø nylon mesh into 2 mL Eppendorf tubes.

3.7.3. Allow the suspension to equilibrate for a few minutes before use.

3.7.4. Stain the filtered suspension with 1 mL of Otto II buffer.

3.7.5. Incubate the stained nuclei on ice in darkness for 15 min prior to analysis.

3.7.6. Mix gently and analyze 50 µL aliquots of stained nuclei using a CyFlow Space flow cytometer. Consider at least 5000 measured events per sample as a replicate (*see* **Note 14**).

4 Notes

1. Choosing an androgenesis induction protocol tailored to different rye F1 lines offers several advantages, particularly in the context of genetic diversity and breeding efficiency. In the case of cross-pollinating rye, partially and fully self-pollinating varieties play an important role in genetic recombination and the restoration of fertility in male-sterile lines. For the initial screening, it is advisable to select at least five F1 lines per type of pollination system. Each line may respond differently to induction methods, allowing for the identification of optimal protocols for various genetic backgrounds.
2. Calculate the total volume needed for 1 day treatment. Prepare approximately 0.3 L of solution per genotype and treatment and store in a cool, dry place at 4°C, out of direct sunlight.
3. Calculate the total volume needed for 1 day treatment. Prepare approximately 0.3 L of solution per genotype and treatment and store it in a cool, dry place out of direct sunlight, or in a refrigerator at 4°C to prevent oxidation.
4. Calculate the total volume of 50 µM SS working solution needed for 1 day treatment, approximately 0.3 L of solution per genotype and treatment and store in a cool, dry place to maintain stability.
5. Setup solutions for autoclaving in advance and use them after cooling. Calculate the appropriate total volume needed for the isolation day. Make approximately 0.5 L of solution per genotype and treatment, and store at 4°C.
6. To prepare 0.1 M citric acid, slowly add the powder to water and stir until thoroughly dissolved. Adjust the pH of 0.1 M citric acid with 5 M NaOH until reaching the desired pH, 2.5 or 3.0. To prepare 100 mL of a 5 M NaOH solution: place approximately 50 mL of distilled water in a 100 mL volumetric flask and carefully add 20 g of NaOH while stirring to dissolve it. Since NaOH dissolves exothermically, allow the solution to cool if it becomes too warm. Once fully dissolved, add more distilled water to bring the volume to 100 mL and mix thoroughly. Remember to wear gloves and goggles when

handling NaOH, as it is highly caustic, and always add NaOH to water (not the reverse) to reduce the risk of splashing. The detergent Tween 20 is weaker than Triton X-100.

7. The pH of Otto II buffer is not adjusted because β-mercaptoethanol suppresses the negative effects of phenols and other cytosolic compounds. The pH of this solution is about 8.5–9.5, depending on concentration and temperature. If the pH is higher than 9.0 or lower than 8.0, add small amounts of 1 M or 0.1 M HCl, or 1 M or 0.1 M NaOH to adjust the pH to 8.0–9.0.
8. Work in a well-ventilated area when preparing PGR solutions as acids and alkalis can be dangerous and harmful. It is essential to wear gloves when working with 1 N NaOH. PGRs are considered stable and can be stored at −20°C.
9. If phytagel is added to a warm or hot medium, it will agglomerate and will not gel properly after autoclaving. The setting temperature of Phytagel is 27°C–32°C.
10. Perlite presoaked in Hoagland's salt solution should be autoclaved beforehand and used after cooling.
11. Any commonly available soil mix suitable for general plant growth should work well for growing donor plants.
12. To avoid contamination, use freshly filtered liquid medium on each isolation day. For filtration, use freshly thawed, portioned 50 mL aliquots of the previously prepared medium stored at −20°C.
13. After 3 weeks, anthers typically turn dark yellow with brown spots. By the fourth week, some anthers begin to break longitudinally, revealing white, compact structures resembling globular embryos—ELS (Fig. 3H, I). ELS can develop into amorphous masses, eventually undergoing somatic embryogenesissomatic embryogenesis on their surface, resulting in the so-called androgenic callus. Both structures, as well as various intermediate forms, coexist within the same anther culture.
14. Gently mix the contents by inverting the tube several times to ensure thorough mixing. Avoid vigorous shaking to prevent damage to the nuclei.

Acknowledgments

The research was supported by the IPP PAS institutional funding, and individual national research project financed by the Ministry of Agriculture and Rural Development in 2017–2018.

References

1. Koistinen VM, Mattila O, Katina K et al (2018) Metabolic profiling of sourdough fermented wheat and rye bread. Sci Rep 8(1):5684. https://www.nature.com/articles/s41598-018-24149-w
2. Tenhola-Roininen T, Immonen S, Tanhuanpää P (2006) Rye doubled haploids as a research and breeding tool—a practical point of view. Plant Breed 125:584–590. https://doi.org/10.1007/s10681-005-8229-6
3. Immonen S, Tenhola-Roininen T (2003) Protocol for rye anther culture. In: Maluszynski M, Kasha KJ, Forster BP, Szarejko I (eds) Doubled haploid production in crop plants: a manual. Kluwer Academic Publisher, Dordrecht, pp 141–149. https://doi.org/10.1007/978-94-017-1293-4_3
4. Tenhola-Roininen T (2009) Rye doubled haploids: production and use in mapping studies. In: Marjomäki V (ed) Jyväskylä studies in biological and environmental science. University of Jyväskylä, Jyväskylä, p 93
5. Rakoczy-Trojanowska M, Śmiech M, Malepszy S (1997) The influence of genotype and medium on rye (*Secale cereale* L.) anther culture. Plant Cell Tissue Organ Cult 48(1):15–21. https://doi.org/10.1023/A:1005792912316
6. Guo YD, Pulli S (2000) Isolated microspore culture and plant regeneration in rye (*Secale cereale* L.). Plant Cell Rep 19(9):875–880. https://doi.org/10.1007/s002990000194
7. Tenhola-Roininen T, Tanhuanpää P, Immonen S (2005) The effect of cold and heat treatments on the anther culture response of diverse rye genotypes. Euphytica 145(1):1–9. https://doi.org/10.1007/s10681-005-8229-6
8. Dubas E, Żur I, Moravčiková J et al (2021) Proteins, small peptides and other signaling molecules identified as inconspicuous but possibly important players in microspores reprogramming toward embryogenesis. Front Sustain Food Syst 5:745865. https://doi.org/10.3389/fsufs.2021.745865
9. Żur I, Dubas E, Krzewska M et al (2014) Antioxidant activity and ROS tolerance in triticale (×*Triticosecale* Wittm.) anthers affect the efficiency of microspore embryogenesis. Plant Cell Tissue Organ Cult 119(1):79–94. https://doi.org/10.1007/s11240-014-0515-3
10. Żur I, Dubas E, Krzewska M et al (2019) Glutathione provides antioxidative defence and promotes microspore-derived embryo development in isolated microspore cultures of triticale (×*Triticosecale* Wittm.). Plant Cell Rep 38(2):195–209. https://doi.org/10.1007/s00299-018-2362-x
11. Zieliński K, Krzewska M, Żur I et al (2020) The effect of glutathione and mannitol on androgenesis in anther and isolated microspore cultures of rye (*Secale cereale* L.). Plant Cell Tissue Organ Cult 140(3):577–592. https://doi.org/10.1007/s11240-019-01754-9
12. Zieliński K, Dubas E, Gerši Z et al (2021) β-1,3-Glucanases and chitinases participate in the stress-related defence mechanisms that are possibly connected with modulation of arabinogalactan proteins (AGP) required for the androgenesis initiation in rye (*Secale cereale L.*). *Plant Sci* 302:110700. https://www.sciencedirect.com/science/article/pii/S016894522030306X
13. Zieliński K, Krzewska M, Nowicka A et al (2018) Redox regulation of androgenesis in rye (*Secale cereale* L.) In: Grzesiak MT, Hura T (eds) Proceedings of the 11th International Conference Plant Functioning under Environmental Stress, Kraków, p 82
14. Zhuang JJ, Xu J (1983) Increasing differentiation frequencies in wheat pollen callus. In: Hu H, Vega MR (eds) Cell and tissue culture techniques for cereal crop improvement. Science Press, Beijing, p 431
15. El-Tantawy AA, Solís MT, Da Costa ML et al (2013) Arabinogalactan protein profiles and distribution patterns during microspore embryogenesis and pollen development in *Brassica napus*. Plant Reprod 26(3):231–243. https://doi.org/10.1007/s00497-013-0217-8
16. Dubas E, Krzewska M, Surówka E et al (2024) New prospects for improving microspore embryogenesis induction in highly recalcitrant winter wheat lines. Plants 13(3):363. https://www.mdpi.com/2223-7747/13/3/363
17. Wędzony M (2003) Protocol for doubled haploid production in hexaploid triticale (x *Triticosecale* Wittm.) by crosses with maize. In: Maluszynski M, Kasha KJ, Forster BP, Szarejko I (eds) Doubled haploid production in crop plants: a manual. Springer Netherlands, Dordrecht, pp 135–140. https://doi.org/10.1007/978-94-017-1293-4_21
18. Murashige T, Skoog F (1962) A revised medium for rapid growth and bio assays with tobacco tissue cultures. Physiol Plant 15(3):473–497. https://onlinelibrary.wiley.com/doi/abs/10.1111/j.1399-3054.1962.tb08052.x

Chapter 18

Targeted Gene Editing in Wheat During Haploid Production via Wide Hybridization with Transgenic Maize Expressing Cas9 and Guide RNA

Shaobin Zhong, Yueqiang Leng and Shengming Yang

Abstract

The clustered regularly interspersed short palindromic repeats (CRISPR)/Cas9 system is an efficient and versatile genome engineering tool, which has been widely used for targeted mutagenesis and gene functional characterization in various organisms. This system is simple because it only requires a Cas9 enzyme serving as a nuclease and guide RNA (gRNA) containing a 20-nt sequence matching the target gene. Delivery of a vector expressing Cas9 and gRNA or the preassembled Cas9/gRNA complex as a ribonucleoprotein (RNP) into plant cells for gene targeting are usually via the biolistic- or *Agrobacterium*-mediated approach. However, most wheat genotypes suffer from low efficiency of callus induction and plant regeneration from explants receiving the vector or RNP delivered by the biolistic- or *Agrobacterium*-mediated transformation method, limiting the application of genome editing systems in many commercially grown wheat varieties. Here, we describe a stepwise protocol for targeted gene editing in wheat via wide hybridization with transgenic maize expressing Cas9 and gRNA. A binary vector expressing Cas9 and gRNA is constructed and used for *Agrobacterium*-mediated transformation to generate transgenic maize plants, which are used to pollinate emasculated spikes of wheat varieties. After fertilization, the maize chromosomes enter the transient hybrid zygote and the transgene (T-DNA) on a maize chromosome expresses the Cas9 enzyme and gRNA, which forms an RNP complex to edit the target gene in wheat genome. After several cell divisions, maize chromosomes in the hybrid zygote are eliminated, resulting in formation of haploid wheat embryos with the target gene edited, which can be rescued by embryo culture technique to produce haploid plants. Doubled haploid (DH) wheat plants with homozygous gene mutations are developed by chromosome doubling through colchicine treatment of the haploid plants. The wheat × maize hybridization combined with the CRISPR/Cas9 system provides a one-step approach for generating DH lines with the target gene edited from any wheat genotypes of interest.

Keywords CRISPR/Cas9, Guide RNA, Genome editing, Wheat, Maize, Wide hybridization, Haploid plants, Doubled haploid lines

1 Introduction

The clustered regularly interspersed short palindromic repeats (CRISPR)/Cas9-mediated genome editing system provides a revolutionary technology that allows precise and targeted changes to

Ana María Castillo et al. (eds.), *Doubled Haploid Methods in Cereals*, Methods in Molecular Biology, vol. 3029,
https://doi.org/10.1007/978-1-0716-5245-9_18,

an organism's DNA [1–3]. The system consists of Cas9, a nuclease enzyme that cuts double-strand DNA and a single guide RNA (sgRNA) sequence, designed to match the target DNA in the organism [1, 2] (Fig. 1). When Cas9 is combined with the sgRNA, it finds the target DNA sequence and introduces a double-strand break (DSB) (Fig. 1). DSB can be repaired by the cell's natural repair mechanisms, including Non-Homologous End Joining (NHEJ) and Homology-Directed Repair (HDR). NHEJ often leads to small insertions or deletions, which can disrupt a gene, while HDR allows for precise edits by inserting a donor DNA template, leading to precise nucleotide correction or gene insertion through homologous recombination (Fig. 2).

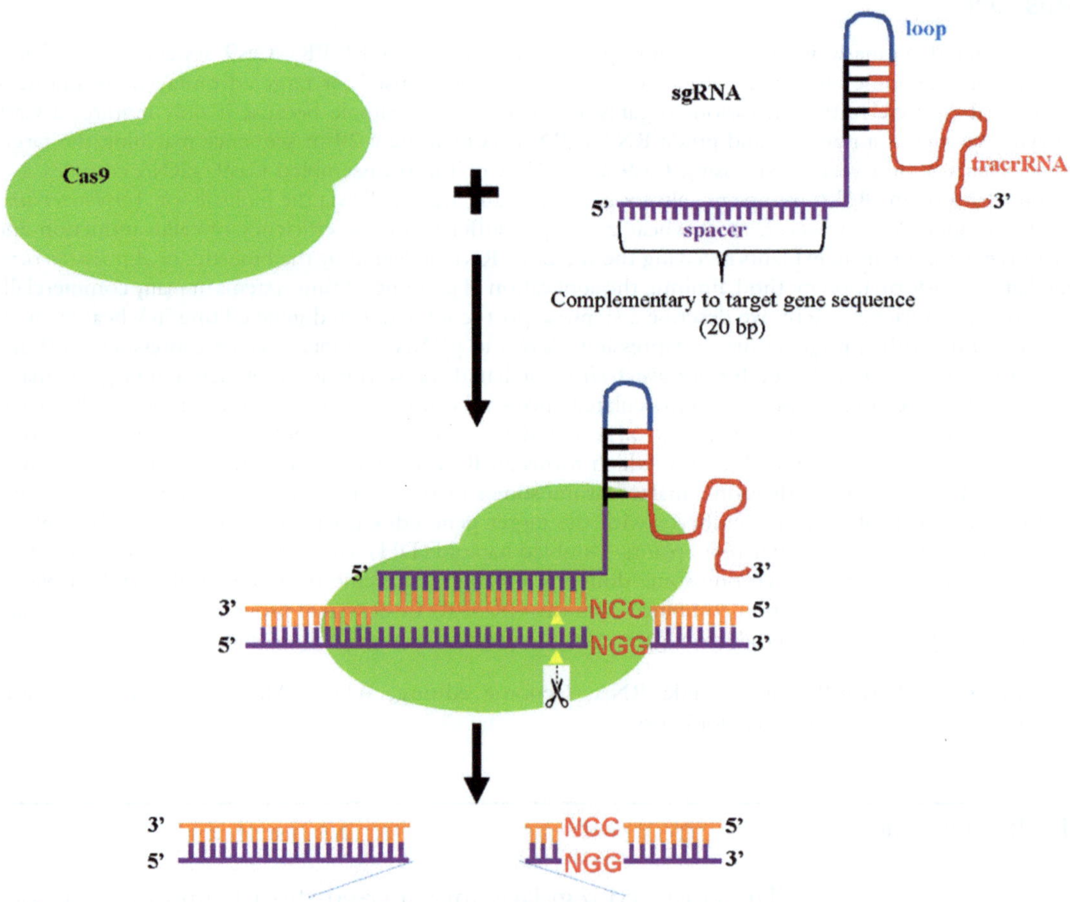

Fig. 1 Schematic illustration of the CRISPR/Cas9-mediated genome editing system for generating of a double-strand DNA break. Cas9 is a nuclease (green) that cuts double-strand DNA. Single guide RNA (sgRNA) consists of a tracrRNA, a loop, and a crRNA containing the 20-bp spacer complementary to the target gene sequence. Cas9 and sgRNA form a complex that cleaves the target double-strand DNA at 3-bp upstream of the protospacer adjacent motif (PAM) (NGG/NCC, red), leading to a double-strand break (DSB) (arrows in yellow)

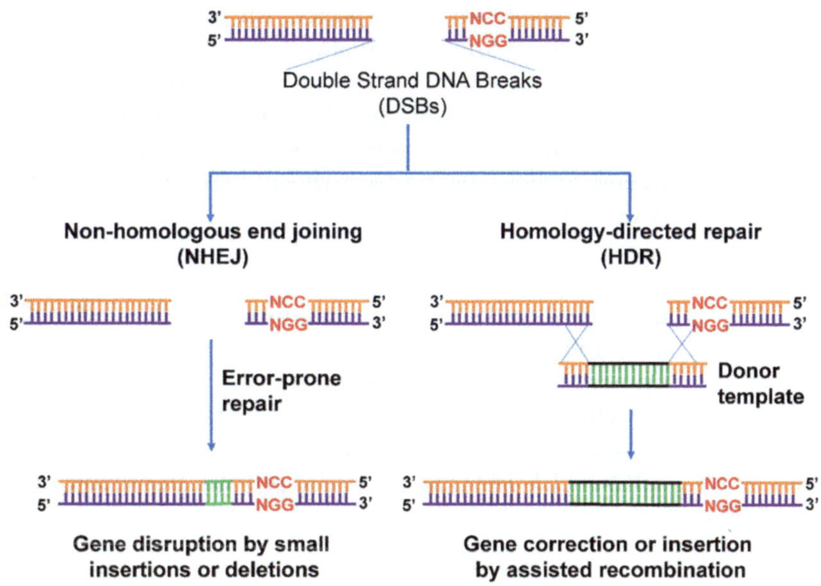

Fig. 2 CRISPR/Cas9 double-strand breaks (DSBs) are repaired by either nonhomologous end joining (NHEJ, left) or homology-directed repair (HDR, right) mechanism in cells. NHEJ results in random small indels (insertions or deletions), while HDR makes a precise nucleotide correction or gene insertion by homologous recombination

To target a gene in plants using the CRISPR/Cas9-mediated genome editing system, researchers need to design a sgRNA specific to the target gene, and then deliver the CRISPR/Cas9 components (vector expressing Cas9 and sgRNA or Cas9/sgRNA RNP complex) into plant cells through methods such as *Agrobacterium*-mediated transformation and particle bombardment [3–7]. Cells that undergo successful editing are selected and grown into mature plants, which carry the edited gene in all cells. Edited plants are then screened to confirm the intended genetic modifications. However, in wheat (*Triticum* ssp.), genome editing through *Agrobacterium*-mediated transformation and particle bombardment has been confronted with difficulties and challenges, including low efficiency of plant transformation and regeneration, time-consuming and genotype-dependent procedures, as well as unintended changes to the genome [3], limiting its applications in many commercially grown wheat varieties. Therefore, alternate approaches for wheat genome editing are needed to overcome the obstacles.

Producing doubled haploid (DH) wheat plants through wide hybridization with maize (*Zea mays* L.) is a useful technique in plant breeding to accelerate the development of homozygous lines [8–10]. This method involves crossing wheat with maize to produce haploid wheat plants, which are later chromosome-doubled to generate fully fertile DH wheat plants [8–10]. In the process, wheat plants are emasculated and pollinated with maize pollen. After fertilization of wheat ovules by maize pollen, the maize chromosomes are often eliminated during the initial cell divisions, leaving a haploid wheat embryo that contains only wheat chromosomes. The developing haploid embryos are "rescued" through embryo culture, which involves excising the embryos and placing them in a suitable nutrient medium to ensure they develop properly [11]. To restore fertility, the haploid plants are treated with colchicine or another chromosome-doubling agent to produce a DH wheat plant. This step ensures that the plants are homozygous and fully fertile. This DH technology is commonly used in wheat breeding programs to rapidly produce homozygous lines, thereby reducing the breeding cycle and accelerating the development of new wheat varieties.

Here, we described a protocol for targeted gene editing in wheat via wide hybridization between wheat and maize expressing Cas9 and sgRNA [12–14] (Fig. 3). In this process, transgenic maize plants are

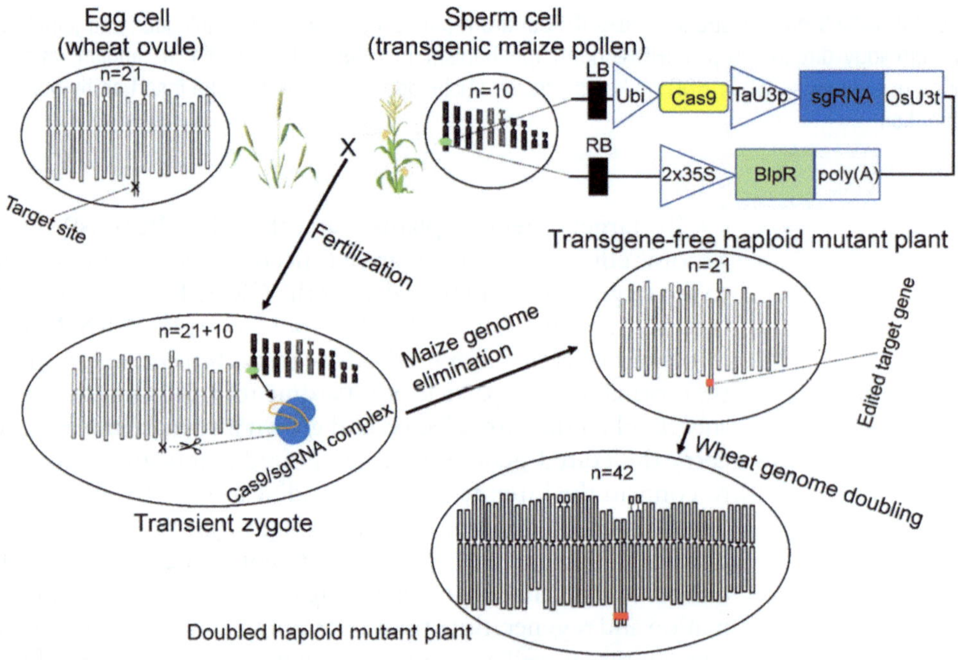

Fig. 3 Schematic illustration of targeted gene editing in wheat via wide hybridization between wheat and maize expressing Cas9 and single guide RNA (sgRNA). In this process, transgenic maize plants are generated using a gene construct that expresses Cas9 and sgRNA and used to pollinate emasculated wheat varieties. After fertilization, the Cas9 enzyme and sgRNA expressed by the transgene (T-DNA) on maize chromosome form a Cas9/sgRNA complex to edit the target gene in wheat embryo cells (transient zygote). After maize genome elimination, transgene-free haploid plants with the target gene edited can be generated by embryo rescue. After chromosome doubling with colchicine treatment of haploid plants, doubled haploid plants with homozygous mutations can be developed. Ubi, maize polyubiquitin-1 (Ubi-1) promoter; TaU3p, wheat U3 promoter; OsU3t, rice U3 terminator; 2x35S, a double-enhanced 35 S promoter from the Cauliflower Mosaic Virus (CaMV); BlpR, phosphinothricin acetyltransferase gene; poly(A), poly(A) signal; LB, left border; RB, right border

generated using a binary vector with the gene construct that expresses Cas9 and sgRNA and used to pollinate emasculated wheat varieties. After fertilization, the maize chromosomes are present in the cells of the transient hybrid zygote and the transgene (T-DNA) on maize chromosome expresses Cas9 enzyme and sgRNA to form a Cas9/sgRNA complex, which edits the target gene in the wheat genome. After maize genome elimination, transgene free haploid embryos with the target gene edited are generated, which can be rescued by the immature embryo culture technique to produce haploid plants. After chromosome doubling with colchicine treatment of haploid plants, DH plants with homozygous mutations are developed (Fig. 3).

2 Materials

2.1 Plant Material

1. Bread wheat genotype: Variety Dayn.
2. Maize genotypes: Inbred line B104 and hybrid variety Hi II.

2.2 Equipment

1. Air pump.
2. Benchtop microcentrifuge.
3. Horizontal Electrophoresis Systems (Catalog No: FB-SB-710, Fisher Scientific).
4. Growth chamber set at 25°C, 350 μmol $m^{-2}s^{-1}$ light intensity provided by Fluence Ray22 lamps and 16/8 h light/dark.
5. Growth chamber set at 16°C, 350 μmol $m^{-2}s^{-1}$ light intensity provided by Fluence Ray22 lamps and 16/8 light/dark.
6. Greenhouse set at 22°C–25°C, 300 μmol $m^{-2}s^{-1}$ light intensity, provided by 600 watt high pressure sodium lamps supplied by PL Light System and 16/8 h light/dark.
7. Greenhouse set to 22°C/18°C day/night, 300 μmol $m^{-2}s^{-1}$ light intensity provided by 600 watt high pressure sodium lamps supplied by PL Light System and 16/8 h light/dark.
8. Incubators set at 25°C or 37°C without light.
9. NanoDrop™ One (Thermo Fisher Scientific).
10. PCR thermal cyclers (Bio-Rad T100 regular and CFX96 Touch Real-time thermal cycler).
11. Laminar air-flow hood.
12. Stereo microscope.
13. pH meter.
14. Shaker.
15. TissueLyser II (Qiagen).
16. Water bath.

2.3 Supplies

1. 96-well DNA extraction plate.
2. Cell culture tubes (14 mL).
3. Conical centrifuge tubes (15 and 50 mL).
4. Fine tip tweezers, scissors, and scalpels.
5. Glassine pollination bags and paper clips.
6. Brown bags.
7. Inoculation loops.
8. Microcentrifuge tubes.
9. Fine, soft paintbrushes.
10. PCR tubes (0.1 mL).
11. Automatic pipettes (200 μL).
12. Petri dishes (100 × 15 mm and 100 × 25 mm).
13. Falcon tubes (50 mL).
14. Sterile plastic cups (200 mL) with lids.
15. Clay pots (15.2 cm diameter).
16. Potting mix (75%–85% Sphagnum Peat Moss and 10%–20% Perlite from Pro-mix BX; Premier Tech Horticulture, Canada).
17. Perlite.
18. Tower paper.
19. Identification tags.
20. Metal wire.
21. Non-Phthalate PVC Tubing (ID: 1/4 in. Thermo Fisher Scientific: 8701-4060).
22. Aluminum foil.
23. Steel beads (4.5 mm).
24. Jiffy-7® Peat Pellets (Greenhouse Megastore: SO-JFPP-0050).

2.4 Media and Reagents

1. 2,4-Dichlorophenoxy acetic acid (2,4-D) solution: 213.05 mg/L, pH 10.36.
2. Autoclaved distilled water.
3. Chemically competent *Escherichia coli* cells (strain DH5α).
4. Sodium hypochlorite solution (1.65%): Take 20 mL of CLOROX Germicidal Bleach (8.25% sodium hypochlorite) and 2 drops of Tween 20 per 100 mL.
5. Colchicine solution: Colchicine (0.5 g/L), DMSO (20 mL/L), gibberellic acid (GA3) (100 mg/L), and one drop of Tween 80 per 100 mL. Adjust pH to 6.0–7.0.

6. Electrophoresis-grade agarose (MidSci: BE-A500).
7. Embryo rescue medium (ERM): 2.2 g/L Murashige and Skoog basal salt mixture (Millipore Sigma: M5524), sucrose (30 g/L), adjust pH to 5.8 with 1 N NaOH, agar (8 g/L). After autoclaving, cool down and pour into Petri dishes (100 × 25 mm) or cups (200 mL).
8. First Strand cDNA Synthesis Kit (Thermo Fisher Scientific: 18080400).
9. Kanamycin (50 mg/mL) dissolved in water and filter-sterilized with a 0.22 μm membrane filter.
10. LB medium: Tryptone (10 g/L), yeast extract (5 g/L), NaCl (10 g/L), agar (15 g/L; for plates only). Adjust pH to 7.0 with 1 N NaOH solution. Sterilize by autoclaving (LB Broth can also be bought commercially).
11. PCR purification kit (PureLink™, ThermoFisher Scientific: K310002).
12. DNeasy Plant Kits (QIAGEN: 69106).
13. Plasmid miniprep kit (NEB: T1110S).
14. Restriction enzyme Bsa I-HFv2 (NEB: R3733S).
15. SYBR Green master mix (Bio-Rad: 1725271).
16. T4 DNA ligase (NEB: M0202S).
17. 70% ethanol.
18. Taq polymerase and reagents for PCR (NEB: M0267S).
19. Total RNA Miniprep Kit (NEB: T2010S).
20. pBUE411 plasmid (Addgene: #62200).

2.5 Software and Databases

1. ClustalW (https://www.genome.jp/tools-bin/clustalw).
2. Multiple sequence alignment (http://multalin.toulouse.inra.fr/multalin/).
3. CRISPRdirect (https://crispr.dbcls.jp/).
4. BLAST analysis (http://blast.ncbi.nlm.nih.gov/).
5. Chinese Spring reference genome sequence v2.1 (https://www.ncbi.nlm.nih.gov/datasets/genome/GCF_018294505.1/).
6. WheatCrispr: https://crispr.bioinfo.nrc.ca/WheatCrispr/.

2.6 Primers

The list of primers used for expression analysis is provided in Table 1.

Table 1
List of primers used for expression analysis of *Cas9* and *sgRNA*

Target gene	Primer sequence (5′–3′)
Cas9	Cas9-RT-F1: CATGATTAAGTTCAGGGGCC
	Cas9-RT-R1: AGGTTATCCAGGTCATCGTC
sgRNA	gRNA-RT-F1: CGACTCGGTGCCACTTTTTCAAGTTG
Actin	Actin-F2: AGAGGTTACTCCTTCACCACC
	Actin-R1: CAGTGATCTCCTTGCTCATACG
CDK	ZmCDK-RT-F2: TCAGTGCTCAGCAGGCTCTA
	ZmCDK-RT-R2: CATCCCAGAAGGATGTTCGT

3 Methods

3.1 Selection of Gene Target Sites for the Single Guide RNA (sgRNA) Design

1. For a target gene, get sequences of all three homologs from a hexaploid wheat genotype with the genome sequence available (*see* **Note 1**).
2. Align sequences of the homologs using a program such as ClustalW (https://www.genome.jp/tools-bin/clustalw) or other sequence alignment programs.
3. Select conserved regions from the sequence alignment for sgRNA design if all three copies of the genes are targeted. Otherwise, select the nonconserved region for sgRNA design if only one specific homolog of the gene is targeted.
4. Identify potential 20-bp oligos in the selected target sequence for sgRNA design using a web-based tool such as CRISPRdirect (https://crispr.dbcls.jp/) (*see* **Note 2**).
5. Choose the 20-bp oligos with around 50% of GC and less off-target sites for the vector construction (*see* **Note 3**).
6. To minimize potential off-target activity, run a BLAST analysis of the preselected oligo sequences (http://blast.ncbi.nlm.nih.gov/) against the reference genome.
7. Use the chosen target oligos for synthesis of the sgRNA-expressing cassettes (see Sect. 3.2).

3.2 Construction of a Binary Vector for Maize Transformation

1. Order a DNA fragment that consists of two cassettes with the first cassette containing a 20-bp target oligo sequence 1, gRNA scaffold, and the rice U3 terminator (OsU3t), and the second cassette containing the wheat U3 promoter (TaU3p), 20-bp target oligo sequence 2 (Fig. 4) (*see* **Note 4**).

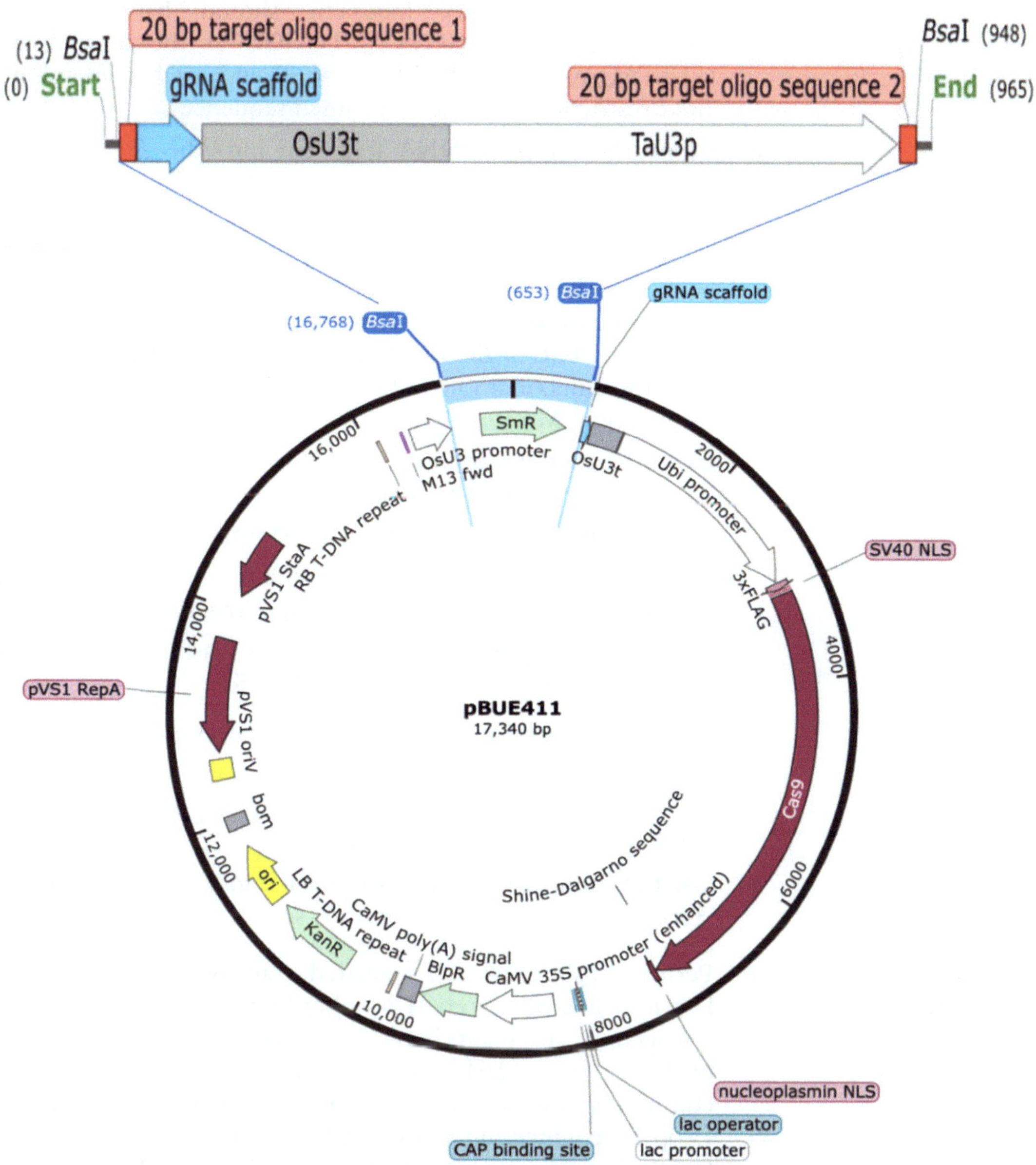

Fig. 4 Assembly of the synthetic DNA fragment expressing gRNAs into the binary vector pBUE411 using the Golden Gate cloning system to construct sgRNA expression cassette. pBUE411 contains rice U3 promoter (OsU3), rice U3 terminator (OsU3t), Ubi-maize polyubiquitin-1 (Ubi-1) promoter, Cas9 gene, 2x35S-a doubled-enhanced 35S promoter from the Cauliflower Mosaic Virus (CaMV 35S promoter), BlpR-phosphinothricin acetyltransferase gene, poly(A)-CaMV poly(A) signal, LB left border (LB T-DNA repeat), and RB right border (LB T-DNA repeat). The synthetic DNA cassette expressing gRNAs has all components indicated except the 5′ and 3′ end sequences (short black lines), ATATATGGTCTCTGGCG and GTTTAGAGACCAATAAT, where the underlined sequences represent the Bsa I restriction site

2. The synthetic DNA fragment should have the tail sequences, ATATATGGTCTCTGGCG at the 5′ end and GTTTAGAGACCAATAAT at the 3′ end, each containing a *Bsa* I restriction site (the underlined sequence).
3. Clone the synthetic DNA fragment into the *Bsa* I sites of the vector pBUE411 (Addgene: #62200) (Fig. 4) using the Golden Gate Cloning method [15]:
   a. Prepare a 15 μL of reaction containing 200 ng of each synthetic DNA fragment, 200 ng of pBUE411 plasmid, 1.5 μL of 10× T4 ligase buffer, 1.5 μL of 10× CutSmart buffer, 1 μL of *Bsa* I (20 units/μL), 1 μL of T4 DNA ligase (400 units/μL) (all buffers and enzymes are from NEB, Ipswich, MA).
   b. Incubate the reaction at 37°C for 5 h followed by 5 min at 50°C and 10 min at 80°C.
   c. Use 5 μL of the reaction for transformation of *E. coli* cells.
   d. Pick four single colonies (*see* **Note 5**) and culture in 5 mL of LB (50 μg/mL of Kanamycin) overnight.
   e. Isolate the plasmids using the Mini Plasmid Isolation Kit following kit instructions (*see* **Note 5**).
4. Digest the plasmids with specific restriction enzymes to confirm the vector structure (*see* **Note 6**).
5. Select two to three plasmids with correct gene structure for whole vector sequencing (*see* **Note 7**).
6. Use the confirmed plasmid for maize transformation using the *Agrobacterium*-mediated method (see Sect. 3.3).

3.3 Maize Transformation

1. Perform *Agrobacterium*-mediated transformation using the hybrid maize variety Hi-II or inbred line B104. We generated transgenic maize plants in our lab using the protocols described by Frame et al. [16] or Kang et al. [17]. We also sent gene editing constructs to the Plant Transformation Facility of the Crop Bioengineering Center (CBC) at Iowa State University for producing transgenic T_0 maize seedlings (*see* **Note 8**).

3.4 Quantify the Relative Expression of Cas9 and sgRNA in Transgenic Maize Plants Using Quantitative Reverse Transcription PCR (qRT-PCR)

1. Collect fresh young leaves from the transgenic maize plants generated from the *Agrobacterium*-mediated transformation.
2. Place the leaves into 96-wells DNA extraction plate and one steel bead (4.5 mm) into each well. After collecting all samples, immediately submerge them in liquid nitrogen.
3. Grind the leaf samples in liquid nitrogen with the TissueLyser II (1 min at oscillation frequency 25 Hz).
4. Extract total RNA using the Monarch® Total RNA Miniprep Kit (NEB, Ipswich, MA), following kit instructions.

5. Measure the concentrations of RNA samples using NanoDrop™ One.
6. Adjust the RNA concentration to 100 ng/μL (*see* **Note 9**).
7. Perform reverse transcription PCR to synthesize complementary DNA (cDNA) using the First Strand cDNA Synthesis Kit (Thermo Fisher Scientific) according to the manufacturer's protocol (*see* **Note 10**).
8. Run qRT-PCR experiments to quantify the relative expression of *Cas9* and *sgRNA*. Genes encoding actin (*Act*) and cyclin dependent kinase (*CDK*) are used as internal references of maize [18] (*see* **Note 11**).
9. Select T_0 plants with higher expression of *Cas9* and *sgRNA* to produce seeds by self-pollination or to be used directly as pollen donor to pollinate emasculated wheat spikes for production of wheat haploid plants (*see* **Note 12**).

3.5 Wheat × Maize Hybridization and Production of Haploid Plants

The protocol of Niu et al. [11] is modified to make crosses between wheat varieties and transgenic maize plants with higher expression of *Cas9* and *sgRNA* to produce haploid plants.

1. Grow selected transgenic maize plants in a greenhouse room maintained at 26°C/22°C day/night and 16/8 h photoperiod.
2. Grow wheat varieties carrying the target gene in another greenhouse room with the conditions set to 22°C/18°C day/night and 16/8 h photoperiod (*see* **Note 13**).
3. When the maize tassel has fully emerged before pollen is shed, emasculate wheat spikes that have fully emerged from the flag leaf with fine tweezers (at 2–3 days before anthesis) (Fig. 5A).
4. Cover the emasculated wheat spikes with glassine bags labeled with the wheat genotype and emasculation date to prevent any cross pollination (Fig. 5A).
5. 2–3 days later, pollinate the emasculated spikes with freshly collected pollen from the transgenic maize plants (Fig. 5B) (*see* **Note 14**).
6. 24 h after the pollination, dip the spikes in the 2,4-D solution (213.05 mg/L) for 5 s and cover the spikes with glassine bags (Fig. 5C).
7. Collect the wheat spikes 13–17 days after the 2,4-D treatment, and harvest green immature kernels for embryo rescue (Fig. 5D, E) (*see* **Note 15**).
8. To isolate the immature embryos, sterilize the green and water-filled kernels with 70% ethanol for 2 min and then with 1.65% of sodium hypochlorite solution for 15 min (Fig. 5F).

Fig. 5 Procedure for wide hybridizationwide hybridization between wheat and transgenic maize expressing Cas9 and sgRNA. (**A**) Emasculation using precise tweezers. (**B**) Pollination with edited maize pollen using a fine soft painting brush. (**C**) Auxin treatment by dipping the pollinated wheat spike in a 2,4-D solution. (**D**) Enlarged caryopses in the spike at 15 days after pollination. (**E**) Enlarged caryopses manually removed from the spike. (**F**) Enlarged caryopses in a 50 mL tube prepared for disinfection. (**G**) Enlarged caryopses after disinfection. (**H**) Haploid embryo isolated from a caryopses. (**I**) Haploid embryos plated on ERM medium. (**J**) Germinated haploid embryos. (**K**) Haploid plantlets growing in plastic cups. (**L**) Acclimation of haploid plant in perlite. (**M**) Developed haploid plants with three or more tillers in soil in the greenhouse. (**N**) Colchicine treatment of haploid plants with three or more tillers. (**O**) Haploid plants transplanted back to soil after colchicine treatment. (**P**) DH seed production from colchicine-treated haploid plants in the greenhouse

9. Rinse kernels with autoclaved distilled water three times, 5 min each.
10. Isolate embryos from sterilized kernels using tweezers and scalpel under a stereo microscope, inside a laminar air-flow hood to maintain aseptic condition (Fig. 5G, H).
11. Culture the excised haploid embryos on ERM medium and maintain the cultures in an incubator at 25°C in the dark for 2 weeks (Fig. 5I).
12. When the coleoptile and small primary roots emerge (Fig. 5J), transfer the small plantlets to new ERM medium in small sterile plastic cups with lids (Fig. 5K).

13. Incubate the cups with the seedlings in a growth chamber under light conditions (16/8 h light/dark) and temperature set at 25°C until green and healthy plantlets develop.
14. Transplant the haploid plantlets into plastic cups with Peat Pellets (soaked thoroughly) and incubate them in a growth chamber with 16/8 h light/dark and 25°C (Fig. 5l).

3.6 Screening of Wheat Haploid Plants for Gene Mutations Using PCR Amplification and Sequencing

1. Collect fresh young leaves from the haploid plantlets into a 96-well DNA extraction plate.
2. After collecting all samples, add one steel bead (4.5 mm) into each well, then immediately submerge them in liquid nitrogen. Grind the leaf samples using the TissueLyser II (1 min at oscillation frequency 25 Hz).
3. Extract the DNA using the QIAGEN DNeasy Plant Mini Kit (Qiagen) following the manufacturer's protocol (*see* **Note 16**).
4. Amplify the fragment of the target gene by PCR using gene specific primers flanking the target oligo sites (see **Note 17**).
5. Check PCR products by gel electrophoresis and purify the amplicons using the PureLink™ PCR Purification Kit following manufacturer's instructions (ThermoFisher Scientific) (*see* **Note 18**).
6. Send the purified PCR products to a sequencing facility for Sanger sequencing (EurofinGenomics, Louisville, KY).
7. Identify the mutations by aligning the sequences (Multiple sequence alignment) from haploid plants against the wild-type gene sequence using Cluster W.
8. Transplant the haploid plants with mutation(s) into 15.2 cm diameter clay pots filled with potting mix and grow them in the greenhouse with 16/8 h light/dark and 22°C–25°C.

3.7 Chromosome Doubling and Doubled Haploid Production

1. At the three to four tillers stage (Fig. 5M), remove the haploid plants from the pots, clean the roots with running water, and dry the plants with a paper towel. If the roots are too long, cut them off to about 1–2 cm length.
2. Tie the identification tag with each individual plant with metal wire and then submerge all plants in a bundle into a colchicine solution (*see* **Note 19**) for 8 h at room temperature (22°C–25°C). Aerate the colchicine solution continuously with a PVC tubing connected to an air pump (Fig. 5N) and cover the container with aluminum foil to prevent the light exposure.
3. After colchicine treatment, wash the lower parts of the plants and roots overnight using running tap water.
4. Then, transfer the individual plants into 15.2 cm diameter clay pots filled with potting mix and incubate them in a growth chamber set at 16°C with a 16/8 h of photoperiod for 2 weeks.

5. Return the plants to the greenhouse under the same conditions as before the colchicine treatment (16/8 h light/dark and 22°C–25°C) for continuous growth until seed setting (Fig. 5O, P).
6. Harvested DH seeds are planted for seed multiplication and phenotyping experiments (*see* **Note 20**).

4 Notes

1. Bread wheatBread wheat (*Triticum aestivum* L., $2n=42$, AABBDD) is an allohexaploid with three genomes (A, B, and D). Thus, most genes have three similar copies (homoeologues). The gene sequences can be obtained from the Chinese Spring reference genome sequence v2.1 (https://www.ncbi.nlm.nih.gov/datasets/genome/GCF_018294505.1/) if no genome sequence is available for the wheat genotype you are interested. To confirm the gene sequences, PCR can be used to amplify the target sequences and confirmed by Sanger sequencing.
2. Other tools for designing sgRNA are also available, like WheatCrispr. (https://crispr.bioinfo.nrc.ca/WheatCrispr/) or any other programs the user prefers. We recommend constructing at least two sgRNAs to disrupt a target gene.
3. We suggest selecting 20 bp oligos with GC% around 50% because the 20-bp oligo is used as one of the primers in qRT-PCR to quantify the sgRNA expression in transgenic maize plants.
4. The first DNA cassette is driven by the rice U3 promoter (OsU3 promoter), which is already present in the binary vector pBUE411. This promoter can be replaced by other promoters for high expression of sgRNA.
5. Normally, most (>95%) of the colonies generated from the Golden Gate Cloning method contain plasmids with the desired gene construct, and therefore, it is not necessary to isolate plasmids from many colonies. From 5 mL of LB culture, we usually can extract 50 μL of plasmid at concentration of 150–200 ng/μL. The plasmid at a concentration of 150–200 ng/μL is good for future *Agrobacterium* transformation.
6. We select different enzymes to digest the plasmids according to the expected vector's map.
7. Based on the restriction enzyme digestion pattern, we can determine if the plasmids contain the potential right gene construct. However, the accuracy of the vector should be

confirmed by whole plasmid sequencing. We usually send our plasmids to Plasmidsaurus (https://www.plasmidsaurus.com/) for sequencing.

8. Maize transformation can be performed by your own lab if you have the personnel and expertise to do the work. We conducted transformation of B104 immature embryos with *Agrobacterium* strain AGL1Thy, which contains the gene editing constructs, following the protocol of Kang et al. [17]. Alternatively, you can send the constructed binary vector to a plant transformation facility, which provides a fee-based service. We have used the Crop Bioengineering Center (CBC) at Iowa State University for maize transformation (https://lab.cropbioengineering.iastate.edu/services-fees), where immature embryos of B104 were used as explants and the *Agrobacterium* strain LBA4404Thy was used for transformation using the protocol described by Kang et al. [17]. Immature embryos of the hybrid maize Hi-II were also used as explants for *Agrobacterium*-mediated transformation using the *Agrobacterium* strain EHA101 and following the protocol of Frame et al. [16]. Other maize varieties can also be used as long as they are transformable and are highly efficient for DH production with different wheat genotypes.
9. We usually treat all RNA samples with DNase I to remove DNA completely, but the DNase I treatment step is included during the isolation of RNA in most of the kits. If the concentration of RNA is lower than 100 ng/μL, new samples can be collected for RNA reisolation. Alternatively, you can add more RNA samples in the next step of reverse transcription PCR.
10. Many commercially available First Strand cDNA Synthesis Kits can be used for this step, and it depends on user's preference.
11. In our lab, qRT-PCR is performed in a Bio-Rad CFX96 Touch Real-time PCR. Each 20 μL of reaction contains 10 μL of SYBR Green master mix (Bio-Rad), 1 μL of each gene specific primer (10 mM), 2 μL of 10× diluted cDNA obtained from Step 4 of Sect. 3.4, and 6 μL of distilled water. The PCR conditions are: 50°C for 2 min, 95°C for 10 min, followed by 40 cycles of 95°C for 15 s and 60°C for 1 min. The relative expression of *CDK*, *Cas9*, and *sgRNA* is calculated with reference to the housekeeping gene of *Act* using the $2^{-\Delta\Delta CT}$ method [19]. Any other established method to quantify the relative expression of genes can be used instead. Primers that we use for qRT-PCR experiment are listed in Table 1.
12. To produce seeds from the T_0 plants, we usually pollinate the ears with pollens collected from their own tassels. However, some T_0 plants have their ears coming out late and cannot

be pollinated by themselves. In this case, we cross them with wild-type pollen to obtain seed from the edited line. We only keep those transgenic maize plants with the expression of *Cas9* and *sgRNA* similar or higher than *CDK* because they are the ones that potentially generate some haploid wheat plants with mutations at the targeted gene. To prevent cross-pollination, cover the emerging ears and tassels with brown bags during plant flowering stage.

13. It is highly recommended to plant multiple sets of wheat lines to match the flowering time of the transgenic maize plants.
14. Pollinate the emasculated wheat spikes when the stigmas are fully developed.
15. It is critical to keep the enlarged immature kernels green. Try to avoid the kernels dry out.
16. DNA can be extracted using any commercially available DNA extraction kit or methods using home-made buffers.
17. We usually design primers 300-bp up and down stream of the targeted sites to amplify the targeted gene sequence. With this size of the PCR fragment, a sequencing reaction can generate good sequences enough to identify the mutations. The regular PCR reaction (20 μL) contains 2 μL of 10× PCR buffer, 1.6 μL of dNTPs (2.5 mM each), 1 μL of each gene specific primer (10 mM), 2 μL of DNA (50 ng/μL), 0.2 μL of Taq polymerase (1 unit), and 12.2 μL of distilled water. The PCR conditions are: 94°C for 2 min, followed by 40 cycles of 94°C for 30 s, 60°C for 30 s and 72°C for 1 min, and final extension at 72°C for 5 min. The PCR products are used for the electrophoresis analysis, future purification and sequencing.
18. PCR products can be purified using any commercially PCR purification kit. Most sequencing companies also provide PCR purification services.
19. Colchicine solution should be freshly made before use. Colchicine is toxic and personal protective equipment should be worn when handling the solution. Colchicine solution should be handled following the hazardous waste disposal policy.
20. Although mutations in the target gene are confirmed by PCR and Sanger sequencing in the haploid plants, DNA samples should be extracted from the DH plants and the target region amplified and sequenced again to ensure that they have the same mutations.

Acknowledgments

The authors gratefully acknowledge the assistance and support of Anil Karmacharya, Shahed Safar, Alireza Poursafar, Olawumi Amusan, and Shirley Zhong during the development of the described protocol. This material is based upon work partially supported by North Dakota Wheat Commission and the U. S. Department of Agriculture under Agreement No. 59-0206-2-156. This is a cooperative project with the U. S. Wheat & Barley Scab Initiative (USWBSI). Any opinions, findings, conclusions, or recommendations expressed in this publication are those of the authors and do not necessarily reflect the view of the U. S. Department of Agriculture.

References

1. Doudna JA, Charpentier E (2014) The new frontier of genome engineering with CRISPR-Cas9. Science 346:1258096
2. Gao C (2021) Genome engineering for crop improvement and future agriculture. Cell 184:1621–1635
3. Cardi T, Murovec J, Bakhsh A et al. (2023) CRISPR/Cas-mediated plant genome editing: outstanding challenges a decade after implementation. Trends Plant Sci 28:1144–1165
4. Shan Q, Wang Y, Li J et al. (2013) Targeted genome modification of crop plants using a CRISPR-Cas system. Nat Biotechnol 31:686–688
5. Shan Q, Wang Y, Li J et al. (2014) Genome editing in rice and wheat using the CRISPR/Cas system. Nat Protoc 9:2395–2410. https://doi.org/10.1038/nprot.2014.157
6. Liang Z, Zong Y, Gao C (2016) An efficient targeted mutagenesis system using CRISPR/cas in monocotyledons. Curr Proto Plant Biol 1:329–344
7. Liang Z, Chen K, Zhang Y et al. (2018) Genome editing of bread wheat using biolistic delivery of CRISPR/Cas9 *in vitro* transcripts or ribonucleoproteins. Nat Protoc 13:413–430. https://doi.org/10.1038/nprot.2017.145
8. Laurie DA, Bennett MD (1986) Wheat × maize hybridization. Can J Genet Cytol 28:313–316
9. Laurie DA, Bennett MD (1988) The production of haploid wheat plants from wheat × maize crosses. Theor Appl Genet 76:393–397
10. Laurie DA, Reymondie S (1991) High frequencies of fertilization and haploid seedling production in crosses between commercial hexaploidy wheat varieties and maize. Plant Breed 106:182–189
11. Niu Z, Jiang A, Hammad WA et al. (2014) Review of doubled haploid production in durum and common wheat through wheat × maize hybridization. Plant Breed 133:313–320
12. Kelliher T, Starr D, Su X et al. (2019) One-step genome editing of elite crop germplasm during haploid induction. Nat Biotechnol 37:287–292
13. Budhagatapalli N, Halbach T, Hieke S et al. (2020) Site-directed mutagenesis in bread and durum wheat via pollination by cas9/guide RNA-transgenic maize used as haploidy inducer. Plant Biotechnol J 18:2376
14. Karmacharya A, Li D, Leng Y et al. (2023) Targeting disease susceptibility genes in wheat through wide hybridization with maize expressing Cas9 and guide RNA. Mol Plant-Microbe Interact 36(9):554–557
15. Xing HL, Dong L, Wang ZP et al. (2014) A CRISPR/Cas9 toolkit for multiplex genome editing in plants. BMC Plant Biol 14: 327
16. Frame BR, Shou H, Chikwamba RK et al. (2002) *Agrobacterium tumefaciens*-mediated transformation of maize embryos using a standard binary vector system. Plant Physiol 129:13–22
17. Kang M, Lee K, Finley T et al. (2022) An improved *Agrobacterium*-mediated transformation and genome-editing method for maize inbred B104 using a ternary vector system and immature embryos. Front Plant Sci 13:860971
18. Lin F, Jiang L, Liu Y et al. (2014) Genome-wide identification of housekeeping genes in maize. Plant Mol Biol 86:543–554
19. Livak KJ, Schmittgen TD (2001) Analysis of relative gene expression data using real-time quantitative PCR and the $2-\Delta\Delta CT$ method. Methods 25:402–408

Chapter 19

In Vivo Rice Haploid-Induction System by an Egg Cell-Specific Peptidase Knockout

Ce Shi, Xuecheng Zhang, Zifu Zhao and Meng-xiang Sun

Abstract

Doubled haploid (DH) technology is a fast and convenient approach for crop breeding and genetic research. Currently, in vitro anther culture is the main method for rice haploid production. However, genotype dependence remains a major problem in the anther culture of most rice subspecies or cultivars for haploid induction. In this chapter, we describe a protocol for in vivo haploid induction in rice using egg cell-specific peptidase (ECS) knockout lines, including mutation of *ECS* using CRISPR-Cas9 system, selection of homozygous *Osecs* mutants, and identification of haploids in the offspring of *Osecs* by flow cytometry. The ECS mediated maternal in vivo haploid-induction system is a convenient, time-saving and labor-saving technique to produce rice DH lines.

Keywords Haploid-induction, Rice, Egg cell, Aspartic proteases

1 Introduction

Rice is the primary source of calories for half of the global population and a crucial staple food for the world's poorest and most undernourished people [1]. Therefore, rice breeding is one of the major challenges for agriculture in different countries, especially in Asia. Homozygous lines are usually used in rice breeding to create hybrid varieties using conventional breeding methods [2]. However, the conventional methods for developing homozygous lines in rice rely on continuous selfing for six to eight generations and are obviously time and labor consuming. Comparatively, doubled haploid (DH)technology only requires two generations, and it is thus becoming a common technology in crop breeding [3]. Currently, DH technology is widely applied by many crop breeders and seed industry companies as a method to quickly obtain homozygous lines [4].

Anther culture in rice has been a valuable biotechnological tool for producing DH plants since 1986 [5], facilitating genetic

Ana María Castillo et al. (eds.), *Doubled Haploid Methods in Cereals*, Methods in Molecular Biology, vol. 3029, https://doi.org/10.1007/978-1-0716-5245-9_19,

studies and breeding programs. However, its effectiveness is often limited by genotype dependency, as not all rice varieties respond equally to the culture conditions [6]. In addition, another significant problem is the occurrence of albinism in regenerated plants, especially in indica genotypes, which hinders the plantlet growth and survival in vivo, thus impacting the overall success of the technique [7]. Currently, DH can be produced in planta by employing haploid inducer stocks. Paternal haploid induction offers several advantages over anther culture in plant breeding. One key benefit is the ability to obtain haploid plants directly from fertilization, leading to faster generation times compared to the multistep process of anther culture [8]. In addition, paternal haploids can capture genetic variation from both parents, resulting in a more diverse gene pool [8].

However, paternal haploid induction also has its challenges. At present, the reported paternal haploid-inducing genes, such as *DMP* (*Domain of Unknown Function 679 Membrane Protein*) [9], *PLA1* (*Phospholipase A1*)/*MTL* (*Matrilineal*)/NLD (*Not like dad*) [10–12], and *PLD3* (*Phospholipase D3*) [13], are all paternally (pollen) expressed genes. While the mutations in these genes can induce haploid offspring and have been used in crop breeding, there are still several problems that need to be solved: (i) These mutants usually lead to abnormal pollen development and germination or failed fertilization (e.g., single fertilization induced-endosperm development failure), and a decreased seed-setting rates, reducing the efficiency of haploid induction. (ii) Male sterility phenotype is a convenient method to easily identify haploids, but these mutants with male fertility defect cause difficulty in quickly distinguishing between haploid and diploid population. (iii) When male-inducing lines are crossed with different crop lines, the resulting haploid offspring may not fully retain the desirable traits of the original genotype. This dilution can lead to variability in agronomic characteristics. (iv) There are hybridization barriers between some genotypes that limit haploid induction. In addition, although the *CENH3* (*centromeric histone H3*) haploid-induction system has a relatively high haploid-induction rate [14], it is not successful in some species, such as soybean [15]. Up to now, only paternal haploid-inducer *MTL/MTL2* has been reported in rice [16, 17]. Therefore, the establishment of new haploid-induced techniques in rice to address problems mentioned above is essential for rice molecular breeding. We recently reported a female haploid-induction factor ECS (egg cell-specific aspartic endopeptidase) [18]. Here, we provide a detailed protocol for in vivo rice haploid induction by the knockout of an egg cell-specific peptidase (Fig. 1).

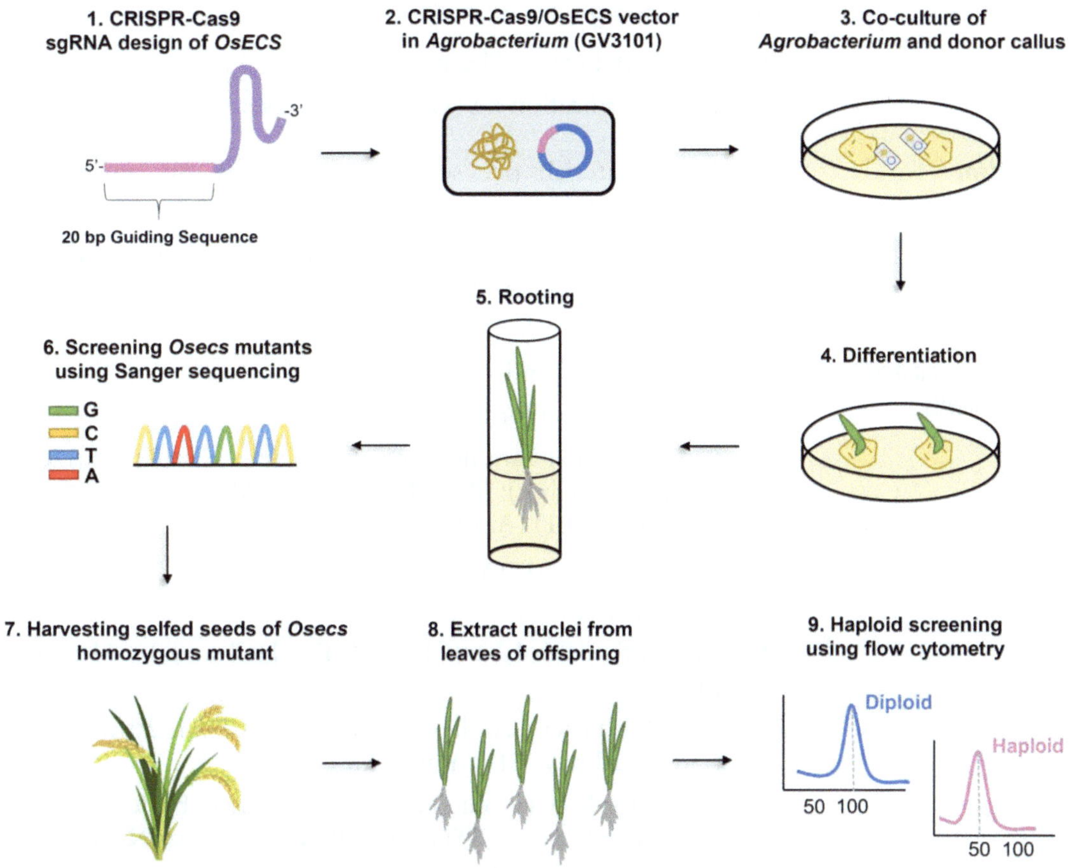

Fig. 1 Diagram of rice haploid-induction system by an egg cell-specific peptidase knockout. Steps 1–6 illustrate the mutation of *OsECS*. Steps 7–9 illustrate the screening of rice haploid plant in *Osecs* offspring

2 Materials

2.1 Vector Construction

1. Vector used for CRISPR-Cas9 in rice: Promoter for sgRNA expression: OsU3; Promoter for hygromycin resistance (*hyg*) gene expression: 35 S cauliflower mosaic virus (CaMV).
2. Competent cells (*Escherichia coli* strain DH5α).
3. Agarose G-10 (Biowest).
4. 2× Phanta UniFi Master Mix (Vazyme).
5. 2× Rapid Taq Master Mix (Vazyme).
6. FastPure Plasmid Mini Kit (Vazyme).
7. FastPure Gel DNA Extraction Mini Kit (Vazyme).

8. BsaI (NEB).
9. T4 DNA Ligase (Thermo).
10. DNA purification kit (Omega Bio-Tek).

2.2 Plant Transformation

1. Plant material: rice seeds (*Oryza sativa* ssp. *japonica* cv. Nipponbare).
2. *Agrobacterium tumefaciens* strain GV3101 (alternative *Agrobacterium* strains may also be used depending on laboratory preference).

2.3 Solutions and Culture Media

1. Ethanol (70% and 75%).
2. Sodium hypochlorite (1.5%).
3. Cetyltrimethyl ammonium bromide (CTAB) extraction buffer (Coolaber): 50 mM Tris-HCl, 4 M NaCl, 1.8% CTAB, and 25 mM EDTA, pH 8.0.
4. Chloroform: isoamyl alcohol (24:1 *v/v*).
5. Isopropanol.
6. TE buffer (Coolaber): 10 mM Tris-HCl and 1 mM EDTA, pH 8.0.
7. 1× TAE buffer (Coolaber): 40 mM Tris-HCl, 20 mM Acetic acid, and 1 mM EDTA, pH 8.3.
8. Buffer for flow cytometry: nuclei extraction buffer and staining buffer (Partec CyStain UV Precise-Kit) (*see* **Note 1**).
9. LB medium: Combine 10 g of tryptone, 5 g of yeast extract, and 10 g of NaCl in 1 L of Milli-Q water. Adjust the pH to 7.0, then autoclave at 120°C for 20 min.
10. Yoshida rice nutrient salts solution. Prepare 1 L of solution with the following final concentration of components: NH_4NO_3, 114.36 mg/L; NaH_2PO_4, 38.75 mg/L; K_2SO_4, 89.22 mg/L; $CaCl_2$, 110.76 mg/L; $MgSO_4$, 197.76 mg/L; $MnCl_2{\cdot}4H_2O$, 1.875 mg/L; $(NH_4)_6Mo_7O_{24}{\cdot}4H_2O$, 0.093 mg/L; H_3BO_3, 1.168 mg/L; $ZnSO_4{\cdot}7H_2O$, 0.044 mg/L; $CuSO_4{\cdot}5H_2O$, 0.039 mg/L; $FeCl_3$, 5.775 mg/L; $C_6H_8O_7{\cdot}H_2O$, 14.875 mg/L. Store at 4°C.
11. 2,4-Dichlorophenoxyaceticacid(2,4-D)stocksolution(100×): Dissolve 100 mg of 2,4-D in 1 mL of 1 N NaOH solution. Then adjust the total volume to 500 mL with Milli-Q water and filter sterilize. Store at 4°C.
12. CHU (N6) basal salt mixture (Coolaber) [19]: $(NH_4)_2SO_4$, 463 mg/L; $CaCl_2$, 125.33 mg/L; $MgSO_4$, 90.27 mg/L; KNO_3, 2830 mg/L; KH_2PO_4, 400 mg/L; H_3BO_3, 1.6 mg/L; FeNaEDTA, 36.7 mg/L; $MnSO_4{\cdot}H_2O$, 3.33 mg/L; KI, 0.8 mg/L; $ZnSO_4{\cdot}7H_2O$, 1.5 mg/L. Store at 4°C.
13. MS plant salt mixture (Coolaber) [20]: NH_4NO_3, 1650 mg/L; $CaCl_2$, 332.2 mg/L; $MgSO_4$, 180.7 mg/L; KNO_3, 1900 mg/L;

KH_2PO_4, 170 mg/L; H_3BO_3, 6.2 mg/L; $CoCl_2{\cdot}6H_2O$, 0.025 mg/L; $CuSO_4{\cdot}5H_2O$, 0.025 mg/L; FeNaEDTA, 36.7 mg/L; $MnSO_4{\cdot}H_2O$, 16.9 mg/L; $Na_2MoO_4{\cdot}2H_2O$, 0.25 mg/L; KI, 0.83 mg/L; $ZnSO_4{\cdot}7H_2O$, 8.6 mg/L. Store at 4°C.

14. MS vitamin stock solution (100×) [20]: Mix and dissolve 100 mg of glycine, 25 mg of nicotinic acid, 25 mg of pyridoxine hydrochloride, and 50 mg of thiamine hydrochloride in 450 mL Milli-Q water and adjust the total volume to 500 mL with Milli-Q water. Store at 4°C.
15. N6 vitamin stock solution (100×) [19]: Mix and dissolve 100 mg of glycine, 25 mg of nicotinic acid, 25 mg of pyridoxine hydrochloride, and 25 mg of thiamine hydrochloride in 450 mL Milli-Q water and adjust the total volume to 500 mL with Milli-Q water. Store at 4°C.
16. Acetosyringone (AS) stock solution (1000×): Dissolve 100 mg AS in 10 mL of 100% DMSO and filter sterilize. Store at −20°C.
17. AAM medium (for infection). Prepare a solution with the following final concentration of components: KH_2PO_4, 170 mg/L; $CaCl_2$, 347 mg/L; $MgSO_4$, 180 mg/L; KCl, 2940 mg/L; H_3BO_3, 3 mg/L; $CoCl_2{\cdot}6H_2O$, 0.025 mg/L; $CuSO_4{\cdot}5H_2O$, 0.0387 mg/L; FeNaEDTA, 36.7 mg/L; $MnSO_4{\cdot}H_2O$, 7.58 mg/L; $Na_2MoO_4{\cdot}2H_2O$, 0.25 mg/L; KI, 0.75 mg/L; $ZnSO_4{\cdot}7H_2O$, 2 mg/L; glycine, 7.5 mg/L; myo-inositol, 100 mg/L; pyridoxine hydrochloride, 0.5 mg/L; thiamine hydrochloride, 0.1 mg/L; L-glutamine, 876 mg/L; casamino acids, 500 mg/L; nicotinic acid, 0.5 mg/L; arginine, 174 mg/L; sucrose 68.5 g/L; glucose 36 g/L. Adjust the pH to 5.2 and the total volume to 1 L with Milli-Q water and autoclave at 120°C for 20 min. Store at 4°C.
18. Carbenicillin (Carb) stock solution (1000×): Dissolve 2.5 g Carb in 10 mL of Milli-Q water and filter sterilize. Store at −20°C.
19. Hygromycin (Hyg) stock solution (1000×): Dissolve 0.5 g Hyg in 10 mL of Milli-Q water and filter sterilize. Store at −20°C.
20. 1-Naphthaleneacetic acid (NAA) stock solution (1000×): Dissolve 40 mg NAA in 3 mL of 0.1 N NaOH solution and adjust the volume to 200 mL with Milli-Q water and filter sterilize. Store at 4°C.
21. Kinetin Stock solution (50×): Dissolve 50 mg kinetin in 20 mL 0.1 N HCl solution and adjust the volume to 500 mL with Milli-Q water and filter sterilize. Store at 4°C.
22. N6D medium (for callus induction): Mix and dissolve 3.98 g CHU (N_6) basal salt mixture, 10 mL of N6 vitamin stock

solution (100×), 0.1 g of myo-inositol, 0.3 g of casamino acid, 2.878 g of proline, 10 mL of 2,4-D stock solution (100×), and 30 g of sucrose in 800 mL of Milli-Q water. Adjust the pH to 5.8 and add 3 g of gellan gum. Adjust the total volume to 1 L with Milli-Q water and autoclave at 120°C for 20 min. Cool to 50°C and pour 40 mL aliquots into Petri dishes (90 × 20 mm). Store in the dark at 25°C.

23. 2N6-AS medium (for cocultivation). Prepare 1 L N6D medium, and adjust the pH to 5.2 and add 3 g of gellan gum. Autoclave at 120°C for 20 min. Cool to 50°C, add 1 mL of AS stock solution (1000×) and pour 40 mL aliquots into Petri dishes (90 × 20 mm). Store in the dark at 25°C.
24. N6D-S medium (for selection). Prepare 1 L of N6D medium, and adjust the pH to 5.8 and add 4 g of gellan gum. Autoclave at 120°C for 20 min. Cool to 50°C, add 1 mL of Carb (1000×) and 1 mL of Hyg (1000×). Pour 40 mL aliquots into Petri dishes (90 × 20 mm). Store in the dark at 25°C.
25. MS medium: Mix and dissolve 4.6 g of MS plant salt mixture, 10 mL of MS vitamin solution (100×), 0.1 g of myo-inositol, 2 g of casamino acid, 30 g of sucrose, and 30 g of sorbitol in 800 mL of Milli-Q water. Adjust the pH to 5.8 and add 3 g of gellan gum. Adjust the total volume to 1 L with Milli-Q water and autoclave at 120°C for 20 min. Cool to 50°C and pour 80 mL aliquots into sterile plastic container (100 × 100 × 110 mm). Store in the dark at 25°C.
26. MS-NK medium (for regeneration). Prepare 1 L of MS medium. Add 1 mL of NAA (1000×) and 20 mL of kinetin (50×). Adjust the pH to 5.8 and add 3 g of gellan gum. Adjust the total volume to 1 L with Milli-Q water and autoclave at 120°C for 20 min. Cool to 50°C, add 1 mL of Carb (1000×) and 1 mL of Hyg (1000×). Pour 40 mL aliquots into Petri dishes (90 × 20 mm). Store in the dark at 25°C.
27. MS-HF medium (for plant growth). Prepare 1 L of MS medium. Adjust the pH to 5.8 and add 3 g of gellan gum. Adjust the total volume to 1 L with Milli-Q water and autoclave at 120°C for 20 min. Cool to 50°C, add 1 mL of Carb (1000×) and 1 mL of Hyg (1000×). Pour 10 mL aliquots into plastic tubes (30 × 200 mm). Store in the dark at 25°C.

2.4 Equipment

1. Conditioned greenhouse with a photoperiod of 16/8 h light/dark with a light intensity of 500 μmol $m^{-2}s^{-1}$ at 30°C for growing donor plants, regeneration and acclimatization of plantlets.
2. Autoclave for medium preparation.
3. Lighting incubators with a photoperiod of 12/12 h light/dark with a light intensity of 120 μmol $m^{-2}s^{-1}$ at 28°C–30°C.

4. Incubators (25°C–37°C).
5. Incubator shakers at 28°C and 37°C.
6. Electrophoresis machine.
7. Electroporator.
8. Laminar flow cabinet.
9. PCR machine.
10. DNA gel imaging systems (UV).
11. NanoDrop (Thermo Fisher Scientific).
12. Flow cytometer with a laser (405 nm).
13. Centrifuge.
14. Refrigerated centrifuge (4°C).
15. Balance.
16. pH meter.
17. Fridge.
18. Freezer at −20°C and −80°C.
19. Stirrer.

2.5 Laboratory Materials

1. Liquid nitrogen.
2. EP tube.
3. Aluminum foil.
4. Filter paper.
5. Pipettes and pipettes tips.
6. Alcohol burner.
7. Tweezers.
8. Scissors.
9. Erlenmeyer flask (250 mL).
10. Syringe filter (0.22 μm pore).
11. Polystyrene Petri dishes (90 × 20 mm).
12. Sterile plastic test tubes (30 × 200 mm).
13. Sterile plastic containers (100 × 100 × 110 mm).
14. Falcon® 5 mL round bottom polystyrene test tube, with cell strainer snap cap (35-μm nylon mesh) (Fig. 2A, B).
15. Razor blade.
16. Beakers.
17. Parafilm.

2.6 Primer Sequences

1. Primer 1: 5′-cagtGGTCTCaTGCACCGTGACTCACCCAG-GTCGC-3′.
2. Primer 2: 5′-cagtGGTCTCaAAACGCGACCTGGGTGAGT-CACGG-3′.

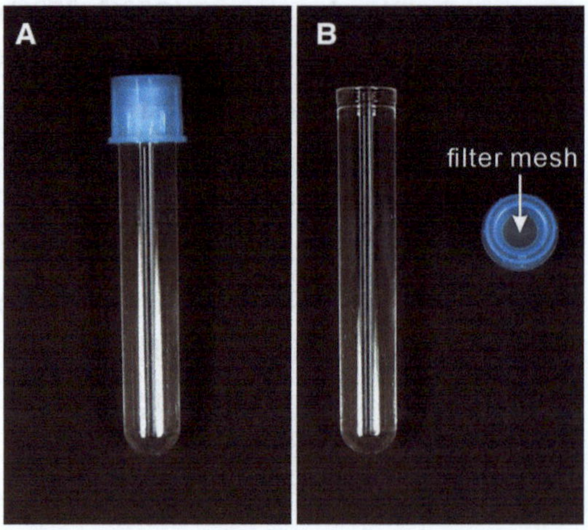

Fig. 2 Tubes used for nucleus extraction and flow cytometry. (**A**) A 5 mL round bottom polystyrene test tube. (**B**) A 35 μm nylon mesh is incorporated into the dual-position snap cap, which can be used to collect the dissociated samples for downstream, processing in flow cytometry instruments

3. Primer 3: 5′-ACCGGTAAGGCGCGCCGTAGT-3′.
4. Primer 4: 5′-GCGATTAAGTTGGGTAACGCCAGGG-3′.
5. Primer 5: 5′-CGCCATTAGCAGCGTGAGGAG-3′.
6. Primer 6: 5′-GCCGCCGATGAGGTGTCG-3′.

3 Methods

3.1 Construction of CRISPR-Cas9 Vector for Mutation of OsECS

1. Select the sgRNA (single guide RNA) sequence for *OsECS* (LOC_Os01g08330) (Nipponbare) mutation using CRISPR-GE (http://skl.scau.edu.cn/): 5′-CCGTGACTCACCCAGGTCGC-3′ (*see* **Notes 2 and 3**).
2. Design the primers based on the target site of LOC_Os01g08330: 5′-CCGTGACTCACCCAGGTCGC-3′. The primer sequences are shown as Primer 1 and Primer 2 (Fig. 3).

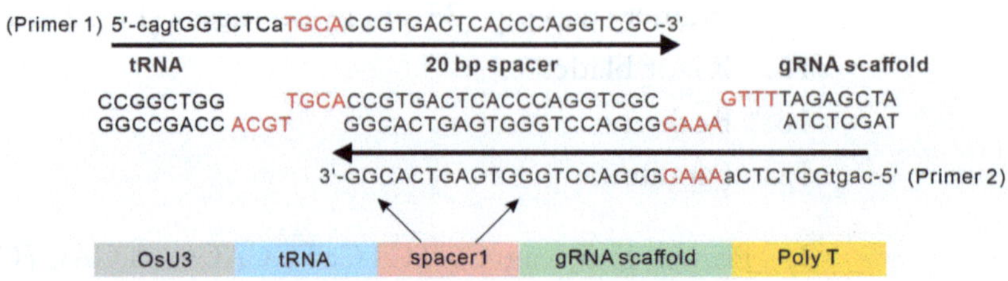

Fig. 3 Schematic image of gRNA oligos to be synthesized in CRISPR-Cas9 vector: cagt, the protective base; GGTCTCa, the BsaI recognition sequence; TGCA/AAAC, the complementary sticky end sequences of the vector

3. Prepare the target fragment: Mix 10 μL of Primer 1 (100 μM) and Primer 2 (100 μM) and incubate at 95°C for 5 min for later use.
4. To perform Golden Gate cloning, prepare the following reaction system: 1 μL of Vector, 1 μL of target fragment, 1 μL of BsaI, 1 μL of T4 DNA Ligase, 2 μL of 10× buffer and adjust the volume to 20 μL with Milli-Q water. Immediately incubate at 37°C for 60 min, then heat at 65°C for 20 min to terminate the reaction and transfer to ice for later use.
5. Thaw competent DH5α cells on ice, add plasmid DNA, and incubate on ice for 30 min. Perform a heat-shock at 42°C for 90 s, then return to ice. Add LB medium, incubate at 37°C for 1 h, and plate onto LB agar with 50 μg/mL kanamycin. Incubate overnight at 37°C to grow colonies.
6. Pick colonies and transfer to PCR tubes. Prepare the PCR reaction system (50 μL) using 2× Rapid Taq Master Mix (Vazyme) and presynthesized primers (2 μL of 10 μM Primer 3 and 2 μL of 10 μM Primer 4, see Sect. 2.6). The PCR amplification program is as follows: 95°C for 5 min for predenaturation, 95°C for 15 s for denaturation, 55°C for 15 s for annealing, and 72°C for 30 s for extension, for a total of 35 cycles, followed by 72°C for 5 min for final extension. Analyze products on a 1.5% agarose gel with 1× TAE buffer. Colonies showing the correct band size are positive.
7. Inoculate a positive colony in LB broth with 50 μg/mL kanamycin and incubate overnight. Use a FastPure Plasmid Mini kit (Vazyme) to extract DNA, following steps for cell lysis, DNA binding, washing, and elution. Quantify the plasmid DNA using a Nanodrop.
8. Dilute the extracted plasmid to 100–500 ng/μL and submit it to a sequencing service. Analyze the results using alignment software (e.g., SnapGene) to confirm the plasmid sequence and verify the insert. This ensures the plasmid, named CRISPR-Cas9/OsECS, is correct and ready for further use.
9. Transfer the CRISPR-Cas9/OsECS vector into *Agrobacterium tumefaciens* GV3101 through electroporation using a 0.1-cm cuvette at 1.8 kV for 5–7 ms to obtain the recombinant strain GV3101/CRISPR-Cas9/OsECS (Fig. 1).
10. Propagate the recombinant *Agrobacterium* in LB medium supplemented with 50 μg/mL kanamycin at 28°C.

3.2 Induction of Calli

1. Dehusk mature healthy seeds using a rice husker. Sterilize the seeds with 75% alcohol for 1 min, then rinse with sterilized water for 1 min. Add 1.5% sodium hypochlorite and incubate for 20 min, then rinse three times with sterilized water.

2. Inoculate about 20 whole seeds on N6D medium in a Petri dish (90×20 mm) and culture at 28°C in the dark for 3–4 weeks. Calli will form from the scutella after this period (Fig. 4A, B).

3.3 Agrobacterium Infection

1. Streak a single colony of GV3101/CRISPR-Cas9/OsECS on 100 mL LB medium, supplemented with 50 μg/mL kanamycin. Incubate at 28°C for 2 days (OD_{600}=0.5–0.6) (Fig. 1).
2. Collect *Agrobacterium* by centrifugation at 5000 × *g* for 10 min at 4°C and resuspend the cells in 30 mL AAM medium containing 30 μL of AS (1000×) (OD_{600}=0.05–0.1).
3. Select and collect the growing calli (yellowish-white, relatively dry and about 1–3 mm in diameter) and subculture on fresh N6D medium at 30°C in the dark for 3 days.
4. Collect about 20 subcultured calli and immerse them in the *Agrobacterium* resuspension in a 250 mL Erlenmeyer flask for 15 min.
5. After infection, pour out the solution, use sterilized filter paper to remove excess *Agrobacterium* from the calli, and air dry for 2 h.
6. Drip 1 mL AAM onto a sheet of filter paper placed on 2N6-AS medium, and then transfer the calli onto the filter paper.
7. Use a sterile spoon to evenly spread the thoroughly dried calli on 2N6-AS medium (do not move them once placed to

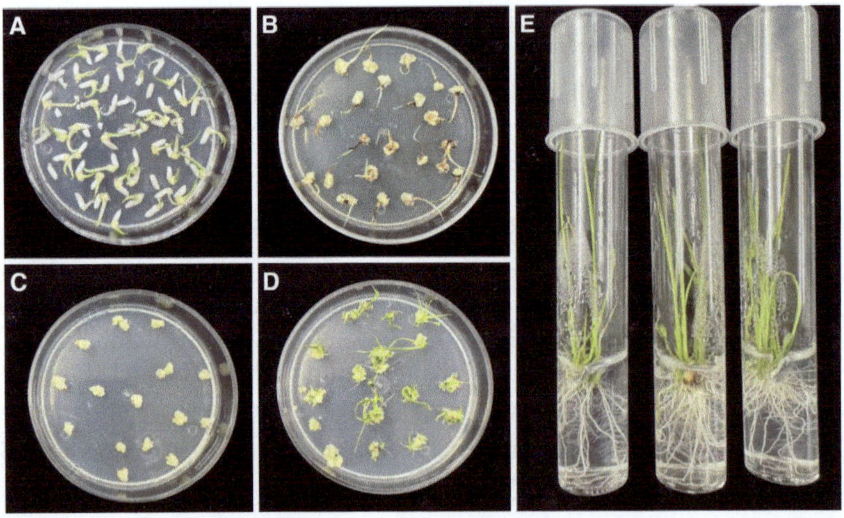

Fig. 4 Procedure for the production of transgenic rice plants. (**A, B**) Calli generated from the scutella of mature seeds on N6D medium. (**C**) Selection of hygromycin-resistant calli on fresh N6D-S medium. (**D**) Shoot regeneration from calli resistant to hygromycin on MS-NK medium. (**E**) Rooting and growth of transgenic rice on MS-HF medium

minimize contact between the medium and the surface of the callus, preventing excessive growth of *Agrobacterium*). Seal with Parafilm (Fig. 1).

8. Wrap the plate with aluminum foil and incubate at 28°C in the dark for 2–4 days.

3.4 Calli Selection and Plant Regeneration

1. Collect the infected calli into a 50 mL sterile tube and wash them with sterilized water for six to eight times until the poured-out water is clear.
2. Rinse the calli with 30 mL of sterilized water containing 30 μL of Carb (1000×) and soak at room temperature for 30 min.
3. Place calli onto sterilized filter paper and dry thoroughly.
4. Transfer about 15 calli onto fresh N6D-S medium in a Petri dish (90 × 20 mm) and incubate in the dark at 28°C for 3–4 weeks. During the callus selection process, be sure to select monoclonal calli (*see* **Note 4**) (Fig. 4C).
5. Inoculate the growing calli on MS-NK medium in Petri dishes (90 × 20 mm) and culture them under light conditions at 30°C for 3–4 weeks (Fig. 4D).
6. Transfer differentiated shoots (2–5 cm in length) to MS-HF medium in a sterile plastictest tube (30 × 200 mm) and culture them under light conditions at 30°C for 3–4 weeks for seedling growth (Fig. 4E).
7. Once the seedlings have developed a few roots (2–3 cm), transplant them into prepared soil. Use well-drained, loamy soil rich in organic matter and adjust pH to 5.5–7.0.

3.5 Identification of Osecs Mutants

1. Grow transgenic rice plants in a greenhouse at 30°C with a 16 h photoperiod and a light intensity of 500 μmol/m^2/s, and add Yoshida rice nutrient salt solution every 2 weeks.
2. Collect the leaves of T0 generation transgenic rice plants, extract genomic DNA (gDNA) using CTAB method. Grind young leaves (~50 mg) in liquid nitrogen, followed by mixing with 600 μL of CTAB extraction buffer and incubating at 60°C to lyse cells. Add 500 μL of Chloroform/isoamyl alcohol to separate DNA from proteins, and centrifuge the mixture at 10,000 × *g* for 10 min. Precipitate the DNA using cold isopropanol, then wash with 1 mL of 70% ethanol and remove impurities. Finally, resuspend the DNA pellet in 50 μL of TE buffer, to obtain gDNA for analysis.
3. Using the 1 μL of gDNA as a template, perform PCR amplification (50 μL reaction system) as follows: Prepare the PCR reaction system using 2× Phanta UniFi Master Mix (Vazyme) and presynthesized primers (2 μL of 10 μM Primer 5 and 2 μL of 10 μM Primer 6, see Sect. 2.6). The PCR amplification

program is as follows: 98°C for 30 s for pre-denaturation, 98°C for 10 s for denaturation, 60°C for 10 s for annealing, and 72°C for 30 s for extension, for a total of 35 cycles, followed by 72°C for 5 min for final extension.

4. Analyze 50 μL of PCR products on a 1.5% agarose gel with 1× TAE buffer. Visualize the PCR amplification products under UV light, excise the target DNA band from the gel, and extract the DNA using a FastPure Gel DNA Extraction Mini Kit (Vazyme). Quantify it and use the purified product for Sanger sequencing (Fig. 1). Compare the sequencing results with the wild-type *OsECS* sequence (*see* **Note 5**).

3.6 Obtaining Selfed Seeds from Homozygous Osecs Mutants

1. Harvest the seeds approximately 5–8 weeks after the natural self-pollination of T0 *Osecs* mutants.
2. Identify homozygous *Osecs* mutations (see Sect. 3.5) in the T1 generation based on Sanger sequencing.
3. Harvest seeds 5–8 weeks after the natural self-pollination of T1 homozygous *Osecs* mutants.
4. Sow the sterilized homozygous *Osecs* mutant seeds on MS medium for subsequent haploid screening (Fig. 1).

3.7 Ploidy Level Analyses

The ploidy level of the plantlets can be determined by flow cytometric analysis (Fig. 1).

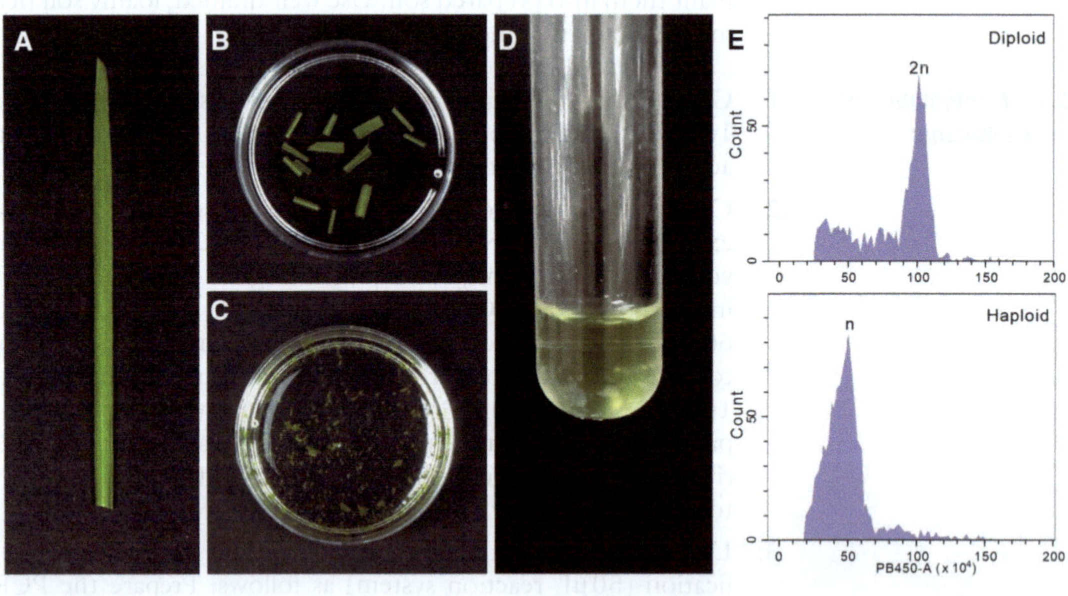

Fig. 5 Extraction of rice leaf nuclei for haploid screening. (**A**) A young leaf of the *Osecs* mutant. (**B**) The leaf is cut into pieces with scissors. (**C**) The leaf is chopped as finely as possible with a razor blade and the nuclei extraction buffer is added. (**D**) The suspension is filtered and prepared for flow cytometry. (**E**) Haploid screening using Flow Cytometry analysis

1. Collect leaf samples (approximately 5–8 cm in length) from plantlets grown in the greenhouse (Fig. 5A). Label and place them on ice.
2. Quickly chop the rice leaves with a razor blade (Fig. 5B, C) (*see* **Note 6**).
3. Add 500 μL of nuclei extraction buffer to the sample to extract nuclei (*see* **Note 7**).
4. Filter the nuclei suspension through a 35-μm nylon mesh into a polystyrene test tube (Fig. 5D) (*see* **Note 8**).
5. Add 1 mL of staining buffer and incubate for 5 min at 25°C to stain the DNA.
6. After sample preparation, insert the tube into the sample port to determine the ploidy level based on the x-axis position of the fluorescence peak compared to the control (Fig. 5E).

4 Notes

1. You can prepare your own nuclei extraction solution, as the amount provided in this kit is insufficient: 45 mM $MgCl_2$, 20 mM MOPS, 30 mM sodium citrate (pH 7.0), and 0.1% Triton X-100.
2. Use the CRISPR-GE website to design several targets. When selecting sgRNAs for rice genome editing, consider the following factors: (i) Specificity: Choose sgRNAs that minimize off-target effects by ensuring high specificity to the target sequence. (ii) Secondary structure: Avoid sgRNAs that form strong secondary structures, which may affect their binding efficiency. (iii) GC content: Aim for a balanced GC content (40%–60%) to enhance the stability and binding of the sgRNA. The sgRNA used in this chapter is just an example for Nipponbare and may not necessarily be the target with the highest actual knockout efficiency. In addition, the sequences of *OsECS* vary among rice superspecies or cultivars. Thus, it is necessary to obtain the exact sequence for the specific rice.
3. Depending on the CRISPR system used in your laboratory, it is recommended to design more than two sgRNA targets to increase the knockout efficiency.
4. Select as many resistant positive calli as possible for subsequent screening of *Osecs* mutants.
5. Test as many different gene-edited lines as possible to obtain homozygous mutant lines in the T0 generation.
6. Use young leaves (14 days after seed germination) to extract nuclei. It is recommended to replace the razor blade after cutting two to three leaves.

7. Add approximately 100 μL of nuclei extraction solution, then add 400 μL after cutting the leaves into a paste and pipette several times.
8. Generally, the nuclei suspension will naturally settle at the bottom of the tube after 2 min. If necessary, use a horizontal centrifuge at 250 × *g* for 30 s.

Acknowledgments

This work was supported by the National Natural Science Foundation of China (32130031; 32400278) and the Funding for Discipline Development of Guizhou Province [XKBF(2025)030].

References

1. Bin Rahman AR, Zhang J (2023) Trends in rice research: 2030 and beyond. Food Energy Secur 12:e390
2. Zheng X, Wei F, Cheng C, Qian Q (2024) A historical review of hybrid rice breeding. J Integrative Plant Biol 66:532–545
3. Li SZ, Wang J, Jia SG et al (2023) Synthetic apomixis: from genetic basis to agricultural application. Seed Biol 2:10
4. Shen K, Qu M, Zhao P (2023) The roads to haploid embryogenesis. Plants 12:243
5. Niizeki H, Oono K (1968) Induction of haploid rice plant from anther culture. Proc Jap Acad Sci 44:554–557
6. Lantos C, Jancsó M, Székely Á et al (2023) Development of *in vitro* anther culture for doubled haploid plant production in indica rice (*Oryza sativa* L.) genotypes. Plants 12:1774
7. Pattnaik SS, Dash B, Bhuyan SS et al (2020) Anther culture efficiency in quality hybrid rice: a comparison between hybrid rice and its ratooned plants. Plants 9:1306
8. Quiroz LF, Gondalia N, Brychkova G et al (2024) Haploid rhapsody: the molecular and cellular orchestra of *in vivo* haploid induction in plants. New Phytol 241:1936–1949
9. Zhong Y, Liu C, Qi X et al (2019) Mutation of *ZmDMP* enhances haploid induction in maize. Nat Plants 5:575–580
10. Liu C, Li X, Meng D et al (2017) A 4-bp insertion at *ZmPLA1* encoding a putative phospholipase a generates haploid induction in maize. Mol Plant 10:520–522
11. Kelliher T, Starr D, Richbourg L et al (2017) MATRILINEAL, a sperm-specific phospholipase, triggers maize haploid induction. Nature 542:105–109
12. Gilles LM, Khaled A, Laffaire JB et al (2017) Loss of pollen-specific phospholipase NOT LIKE DAD triggers gynogenesis in maize. EMBO J 36:707–717
13. Li Y, Lin Z, Yue Y et al (2021) Loss-of-function alleles of *ZmPLD3* cause haploid induction in maize. Nat Plants 7:1579–1588
14. Ravi M, Chan S (2010) Haploid plants produced by centromere-mediated genome elimination. Nature 464:615–618
15. Wang J, Wang XF, Yang WC, Li HJ (2023) Loss of function of *CENH3* causes genome instability in soybean. Seed Biol 2:24
16. Yao L, Zhang Y, Liu C et al (2018) *OsMATL* mutation induces haploid seed formation in indica rice. Nat Plants 4:530–533
17. Jang JH, Noh G, Seo HS et al (2023) Loss of function of pollen-expressed phospholipase *OsMATL2* triggers haploid induction in japonica rice. Plant Physiol 193:1749–1752
18. Zhang X, Shi C, Li S et al (2023) A female in vivo haploid-induction system via mutagenesis of egg cell-specific peptidases. Mol Plant 16:471–480
19. Chu C (1978) The N6 medium and its applications to anther culture of cereal crops. In: Proc symp plant tissue culture. Science Press, Peking, pp 43–50
20. Murashige T, Skoog F (1962) A revised medium for rapid growth and bio assays with tobacco tissue cultures. Physiol Plant 15:473–497

Index

Ana María Castillo et al. (eds.), *Doubled Haploid Methods in Cereals*, Methods in Molecular Biology, vol. 3029, https://doi.org/10.1007/978-1-0716-5245-9,

MIX
Papier aus verantwortungsvollen Quellen
Paper from responsible sources
FSC® C105338

If you have any concerns about our products,
you can contact us on
ProductSafety@springernature.com

In case Publisher is established outside the EU,
the EU authorized representative is:
Springer Nature Customer Service Center GmbH
Europaplatz 3, 69115 Heidelberg, Germany

Printed by Libri Plureos GmbH
in Hamburg, Germany